Pavel Vladimirov, Christopher Dorn (eds.)

Proceedings of the 16th International Workshop on Beryllium Technology (BeWS-16)

October 19-20, 2023, Santander, Spain

Karlsruhe Institute of Technology
KIT SCIENTIFIC REPORTS 7766

Proceedings of the 16th International Workshop on Beryllium Technology (BeWS-16)

October 19-20, 2023, Santander, Spain

by
Pavel Vladimirov, Christopher Dorn (eds.)

BeWS-16
2023
16th International Workshop
On Beryllium Technology
Santander, Cantabria, Spain
19-20 October 2023

Organized under the auspices of
The IEA Technology Collaboration Program on Fusion Materials (FM TCP)

KIT Scientific Publishing

Report-Nr. KIT-SR 7766

Impressum

KIT Scientific Publishing

Karlsruher Institut für Technologie (KIT)
KIT Scientific Publishing
Straße am Forum 2
D-76131 Karlsruhe

KIT Scientific Publishing is a registered trademark
of Karlsruhe Institute of Technology.
Reprint using the book cover is not allowed.

www.bibliothek.kit.edu/ksp.php | E-Mail: info@ksp.kit.edu | Shop: www.ksp.kit.edu

Print on Demand 2025 – Gedruckt auf FSC-zertifiziertem Papier

ISSN 1869-9669
ISBN 978-3-7315-1407-7
DOI 10.5445/KSP/1000178349

Table of contents

Executive Summary

The 16th International Workshop on Beryllium Technology (BeWS-16) took place on 19-20 October 2023 in Santander, Spain, in conjunction with ICFRM-21 (International Conference on Fusion Reactor Materials) in Granada. Local organization was led by "Leading Metal Mechanic Solutions" in Cantabria, Spain, a commercial enterprise with a history of beryllium fabrication work for fusion applications, specifically ITER First Wall tiles. Overall technical support for the workshop came from the Karlsruhe Institute of Technology (KIT) in Germany.

The workshop venue was kindly provided free of charge by the Cantabria Government and the local organizers at Leading. The workshop was held in a hybrid mode, allowing for both in-person and online presentations and participation. About 50 participants represented academia, research institutes, and commercial industry from Spain, Germany, the USA, Japan, the UK, Kazakhstan, Latvia, and Sweden.

The BeWS-16 program included two full working days with eight technical sessions. The workshop's first day featured an engaging opening session, which included a welcome and introduction from the official representative of the Society for the Regional Development of Cantabria (SOREDCAN), Mr. Angel Gonzalez Madrazo, and the head of the "Leading," Mr. Marcos Perez.

The first session started with invited presentations by Dr. Jae-Hwan Kim of QST in Japan on the current status of advanced neutron multipliers for JA DEMO (Demonstration Power Plant), followed by an overview on analysis methods of beryllium used in magnetic-confined fusion devices delivered by Professor Marek Rubel of KTH in Sweden. The "Fabrication & Recycling of Beryllium and Beryllides" session showcased innovative approaches, sharing new process ideas on beryllium extraction from beryl ore, which have the potential to provide a new commercial source for lower cost fusion-grade beryllium, also with ultra-low uranium content. New concepts on recycling for both irradiated and residual beryllium and beryllides were also presented.

The next session addressed the challenges of molten salt breeders, specifically FLiBe (lithium-beryllium-fluoride), being actively considered by multiple fusion start-up companies. The last session of the first day on "Mechanical Properties & Irradiation Damage" provided valuable insights into screening tests for thermo-mechanical properties of beryllium pebble beds, a form of the material that would be used as a neutron multiplier in a solid-breeder blanket design. That information was presented by Dr. Jörg Reimann of KIT in Germany and also included data on the mechanical properties of beryllide intermetallic compounds. The day was concluded with the BeWS-16 conference dinner at the Eurostars Hotel Real in Santander.

The second day of the workshop kicked off with a presentation on the capabilities of the local organizers and host, "Leading Metal Mechanic Solutions", and it was given by Mr. Ignacio Cobo, a senior manager at Leading. Talks in the "Irradiation Damage & He-Implantation" session included those by Drs. Slava Kuksenko of UKAEA and Yutaka Sugimoto of QST. Their presentations focused on the investigation of radiation-damage effects in beryllium.

The Modeling session, featuring *ab initio* simulations of helium and hydrogen behavior in beryllides, provided valuable modeling insights. The closing session was devoted to "Health & Safety," specifically the design of beryllium-handling and fabrication facilities, surface sampling techniques, methods for sample analysis, and screening tests. Mr. Aniceto Goraieb of KBHF (Karlsruhe Beryllium Handling Facility) in Germany, concluded the workshop with a presentation on the BeYOND Industrial Forum, announcing that the focus of future meetings would be expanded to include discussion on public-private partnerships (PPP) in fusion.

The BeWS-16 International Organizing Committee (IOC) members voted to recognize three outstanding contributors to the workshop series. Lifetime achievement awards were presented to Dr. Gunte Kizane of the University of Latvia and to Prof. Marek Rubel of KTH in Sweden for their many contributions to beryllium research and their support of the Beryllium Workshops with regular attendance over the span of their long careers in the field. The Mario Dalle Donne Memorial Award (MDDMA) was bestowed upon Dr. Ramil Gaisin of KIT in Germany in recognition of his remarkable achievements in advancing the development and characterization of beryllide intermetallic compounds in the form of full-sized industrially fabricated blocks.

The next Beryllium Workshop (BeWS-17) is planned to be held in the autumn of 2025, most likely in conjunction with ICFRM-22 in Japan, but details are yet to be determined. An announcement confirming dates and location will be made sometime in the latter half of 2024.

SESSION 1

Welcome

1.1 Presentation of the Leading Group Facilities in Spain

Marcos Perez

The presentation of the Leading Group will highlight their key facilities "Leading - Metal Mechanics Solutions (LMMS)" in Spain. This includes the Engineering and R&D Center, Precision Machining Workshop, Special Materials Workshop, and Welding & Assembly Workshop. The Leading Group excels in precision machining, special welding, dimensional control, mold development, and assembly. It also maintains a dedicated clean assembly room. Key markets served by Leading encompass Oil & Gas Meter (30%), Nuclear Industry (30%), Defense Industry (20%), and other areas.

The Beryllium Machining & Assembly facility stands out as the most modern in Europe, leveraging extensive Be-experience since 2010 and specializing in the manufacturing of UHV and Be-VC. The workshop boasts large-sized machining capabilities and prioritizes safety with three levels of protection. The workshop is fully isolated, and sealed. All operators are required to wear respirators, ensuring a secure working environment.

Success cases highlighted in the presentation include:

- the production of beryllium tiles for the ITER First Wall panels (440 shield blanket modules with a total surface of 680 m^2),
- the creation of a beryllium reflector for the ISIS Neutron and Muon Source at the Rutherford Appleton Laboratory, United Kingdom and
- cassettes for the target wheel of the European Spallation Source (ESS) in Lund, Sweden.

Presenting Author: Marcos Perez
Email: mperez@leading.es

LEADING
METAL-MECHANIC SOLUTIONS

LEADING GROUP

16th INTERNATIONAL WORKSHOP ON BERYLLIUM TECHNOLOGY
Santander, Cantabria, Spain
19-20 October 2023

LEADING
METAL-MECHANIC SOLUTIONS

The Leading Group
Presentation

LEADING GROUP
Facilities
5
Leading
Services
3
Special
Materials
Workshop
4
Welding &
Assembly
Workshop
2
Precision
Machining
Workshop
1
Engineering
and R&D
Center

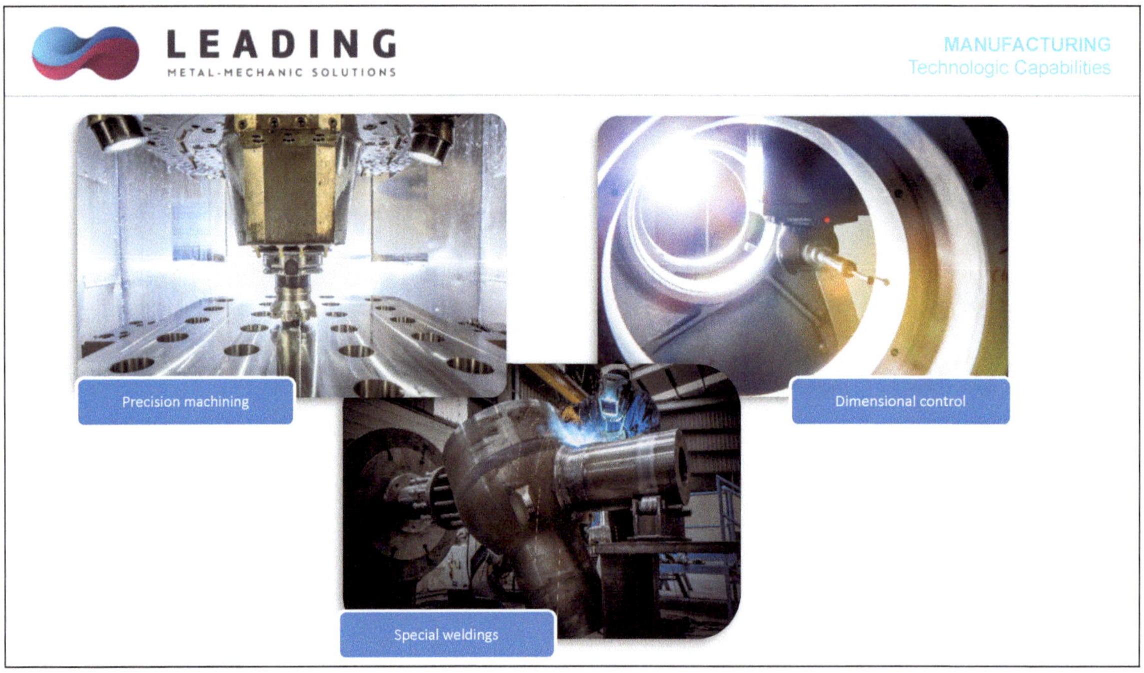
LEADING
METAL-MECHANIC SOLUTIONS
MANUFACTURING
Technologic Capabilities
Precision machining
Dimensional control
Special weldings

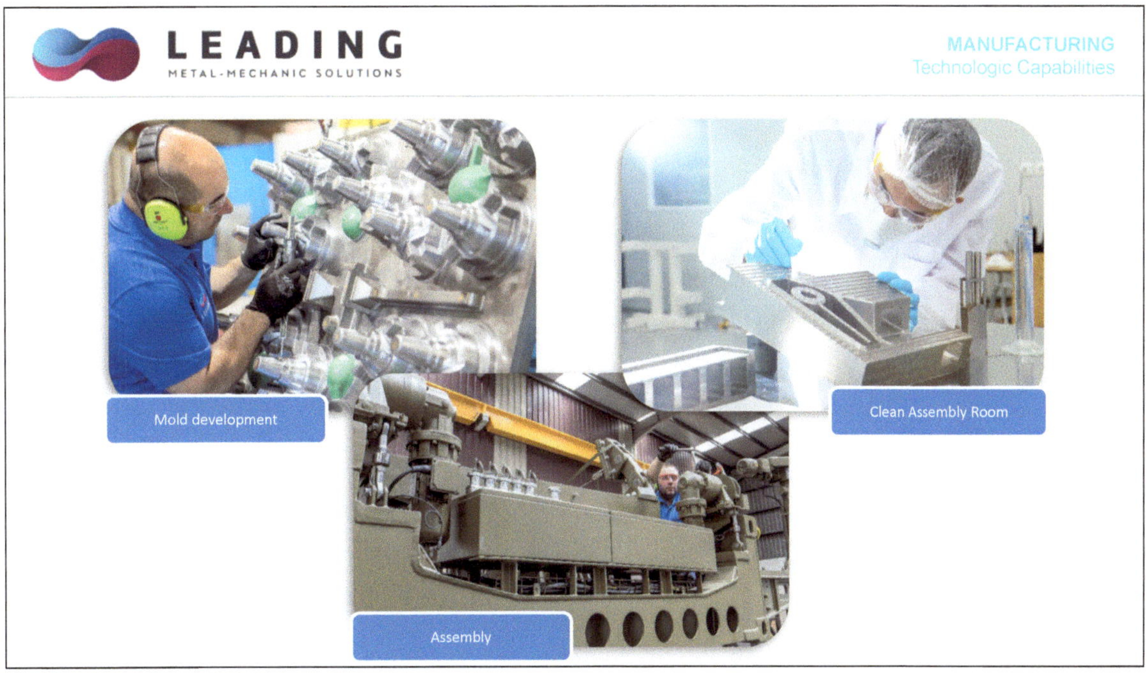
LEADING
METAL-MECHANIC SOLUTIONS
MANUFACTURING
Technologic Capabilities
Mold development
Clean Assembly Room
Assembly

LEADING
METAL-MECHANIC SOLUTIONS
LEADING´s KEY MARKETS
Oil & Gas (Gas meters)
Single chamber
Dual chamber
Meter runs (meter and piping)
30% ( )
Pumps & Valves
Pumps and pump components
5%
Railway Industry
Shaft
Manufactured gear
Engine casing
5%
Defense Industry
Navy components
20% ( )
6

LEADING
METAL-MECHANIC SOLUTIONS
LEADING´s KEY MARKETS
Nuclear Industry
(Fusion and Fission)
Special ring
Shaft
Subcomponent
30% (⬆)
Elevator Industry
Pulley
Gear box
Lift brakes
5%
Heavy Vehicles & Automotion
Chassis
Collector for power engines
Complete fabrication
5%
7

LEADING
METAL-MECHANIC SOLUTIONS
LEADING´s REFERENCES
Relevant clients
Haulotte
CATERPILLAR
HAYWARD TYLER
FMC Technologies
Schindler
DRESSER-RAND
Navantia
SAPA PLACENCIA
Atlas Copco
FUSION FOR ENERGY
iter
VECTOR
SIEMENS
ensa
PSA GROUPE
ESS bilbao
ALSTOM
AREVA
AIRBUS
ess EUROPEAN SPALLATION SOURCE
nexter SYSTEMS
FAURE HERMAN
Snecma SAFRAN Group
BRIDGESTONE
renfe
RENAULT
NISSAN

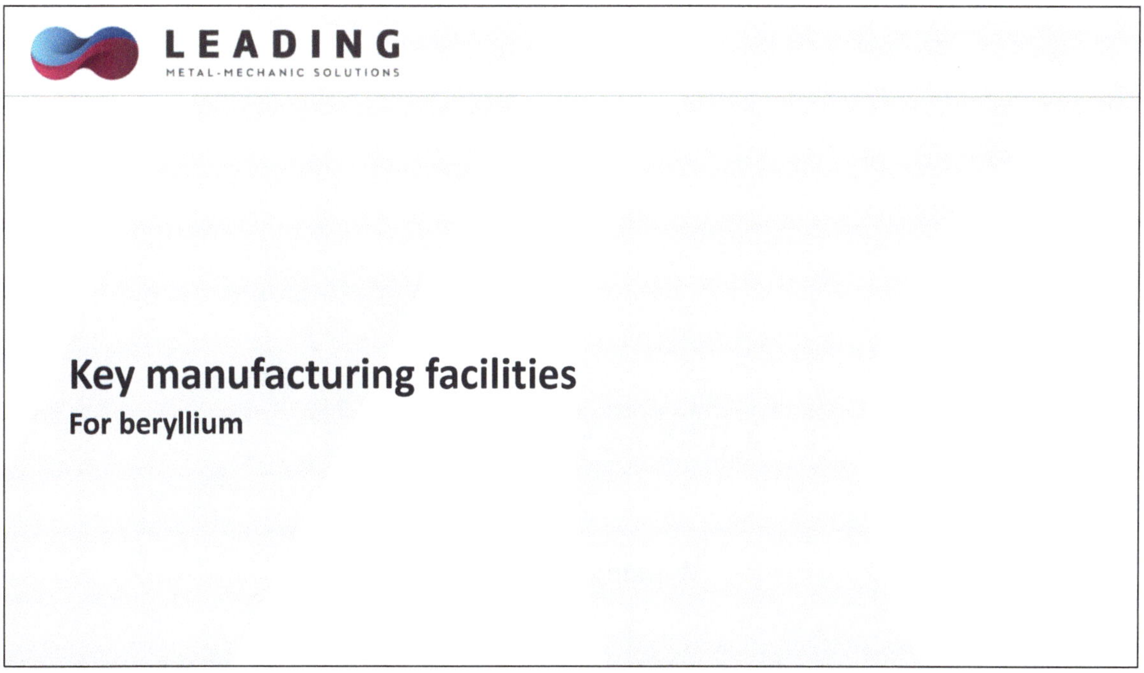
LEADING
METAL-MECHANIC SOLUTIONS
Key manufacturing facilities
For beryllium

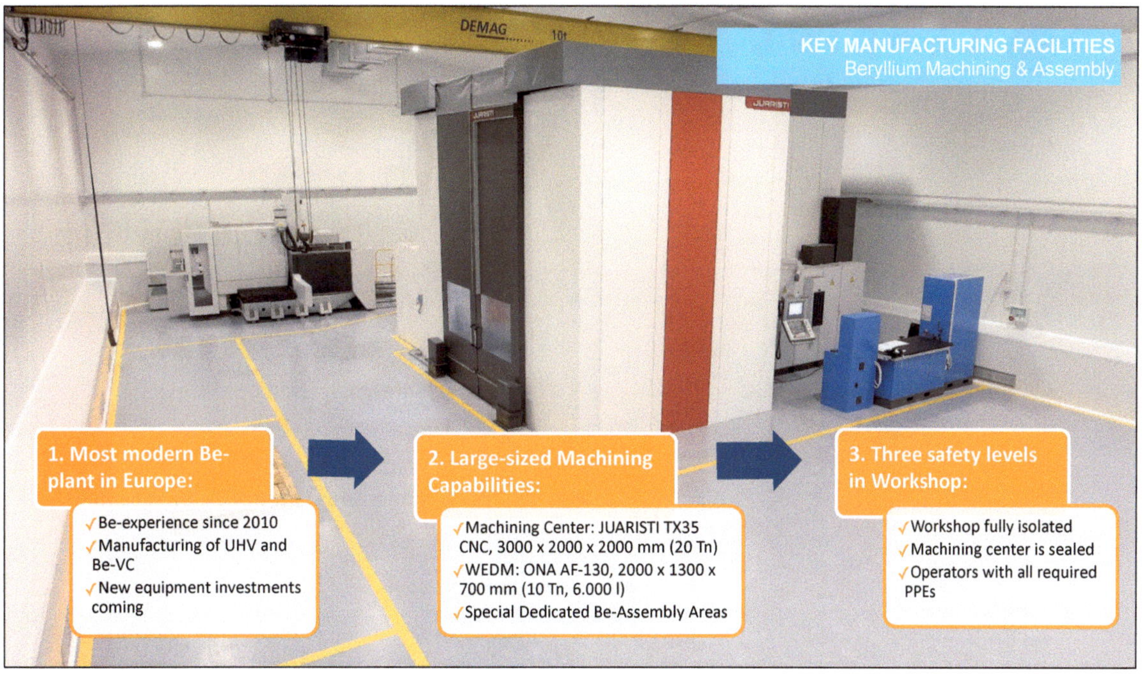
KEY MANUFACTURING FACILITIES
Beryllium Machining & Assembly
1. Most modern Be-plant in Europe:
✓Be-experience since 2010
✓Manufacturing of UHV and Be-VC
✓New equipment investments coming
2. Large-sized Machining Capabilities:
✓Machining Center: JUARISTI TX35 CNC, 3000 x 2000 x 2000 mm (20 Tn)
✓WEDM: ONA AF-130, 2000 x 1300 x 700 mm (10 Tn, 6.000 l)
✓Special Dedicated Be-Assembly Areas
3. Three safety levels in Workshop:
✓Workshop fully isolated
✓Machining center is sealed
✓Operators with all required PPEs
DEMAG
10t
JUARISTI

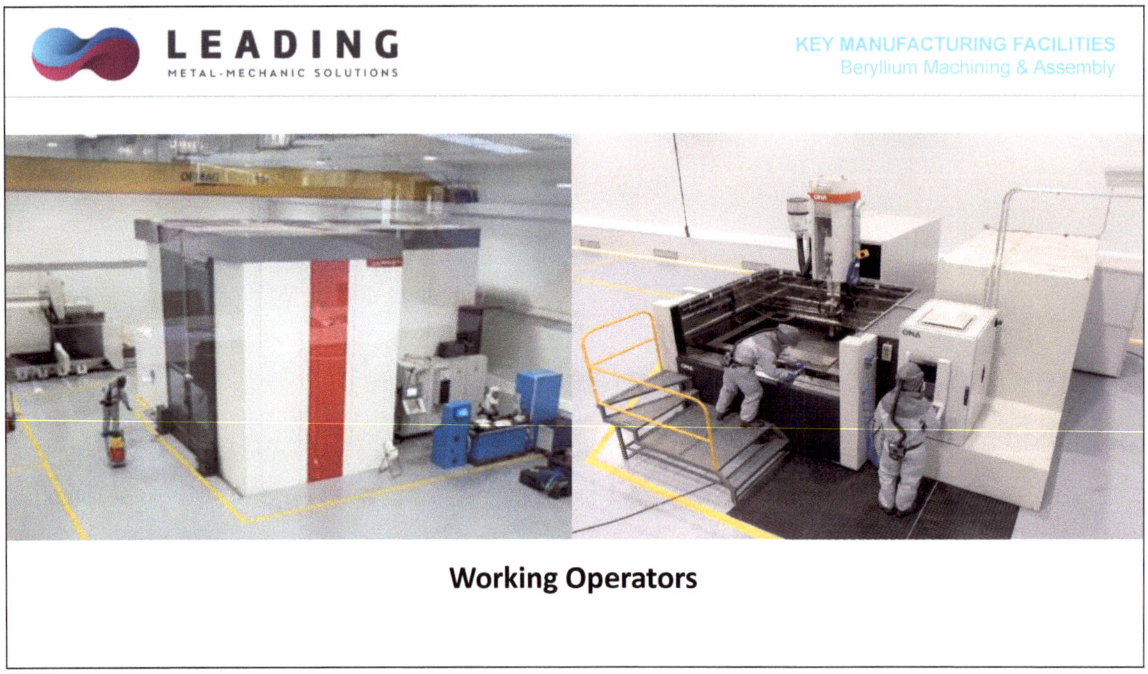
LEADING
METAL-MECHANIC SOLUTIONS
KEY MANUFACTURING FACILITIES
Beryllium Machining & Assembly
Working Operators

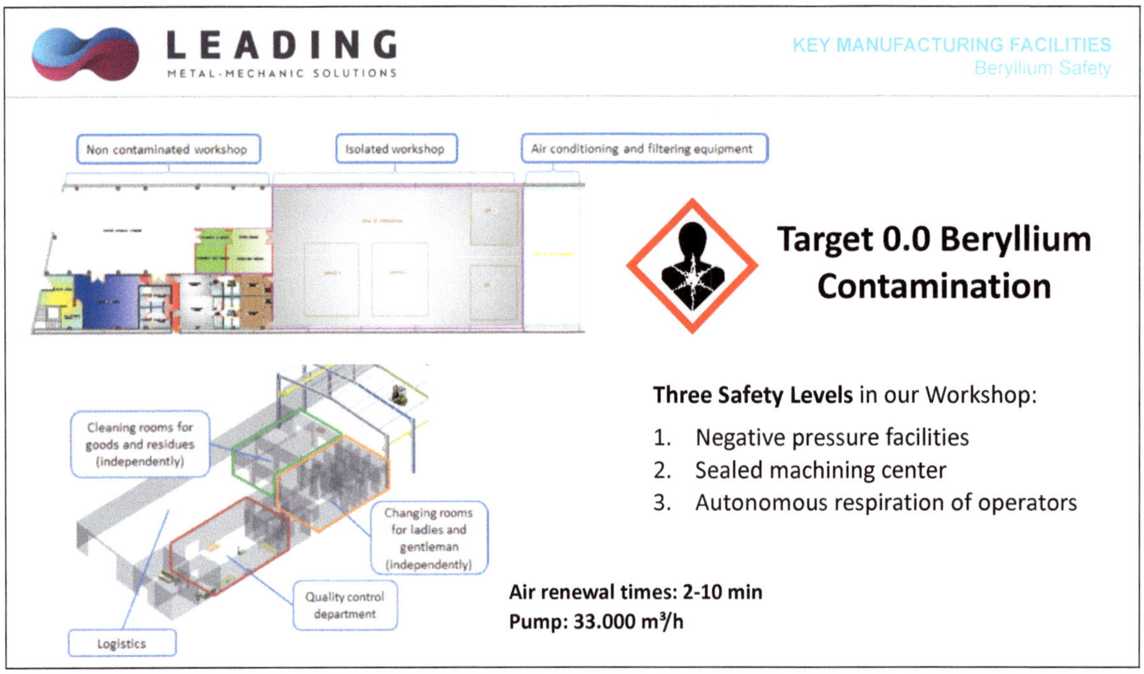
LEADING
METAL-MECHANIC SOLUTIONS
KEY MANUFACTURING FACILITIES
Beryllium Safety
Non contaminated workshop
Isolated workshop
Air conditioning and filtering equipment
Target 0.0 Beryllium Contamination
Cleaning rooms for goods and residues (independently)
Changing rooms for ladies and gentleman (independently)
Quality control department
Logistics
Three Safety Levels in our Workshop:
1. Negative pressure facilities
2. Sealed machining center
3. Autonomous respiration of operators
Air renewal times: 2-10 min
Pump: 33.000 m³/h

LEADING
METAL-MECHANIC SOLUTIONS

KEY MANUFACTURING FACILITIES
Clean Assembly Rooms

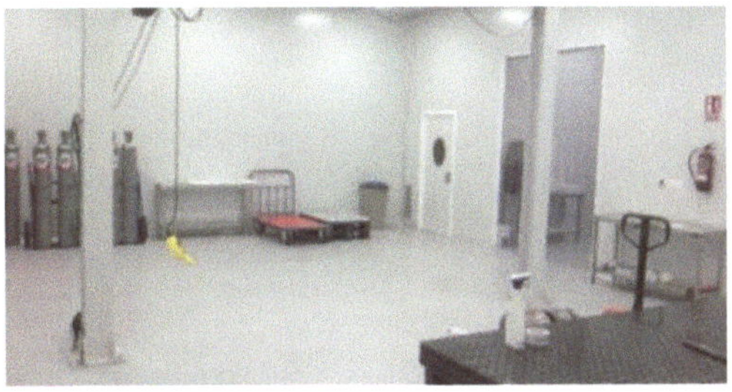

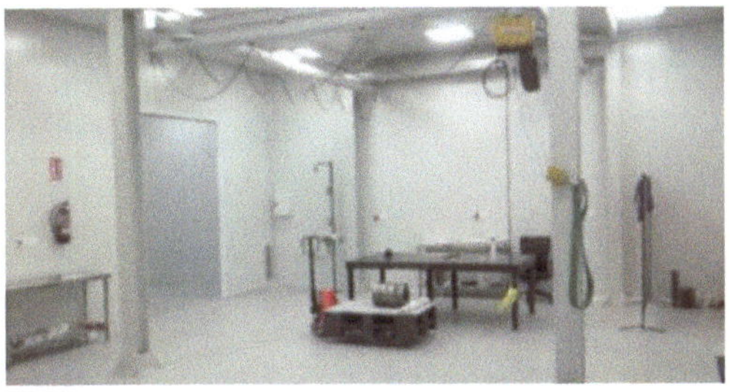

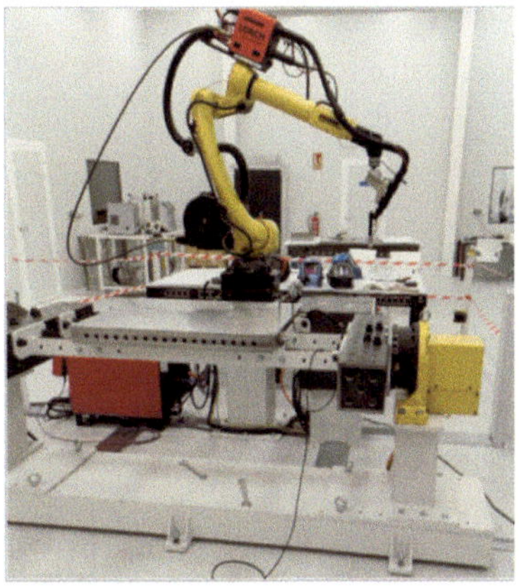

CLEAN ASSEMBLY ROOMs

- ✓ 3 x 240 m2 clean rooms
- ✓ Class 8 under ISO 146441
- ✓ Class 100.000 under Federal Standard 209D
- ✓ Automated welding equipment

LEADING
METAL-MECHANIC SOLUTIONS

BIG SCIENCE PROJECTS

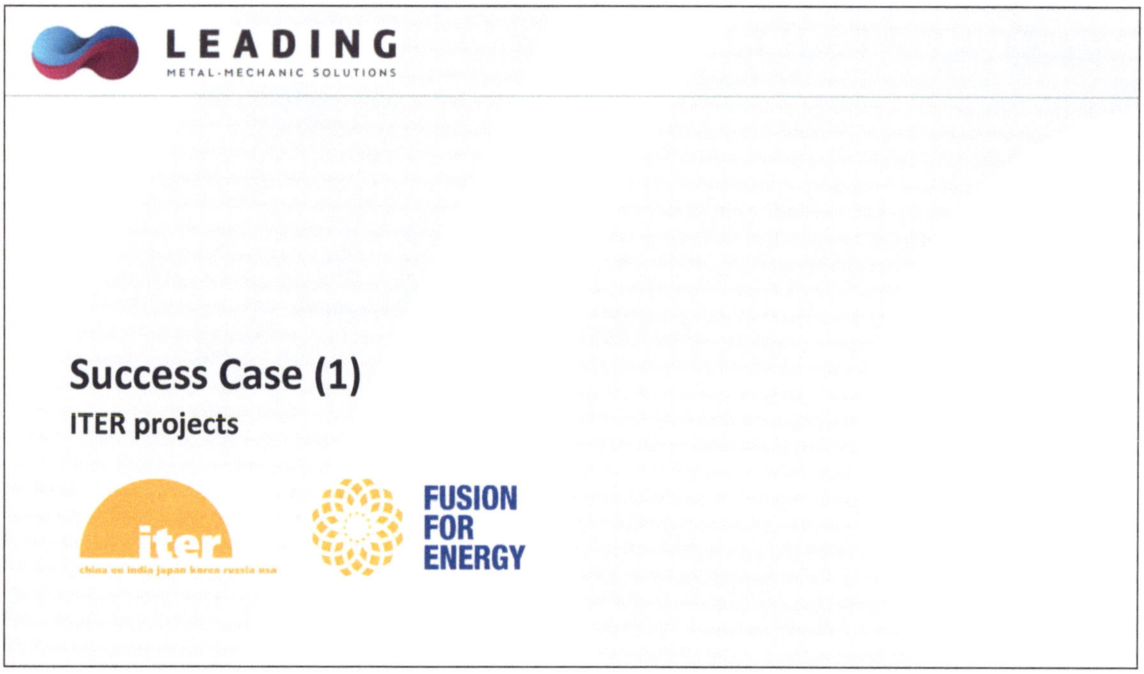
LEADING
METAL-MECHANIC SOLUTIONS
Success Case (1)
ITER projects
iter
china eu india japan korea russia usa
FUSION
FOR
ENERGY

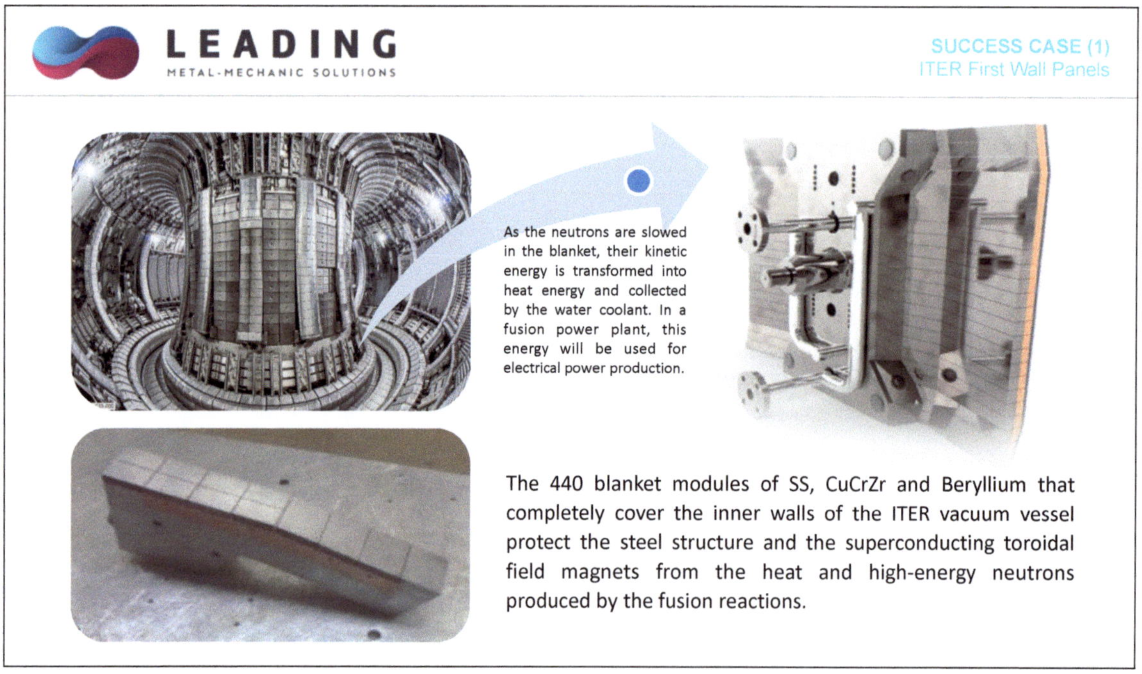
LEADING
METAL-MECHANIC SOLUTIONS
SUCCESS CASE (1)
ITER First Wall Panels
As the neutrons are slowed in the blanket, their kinetic energy is transformed into heat energy and collected by the water coolant. In a fusion power plant, this energy will be used for electrical power production.
The 440 blanket modules of SS, CuCrZr and Beryllium that completely cover the inner walls of the ITER vacuum vessel protect the steel structure and the superconducting toroidal field magnets from the heat and high-energy neutrons produced by the fusion reactions.

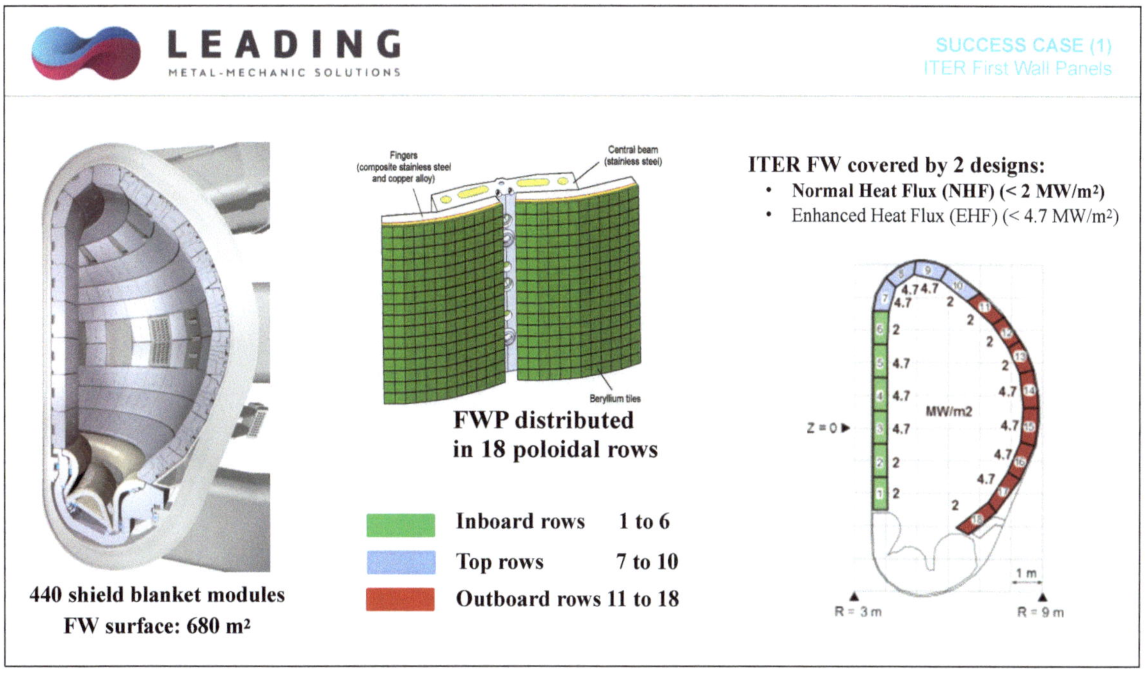
LEADING
METAL-MECHANIC SOLUTIONS
SUCCESS CASE (1)
ITER First Wall Panels
Fingers
(composite stainless steel
and copper alloy)
Central beam
(stainless steel)
Beryllium tiles
FWP distributed
in 18 poloidal rows
Inboard rows 1 to 6
Top rows 7 to 10
Outboard rows 11 to 18
440 shield blanket modules
FW surface: 680 m²
ITER FW covered by 2 designs:
• Normal Heat Flux (NHF) (< 2 MW/m²)
• Enhanced Heat Flux (EHF) (< 4.7 MW/m²)
MW/m2
Z = 0
1 m
R = 3 m
R = 9 m

LEADING
METAL-MECHANIC SOLUTIONS
SUCCESS CASE (1)
ITER First Wall Panels
Machining components of structure/canning and cooling system.
Assembly/Welding the components in the cleaning room.
Canning preparation for Hot Isostatic Pressure Diffusion Bonding of components.
Machining the canning; Semi-prototype Be/ SS / CuCrZr
Manufacturing of the ITER First Wall Panel Semi-Prototype and Full-Scale Prototype
The ITER FIRST WALL PANELS (FWP) are the mechanical components in the ITER Fusion Reactor which aim is confining the plasma in the TOKAMAK (torus shape). They are the first barrier against the fusion nuclear reaction and stand very high temperatures and intensive levels of radiation.
EU must deliver to IO 240 FWP consisting of metal plates of ITER grade stainless steel, copper alloy and beryllium.

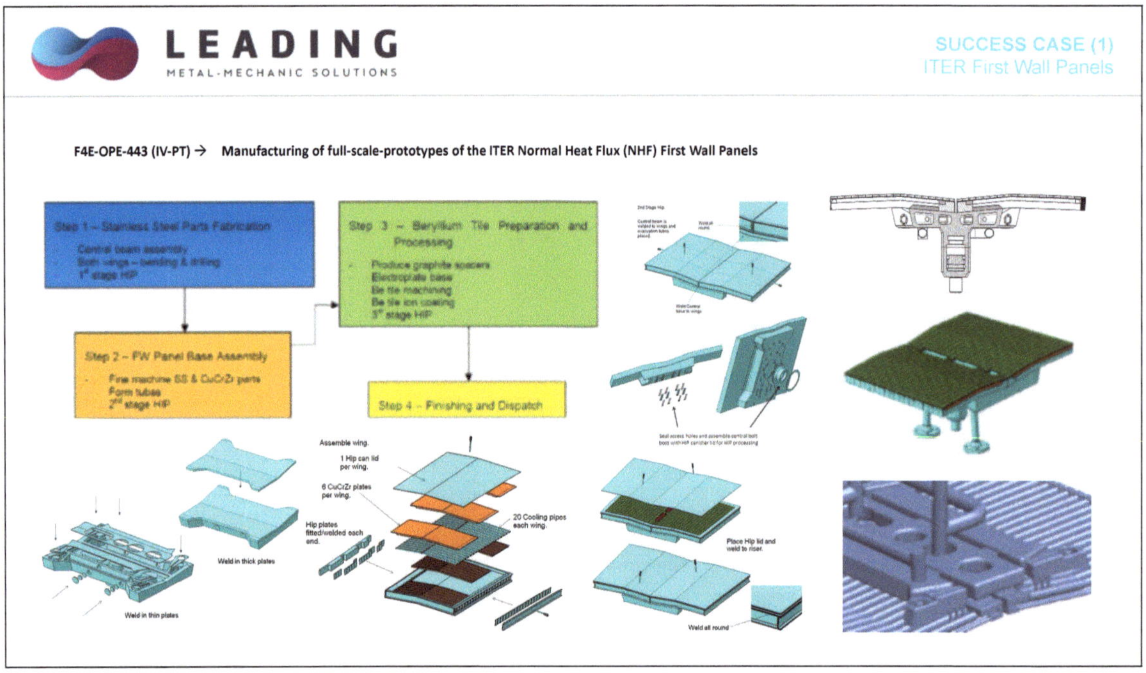
LEADING
METAL-MECHANIC SOLUTIONS
SUCCESS CASE (1)
ITER First Wall Panels
F4E-OPE-443 (IV-PT) → Manufacturing of full-scale-prototypes of the ITER Normal Heat Flux (NHF) First Wall Panels
Step 1 – Stainless Steel Parts Fabrication
Step 2 – FW Panel Base Assembly
Fine machine SS & CuCrZr parts
Form tubes
2nd stage HIP
Step 3 – Beryllium Tile Preparation and Processing
Produce graphite spacers
Electroplate base
Be tile machining
Be tile ion coating
3rd stage HIP
Step 4 – Finishing and Dispatch
Assemble wing.
1 Hip can lid per wing.
6 CuCrZr plates per wing.
Hip plates fitted/welded each end.
20 Cooling pipes each wing.
Weld in thick plates
Weld in thin plates
Place Hip lid and weld to riser.
Weld all round

LEADING
METAL-MECHANIC SOLUTIONS
SUCCESS CASE (1)
ITER First Wall Panels
SS 316L forging
Initial weight: 930kg
Final weight: 350kg
5 axis machine
CAD CAM
programming
IBERDROLA wood.

LEADING
METAL-MECHANIC SOLUTIONS

Success Case (2)

ISIS Reflector

Science & Technology Facilities Council
ISIS Neutron and Muon Source

LEADING
METAL-MECHANIC SOLUTIONS

SUCCESS CASE (2)
ISIS Neutron and Muon source

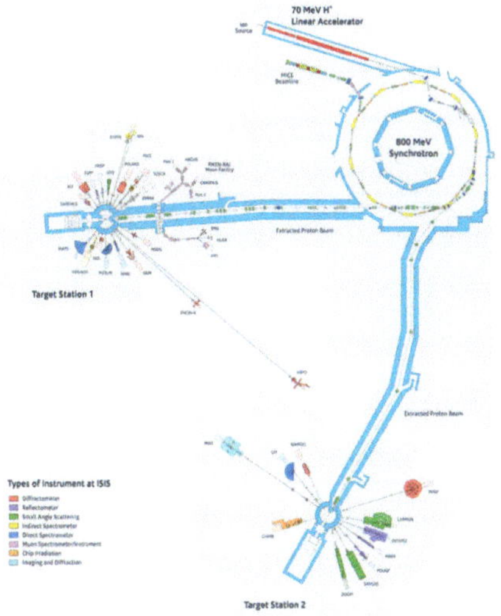

ISIS Target Station 1 Project

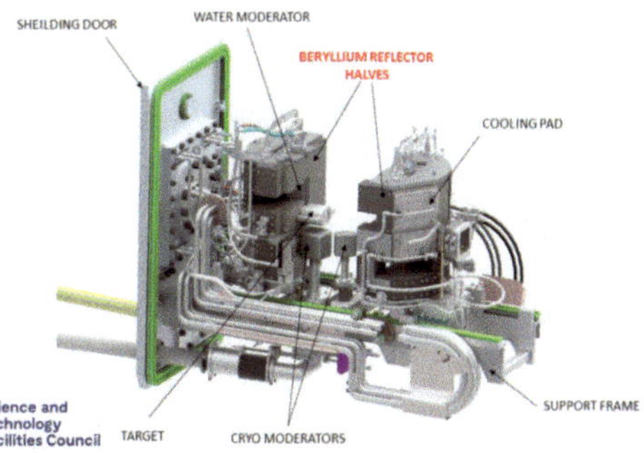

Science and Technology Facilities Council

Target, reflector and moderator structure (TRAM)

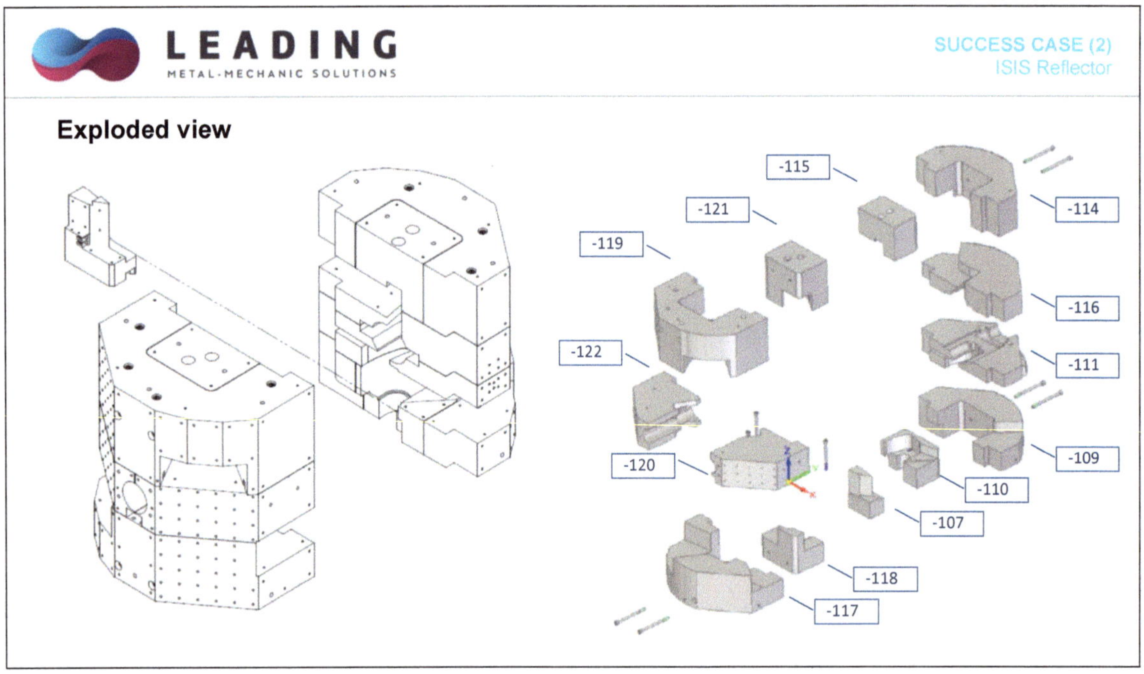
LEADING
METAL-MECHANIC SOLUTIONS
SUCCESS CASE (2)
ISIS Reflector
Exploded view
-115
-121
-119
-114
-116
-122
-111
-109
-120
-110
-107
-118
-117

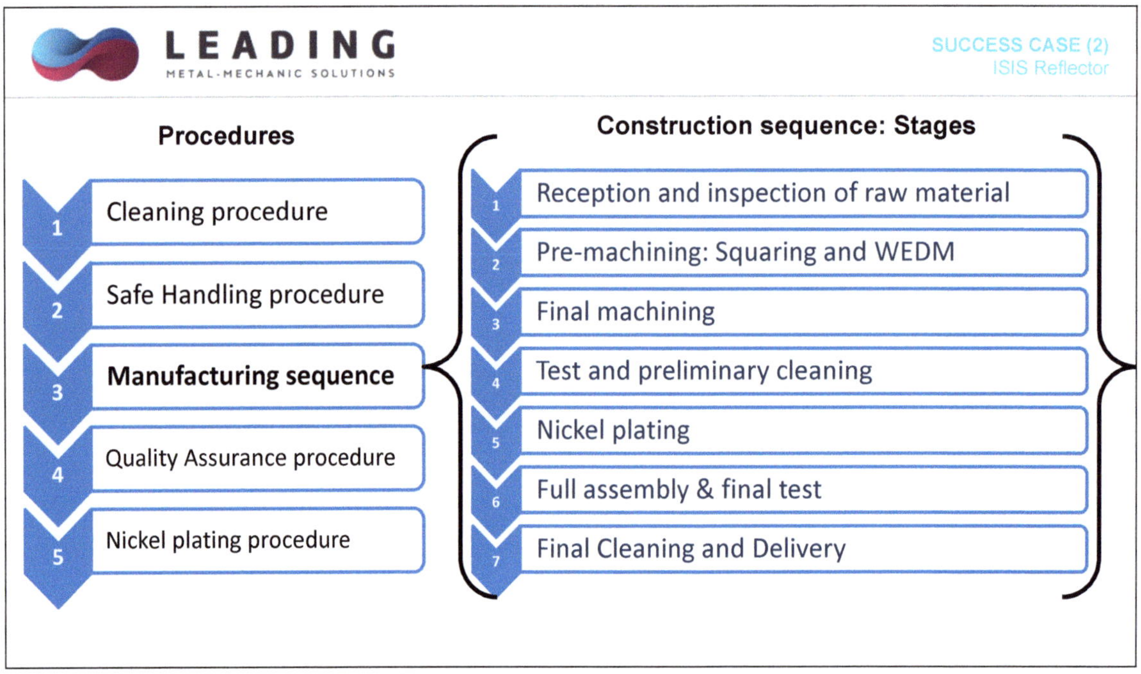
LEADING
METAL-MECHANIC SOLUTIONS
SUCCESS CASE (2)
ISIS Reflector
Procedures
1 Cleaning procedure
2 Safe Handling procedure
3 Manufacturing sequence
4 Quality Assurance procedure
5 Nickel plating procedure
Construction sequence: Stages
1 Reception and inspection of raw material
2 Pre-machining: Squaring and WEDM
3 Final machining
4 Test and preliminary cleaning
5 Nickel plating
6 Full assembly & final test
7 Final Cleaning and Delivery

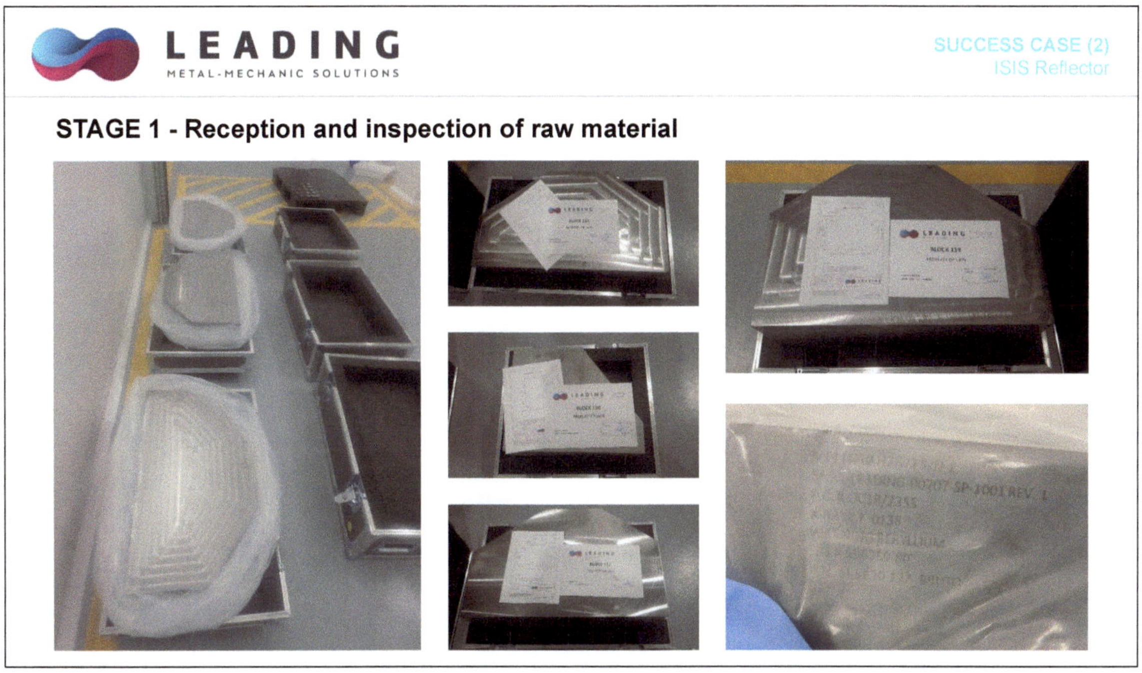
LEADING
METAL-MECHANIC SOLUTIONS
SUCCESS CASE (2)
ISIS Reflector
STAGE 1 - Reception and inspection of raw material

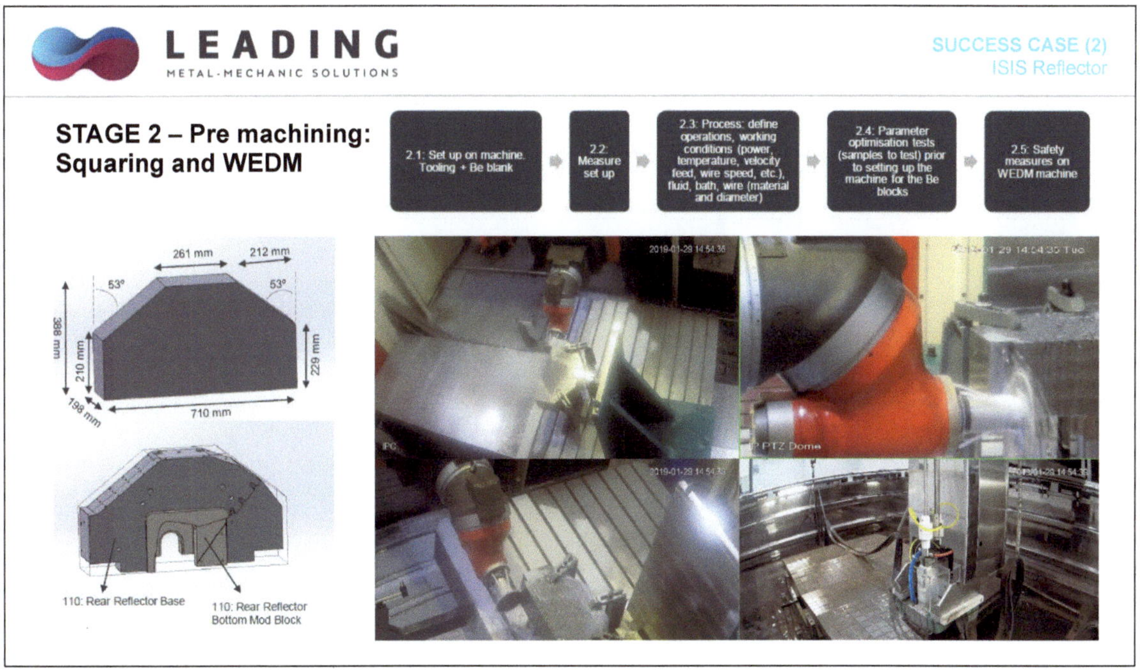
LEADING
METAL-MECHANIC SOLUTIONS
SUCCESS CASE (2)
ISIS Reflector
STAGE 2 – Pre machining: Squaring and WEDM
2.1: Set up on machine. Tooling + Be blank
2.2: Measure set up
2.3: Process: define operations, working conditions (power, temperature, velocity feed, wire speed, etc.), fluid, bath, wire (material and diameter)
2.4: Parameter optimisation tests (samples to test) prior to setting up the machine for the Be blocks
2.5: Safety measures on WEDM machine
261 mm
212 mm
53°
53°
388 mm
210 mm
229 mm
198 mm
710 mm
110: Rear Reflector Base
110: Rear Reflector Bottom Mod Block
2019-01-29 14:54:36

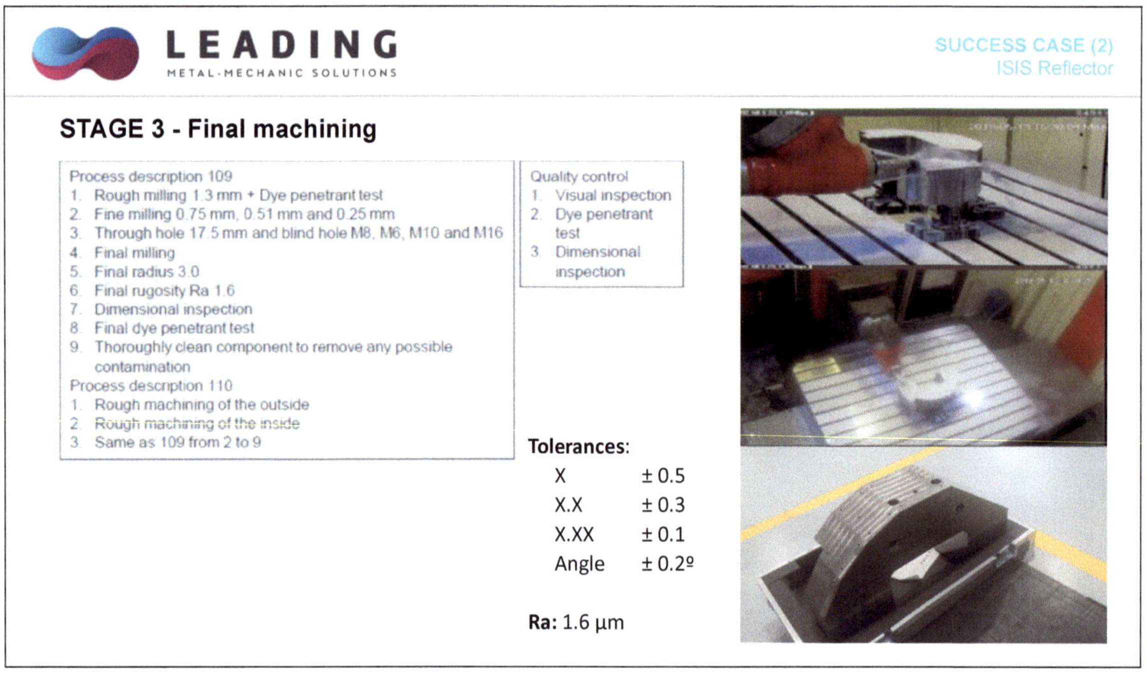
LEADING
METAL-MECHANIC SOLUTIONS
SUCCESS CASE (2)
ISIS Reflector
STAGE 3 - Final machining
Process description 109
1. Rough milling 1.3 mm + Dye penetrant test
2. Fine milling 0.75 mm, 0.51 mm and 0.25 mm
3. Through hole 17.5 mm and blind hole M8, M6, M10 and M16
4. Final milling
5. Final radius 3.0
6. Final rugosity Ra 1.6
7. Dimensional inspection
8. Final dye penetrant test
9. Thoroughly clean component to remove any possible contamination
Process description 110
1. Rough machining of the outside
2. Rough machining of the inside
3. Same as 109 from 2 to 9
Quality control
1. Visual inspection
2. Dye penetrant test
3. Dimensional inspection
Tolerances:
X ± 0.5
X.X ± 0.3
X.XX ± 0.1
Angle ± 0.2º
Ra: 1.6 µm

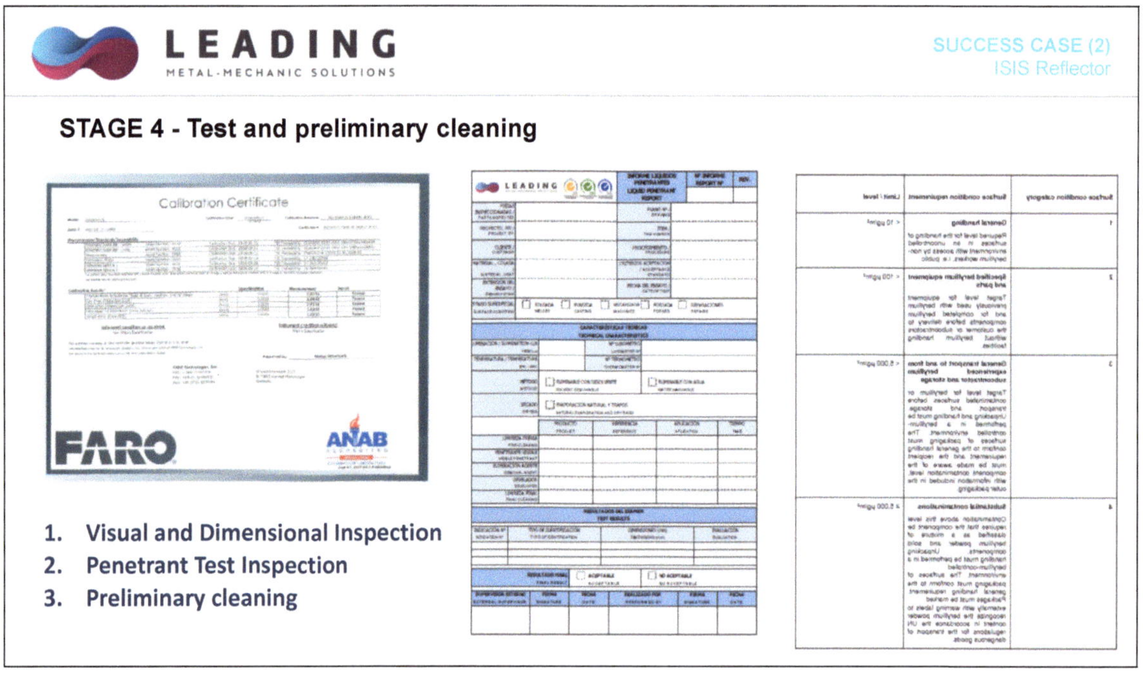
LEADING
METAL-MECHANIC SOLUTIONS
SUCCESS CASE (2)
ISIS Reflector
STAGE 4 - Test and preliminary cleaning
Calibration Certificate
FARO
ANAB
1. Visual and Dimensional Inspection
2. Penetrant Test Inspection
3. Preliminary cleaning

LEADING
METAL-MECHANIC SOLUTIONS

SUCCESS CASE (2)
ISIS Reflector

STAGE 6 – Full assembly & final test

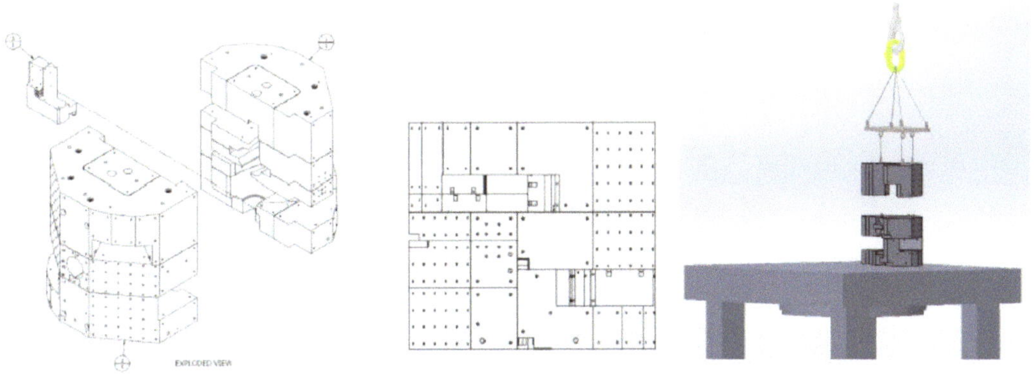

LEADING
METAL-MECHANIC SOLUTIONS

SUCCESS CASE (2)
ISIS Reflector

STAGE 7 – Final cleaning and Delivery

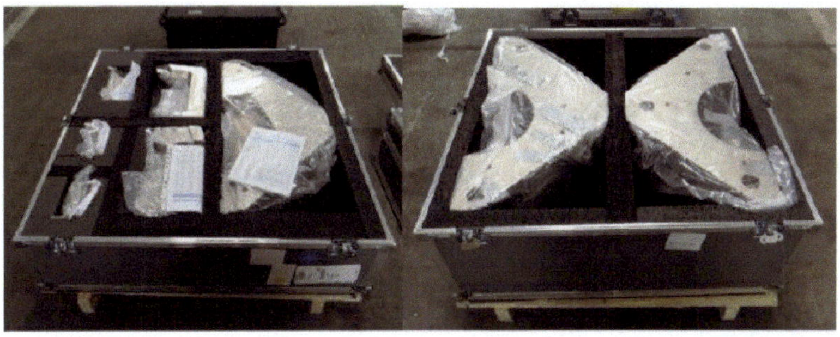

LEADING
METAL-MECHANIC SOLUTIONS

SUCCESS CASE (2)
ISIS Reflector

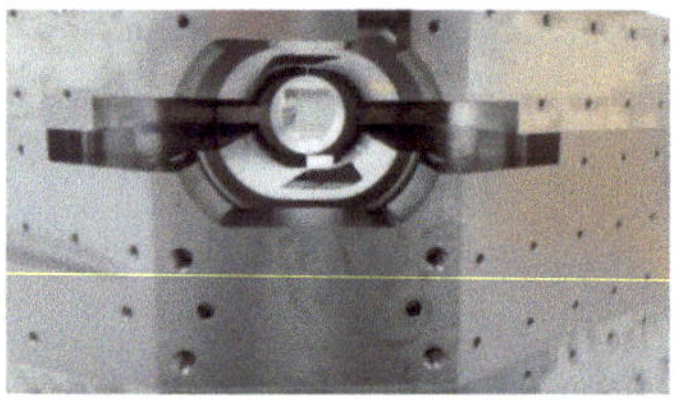

LEADING
METAL-MECHANIC SOLUTIONS

Success Case (3)

ESS Lund

EUROPEAN SPALLATION SOURCE

ESS bilbao

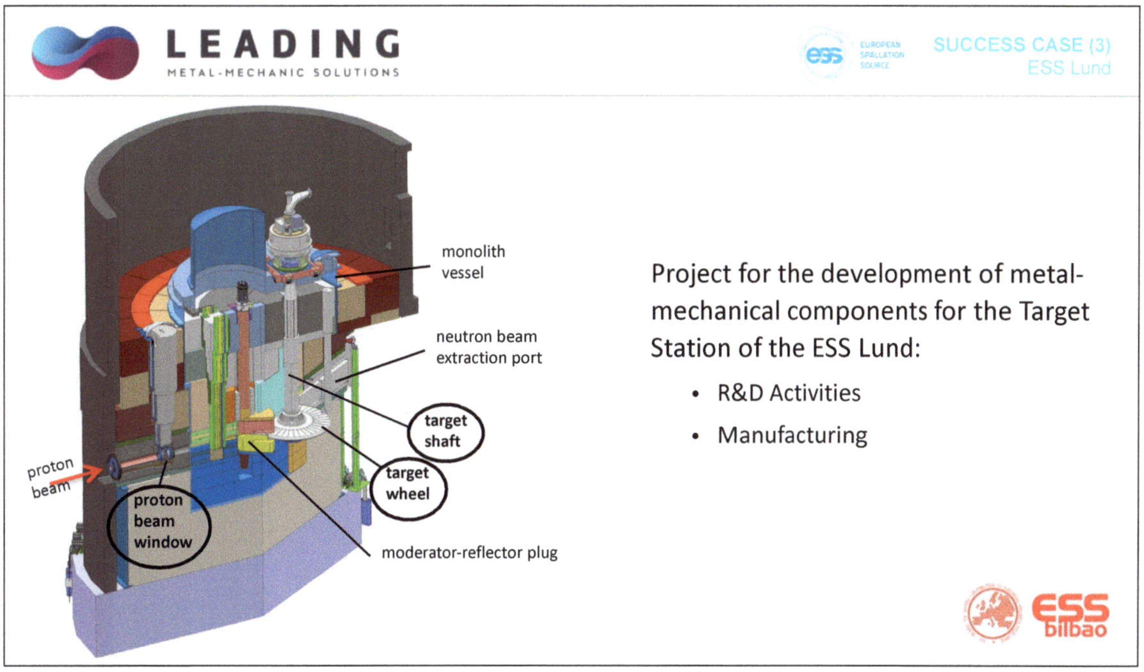
LEADING
METAL-MECHANIC SOLUTIONS
SUCCESS CASE (3)
ESS Lund
monolith vessel
neutron beam extraction port
target shaft
target wheel
proton beam
proton beam window
moderator-reflector plug
Project for the development of metal-mechanical components for the Target Station of the ESS Lund:
• R&D Activities
• Manufacturing
ESS bilbao

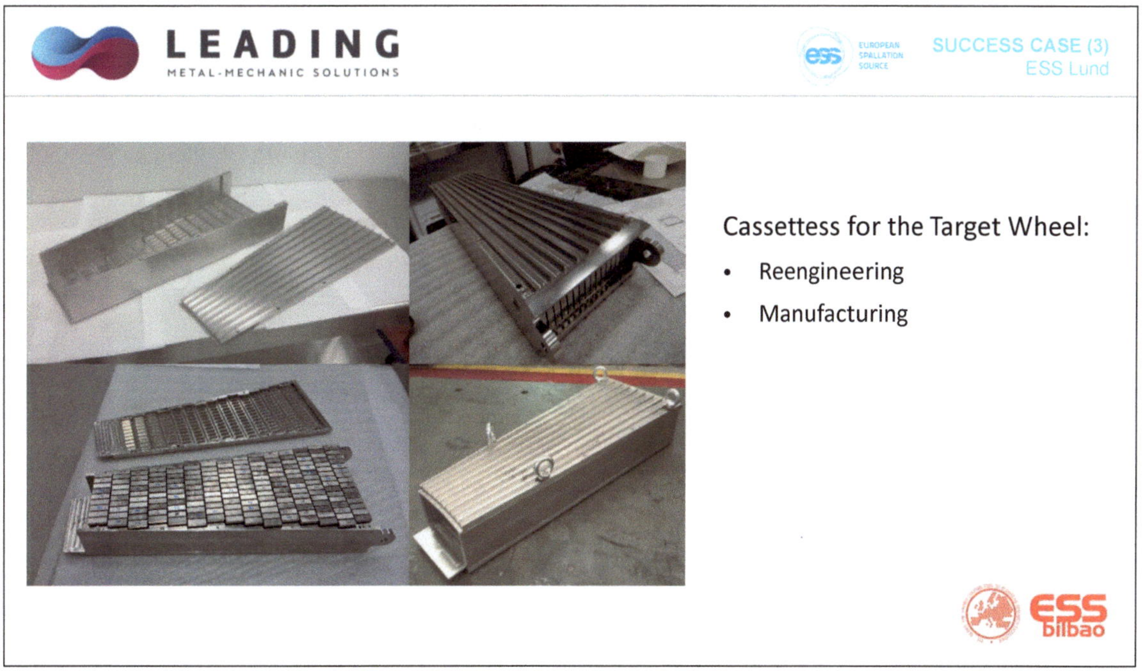
LEADING
METAL-MECHANIC SOLUTIONS
SUCCESS CASE (3)
ESS Lund
Cassettess for the Target Wheel:
• Reengineering
• Manufacturing
ESS bilbao

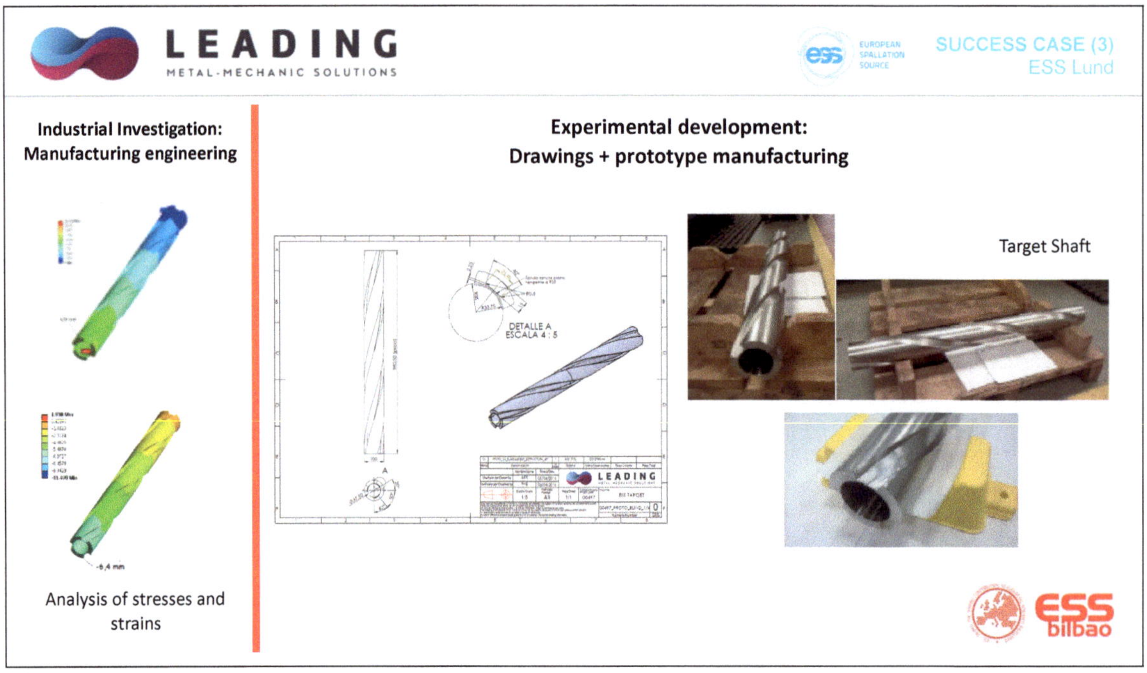
LEADING
METAL-MECHANIC SOLUTIONS
EUROPEAN SPALLATION SOURCE
SUCCESS CASE (3)
ESS Lund
Industrial Investigation:
Manufacturing engineering
Analysis of stresses and strains
Experimental development:
Drawings + prototype manufacturing
DETALLE A
ESCALA 4 : 5
Target Shaft
ESS bilbao

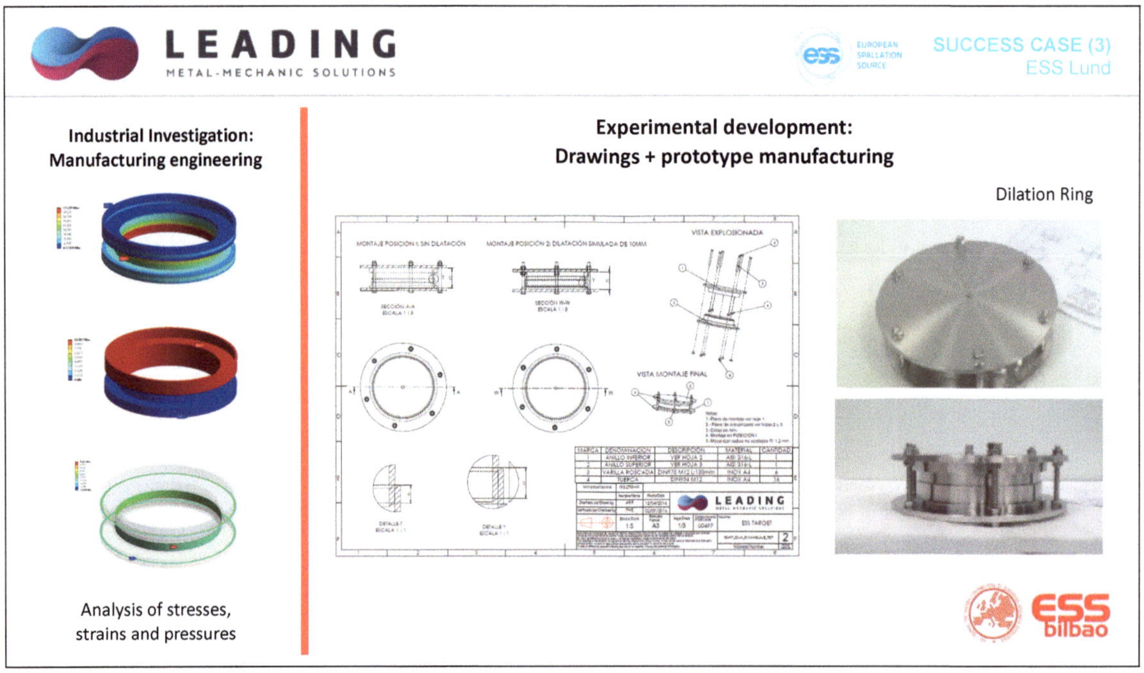
LEADING
METAL-MECHANIC SOLUTIONS
EUROPEAN SPALLATION SOURCE
SUCCESS CASE (3)
ESS Lund
Industrial Investigation:
Manufacturing engineering
Analysis of stresses, strains and pressures
Experimental development:
Drawings + prototype manufacturing
Dilation Ring
VISTA EXPLOSIONADA
VISTA MONTAJE FINAL
ESS bilbao

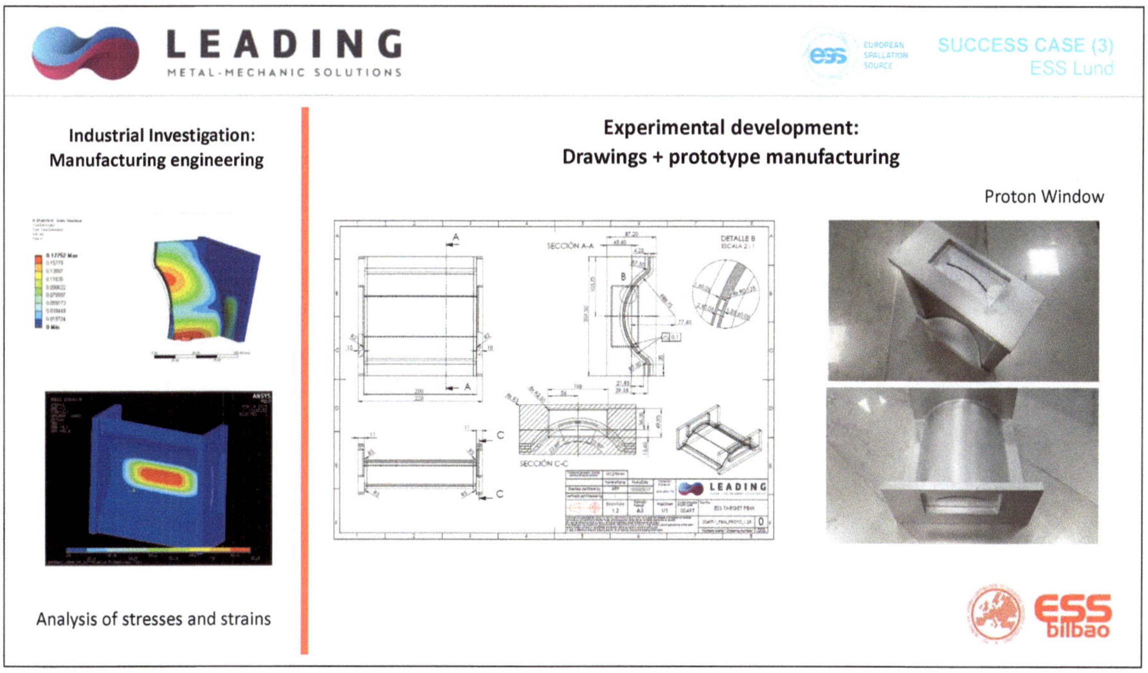
LEADING
METAL-MECHANIC SOLUTIONS
EUROPEAN SPALLATION SOURCE
SUCCESS CASE (3)
ESS Lund
Industrial Investigation:
Manufacturing engineering
Analysis of stresses and strains
Experimental development:
Drawings + prototype manufacturing
Proton Window
ESS bilbao

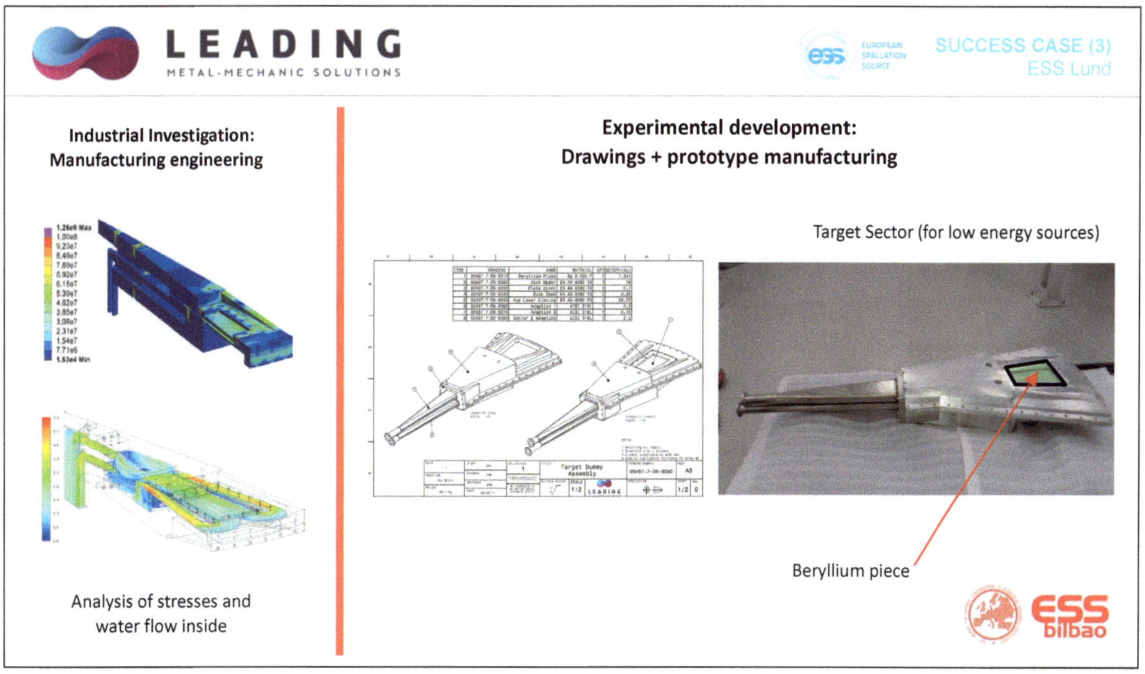
LEADING
METAL-MECHANIC SOLUTIONS
EUROPEAN SPALLATION SOURCE
SUCCESS CASE (3)
ESS Lund
Industrial Investigation:
Manufacturing engineering
Analysis of stresses and
water flow inside
Experimental development:
Drawings + prototype manufacturing
Target Sector (for low energy sources)
Beryllium piece
ESS bilbao

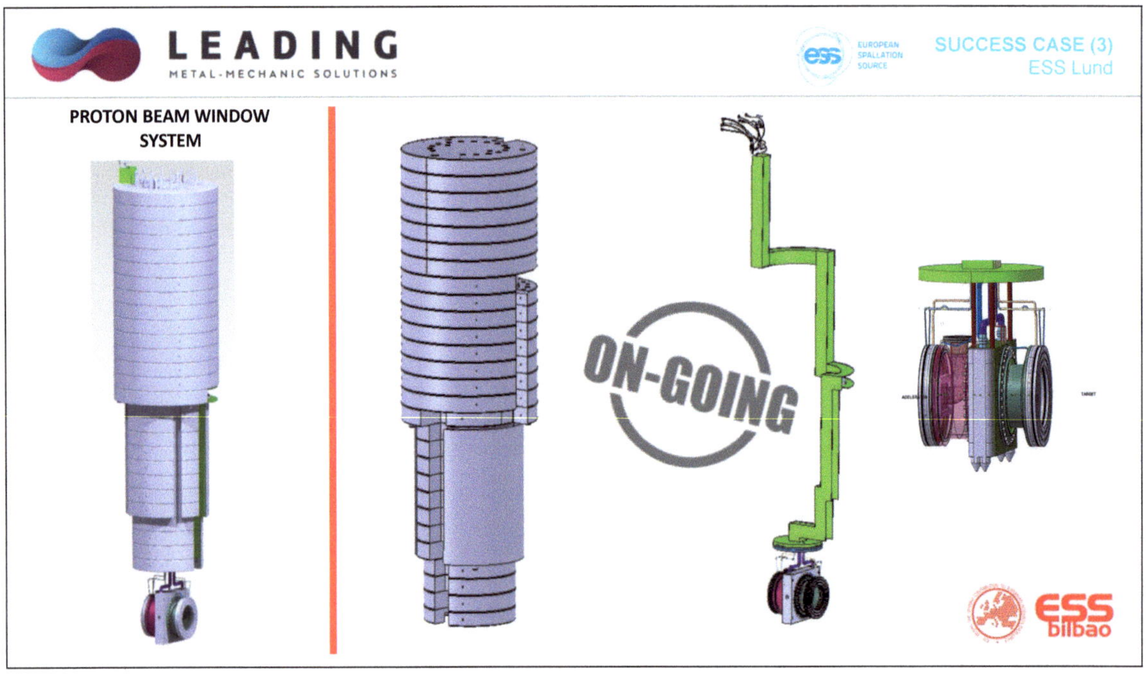
LEADING
METAL-MECHANIC SOLUTIONS
EUROPEAN SPALLATION SOURCE
SUCCESS CASE (3)
ESS Lund
PROTON BEAM WINDOW SYSTEM
ON-GOING
ESS bilbao

LEADING
METAL-MECHANIC SOLUTIONS
Some other projects in Beryllium launched in 2023 for…
INFN Istituto Nazionale di Fisica Nucleare
Imperial College London
ON-GOING

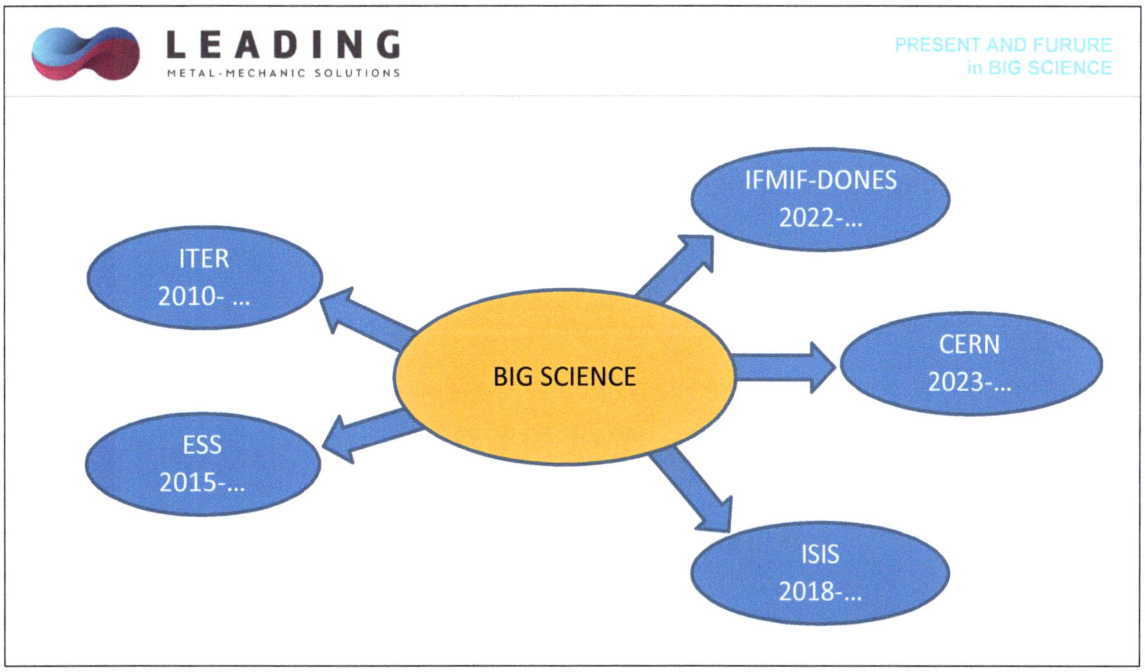
LEADING
METAL-MECHANIC SOLUTIONS
PRESENT AND FURURE
in BIG SCIENCE
IFMIF-DONES
2022-...
ITER
2010- ...
BIG SCIENCE
CERN
2023-...
ESS
2015-...
ISIS
2018-...

LEADING
METAL-MECHANIC SOLUTIONS
INNOVATION AWARD in 2017
Premio QUALITY INNOVATION AWARD 2017
Categoría: Innovación Potencial
cex CENTROS DE EXCELENCIA
QUALITY INNOVATION AWARD
https://www.leading.es/beryllium-leading/presentation.html
SPECIAL MATERIALS MACHINING, S.L.

LEADING
METAL-MECHANIC SOLUTIONS

“ Innovation distinguishes between leaeders and followers”

Steve Jobs

THANKS FOR YOUR ATTENTION

Dr. Ignacio Cobo – R&D Manager- LEADING
icobo@leading.es
+34 647 674 940

SESSION 2

Overview

2.1 Present Status of Advanced Neutron Multipliers for JA DEMO activities

Jae-Hwan Kim, Taehyun Hwang, Yutaka Sugimoto, Shota Yokohama, Yoshiaki Akatsu, Suguru Nakano, Masaru Nakamichi

We have dedicated ourselves to researching and developing granulation process of breeding functional materials for tritium breeders and neutron multipliers over the past several years. Through trial and error, beryllium intermetallic compounds (beryllides) as 1 mm-pebble, have been successfully fabricated by the combination of a plasma sintering process and a rotating electrode process. In addition, the optimization for the granulation process has been carried out to improve the yield and the specific properties.

Many kinds of beryllide pebbles have been successfully fabricated by the same process to search for the best chemical composition of beryllides and diverse characteristics, reactivity to water vapor, swelling properties, hydrogen retention, compatibility with other materials, and pebble bed properties, have been clarified. Those diverse experimental and comparative results have shown that beryllides pebbles are much better properties than beryllium pebbles and the final candidates of beryllides are $Be_{12}V$ and Be-V-Ti ternary ($Be_{12}V_{0.9}Ti_{0.1}$) pebbles.

For the optimal design of solid breeding blanket, blanket designs using beryllide block were suggested. To make this design a reality, the strength of beryllides needs to be clarified because there is limited data available on beryllides. In this study, current R&Ds on breeding functional materials in JA DEMO activities, regarding not only pebble fabrication process but also beryllide blocks' mechanical strength. Additionally, the recycling process of these materials will be addressed by applying the innovative refining process of beryllium ore.

Presenting Author: Jae-Hwan Kim
Presenting Author's Affiliation: National Institutes for Quantum and Radiological Science and Technology, Japan
Email: kim.jaehwan@qst.go.jp

The 16th International Workshop on Beryllium Technology (BeWS-16)
2023.10.19-20 Santander,**Spain**

Present Status of Advanced Breeding Functional Materials (QST activities)

Jae-Hwan Kim, Taehyun Hwang, Yutaka Sugimoto, Suguru Nakano, Yoshiaki Akatsu, Shota Yokohama, Masaru Nakamichi

National Institutes for Quantum Science and Technology

QST

Jae-Hwan Kim | Santander, Spain| 19th Oct. 2023 | page (1/29)

International Fusion Energy Research Centre (IFERC)

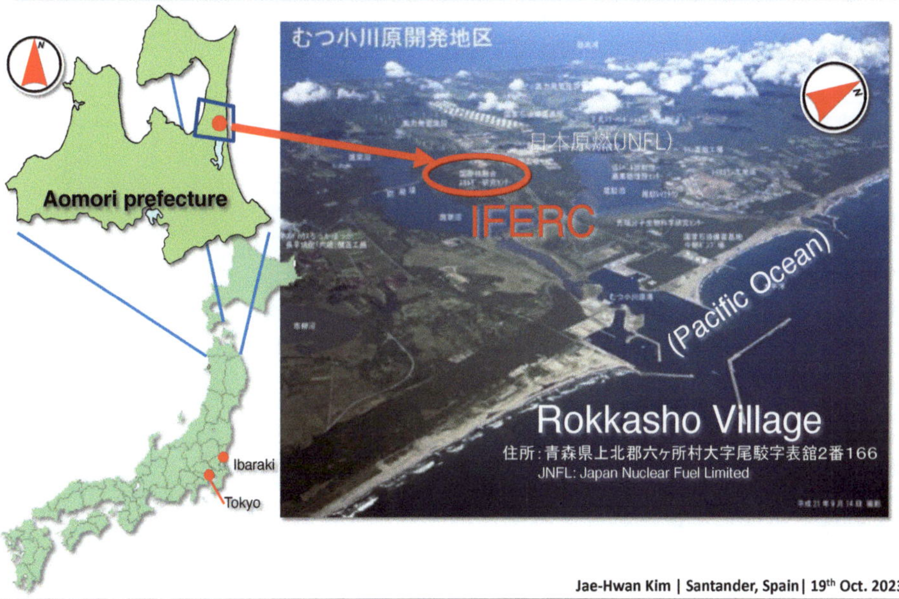

Jae-Hwan Kim | Santander, Spain| 19th Oct. 2023 | page (2/29)

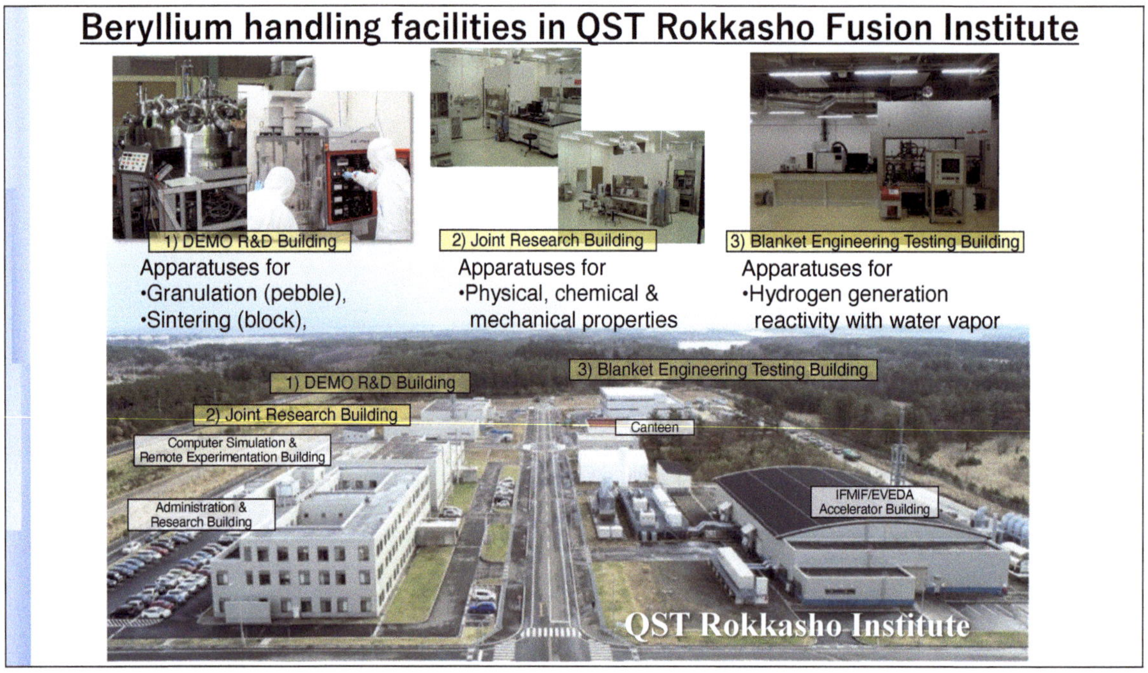
Beryllium handling facilities in QST Rokkasho Fusion Institute
1) DEMO R&D Building
Apparatuses for
•Granulation (pebble),
•Sintering (block),
2) Joint Research Building
Apparatuses for
•Physical, chemical &
mechanical properties
3) Blanket Engineering Testing Building
Apparatuses for
•Hydrogen generation
reactivity with water vapor
1) DEMO R&D Building
3) Blanket Engineering Testing Building
2) Joint Research Building
Canteen
Computer Simulation &
Remote Experimentation Building
Administration &
Research Building
IFMIF/EVEDA
Accelerator Building
QST Rokkasho Institute

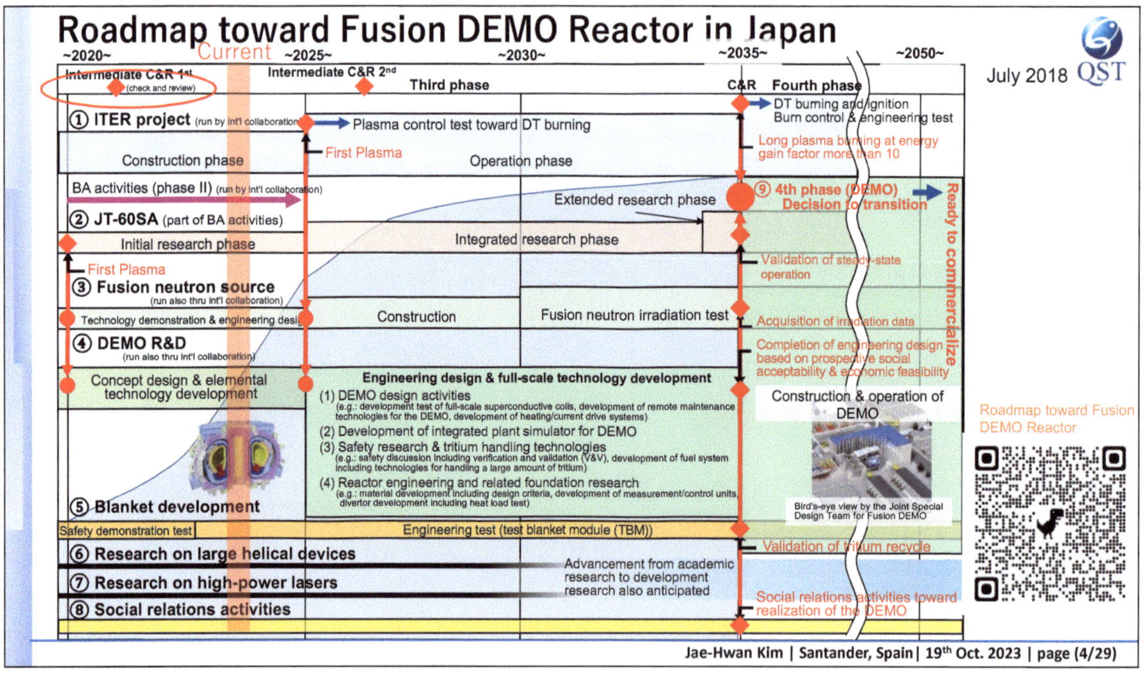
Roadmap toward Fusion DEMO Reactor in Japan
July 2018
QST
~2020~
Current
~2025~
~2030~
~2035~
~2050~
Intermediate C&R 1st
(check and review)
Intermediate C&R 2nd
Third phase
C&R
Fourth phase
① ITER project (run by int'l collaboration)
Plasma control test toward DT burning
DT burning and ignition
Burn control & engineering test
First Plasma
Construction phase
Operation phase
Long plasma burning at energy gain factor more than 10
BA activities (phase II) (run by int'l collaboration)
Extended research phase
⑨ 4th phase (DEMO)
Decision to transition
② JT-60SA (part of BA activities)
Initial research phase
Integrated research phase
Ready to commercialize
First Plasma
③ Fusion neutron source
(run also thru int'l collaboration)
Validation of steady-state operation
Technology demonstration & engineering design
Construction
Fusion neutron irradiation test
Acquisition of irradiation data
④ DEMO R&D
(run also thru int'l collaboration)
Completion of engineering design based on prospective social acceptability & economic feasibility
Concept design & elemental technology development
Engineering design & full-scale technology development
(1) DEMO design activities
(2) Development of integrated plant simulator for DEMO
(3) Safety research & tritium handling technologies
(4) Reactor engineering and related foundation research
Construction & operation of DEMO
Bird's-eye view by the Joint Special Design Team for Fusion DEMO
Roadmap toward Fusion DEMO Reactor
⑤ Blanket development
Safety demonstration test
Engineering test (test blanket module (TBM))
Validation of tritium recycle
⑥ Research on large helical devices
Advancement from academic research to development research also anticipated
⑦ Research on high-power lasers
⑧ Social relations activities
Social relations activities toward realization of the DEMO
Jae-Hwan Kim | Santander, Spain | 19th Oct. 2023 | page (4/29)

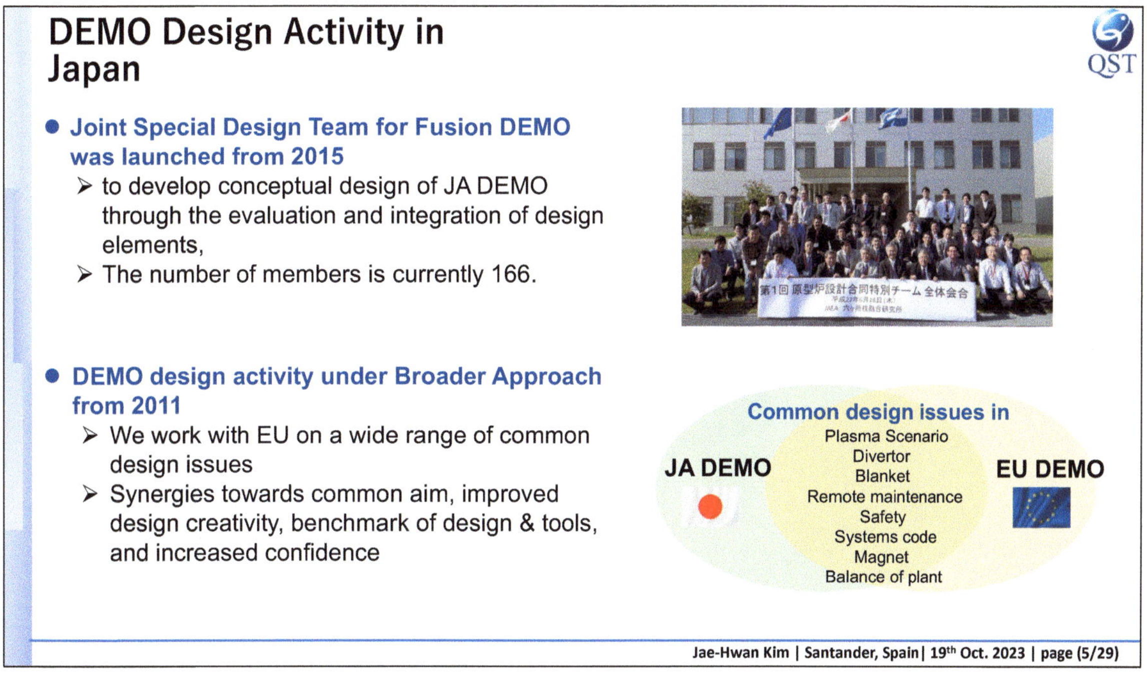
DEMO Design Activity in Japan
QST
Joint Special Design Team for Fusion DEMO was launched from 2015
to develop conceptual design of JA DEMO through the evaluation and integration of design elements,
The number of members is currently 166.
DEMO design activity under Broader Approach from 2011
We work with EU on a wide range of common design issues
Synergies towards common aim, improved design creativity, benchmark of design & tools, and increased confidence
Common design issues in
JA DEMO
EU DEMO
Plasma Scenario
Divertor
Blanket
Remote maintenance
Safety
Systems code
Magnet
Balance of plant
Jae-Hwan Kim | Santander, Spain | 19th Oct. 2023 | page (5/29)

Participating from Industries and Universities
Current situation as of 6th Sept.,2023
QST
30 Industries
166 members
25 Univ. & 4 Inst.
アライドマテリアル
株式会社有沢製作所
KYOTO FUSIONEERING
Helical Fusion
MiRESSO
JX金属
SMO 新むつ小川原株式会社
SUGINO
住友電工
Connect with Innovation
OBAYASHI
住友商事
TOSHIBA
東芝エネルギーシステムズ株式会社
TEPSCO 東電設計
東北環境科学サービス株式会社
ニッコーテクノ株式会社
NIKKO TECHNO Co.,Ltd
JGC 日揮株式会社
日本ガイシ
JANUS
日本エヌ・ユー・エス株式会社
HITACHI
Inspire the Next
古河電工
MIZUHO みずほリサーチ&テクノロジーズ
MITSUI & CO.
三菱重工
三菱電機株式会社
MUFG
三菱UFJリサーチ&コンサルティング
Y-S-E
NAT 株式会社 NAT
KURIHALANT
Skill with Spirit
茨城大学
Ibaraki University
筑波大学
University of Tsukuba
大阪大学
OSAKA UNIVERSITY
核融合科学研究所
九州大学
KYUSHU UNIVERSITY
京都大学
KYOTO UNIVERSITY
近畿大学
KINDAI UNIVERSITY
慶應義塾
Keio University
神戸大学
上智大学
SOPHIA UNIVERSITY
信州大学
静岡大学
摂南大学
IGES
東京大学
THE UNIVERSITY OF TOKYO
東北大学
Tohoku University
鳥取大学
Tottori University
同志社大学
Doshisha University
富山大学
名古屋大学
NAGOYA UNIVERSITY
光産業創成大学院大学
福井大学
物質・材料研究機構
NIMS
QST 量子科学技術研究開発機構
早稲田大学
WASEDA University
島根大学
北海道大学
6

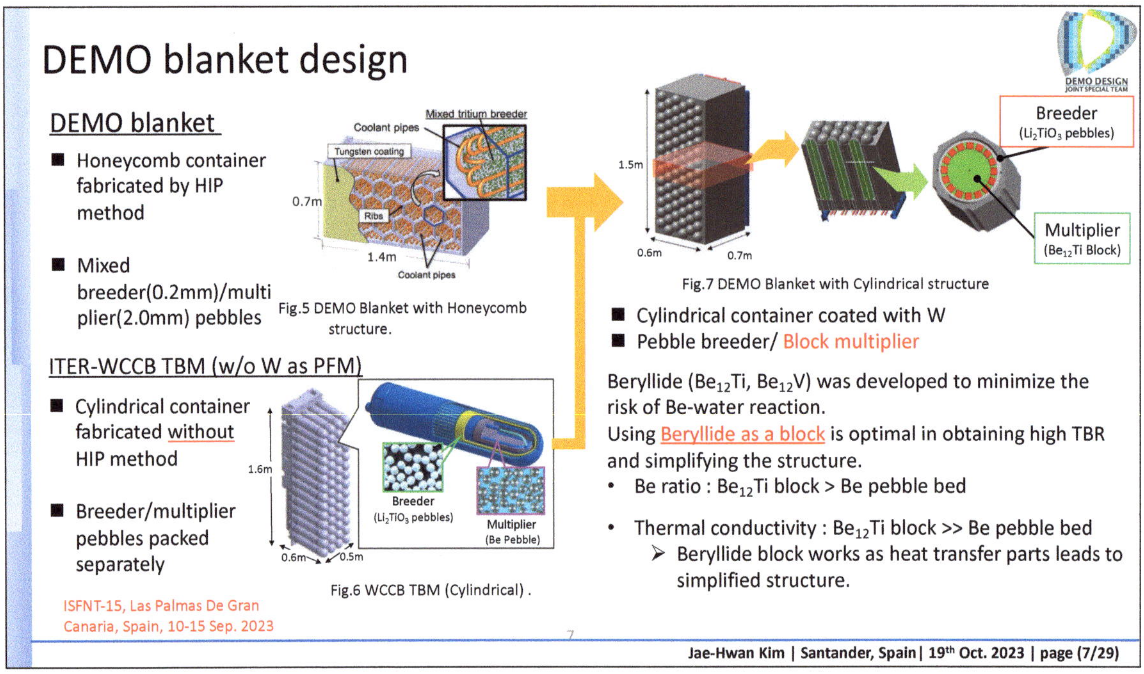
DEMO blanket design
DEMO DESIGN JOINT SPECIAL TEAM
DEMO blanket
■ Honeycomb container fabricated by HIP method
■ Mixed breeder(0.2mm)/multiplier(2.0mm) pebbles
Mixed tritium breeder
Coolant pipes
Tungsten coating
Ribs
0.7m
1.4m
Coolant pipes
Fig.5 DEMO Blanket with Honeycomb structure.
ITER-WCCB TBM (w/o W as PFM)
■ Cylindrical container fabricated without HIP method
■ Breeder/multiplier pebbles packed separately
1.6m
0.6m
0.5m
Breeder (Li_2TiO_3 pebbles)
Multiplier (Be Pebble)
Fig.6 WCCB TBM (Cylindrical) .
ISFNT-15, Las Palmas De Gran Canaria, Spain, 10-15 Sep. 2023
1.5m
0.6m
0.7m
Breeder (Li_2TiO_3 pebbles)
Multiplier ($Be_{12}Ti$ Block)
Fig.7 DEMO Blanket with Cylindrical structure
■ Cylindrical container coated with W
■ Pebble breeder/ Block multiplier
Beryllide ($Be_{12}Ti$, $Be_{12}V$) was developed to minimize the risk of Be-water reaction.
Using Beryllide as a block is optimal in obtaining high TBR and simplifying the structure.
• Be ratio : $Be_{12}Ti$ block > Be pebble bed
• Thermal conductivity : $Be_{12}Ti$ block >> Be pebble bed
➢ Beryllide block works as heat transfer parts leads to simplified structure.
7
Jae-Hwan Kim | Santander, Spain| 19th Oct. 2023 | page (7/29)

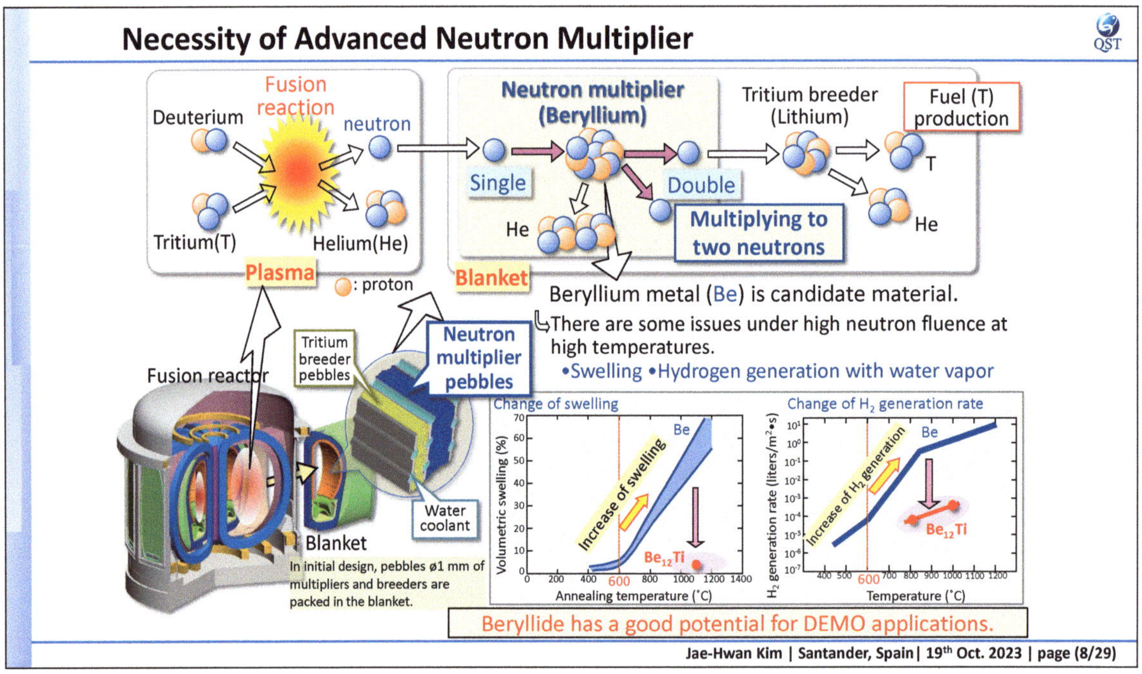
Necessity of Advanced Neutron Multiplier
QST
Fusion reaction
Deuterium
neutron
Tritium(T)
Helium(He)
Plasma
: proton
Neutron multiplier (Beryllium)
Single
Double
He
Multiplying to two neutrons
Tritium breeder (Lithium)
Fuel (T) production
T
He
Blanket
Beryllium metal (Be) is candidate material.
There are some issues under high neutron fluence at high temperatures.
•Swelling •Hydrogen generation with water vapor
Fusion reactor
Tritium breeder pebbles
Neutron multiplier pebbles
Water coolant
Blanket
In initial design, pebbles ø1 mm of multipliers and breeders are packed in the blanket.
Change of swelling
Volumetric swelling (%)
Increase of swelling
Be
$Be_{12}Ti$
Annealing temperature (°C)
Change of H_2 generation rate
H_2 generation rate (liters/m²•s)
Increase of H_2 generation
Be
$Be_{12}Ti$
Temperature (°C)
Beryllide has a good potential for DEMO applications.
Jae-Hwan Kim | Santander, Spain| 19th Oct. 2023 | page (8/29)

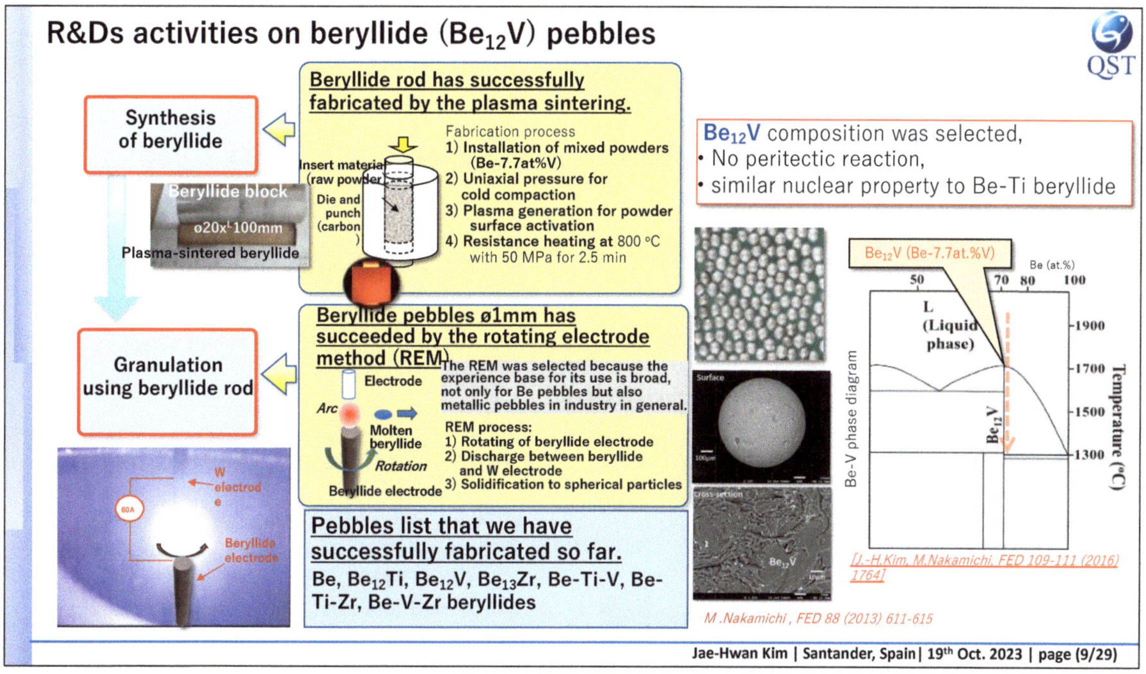

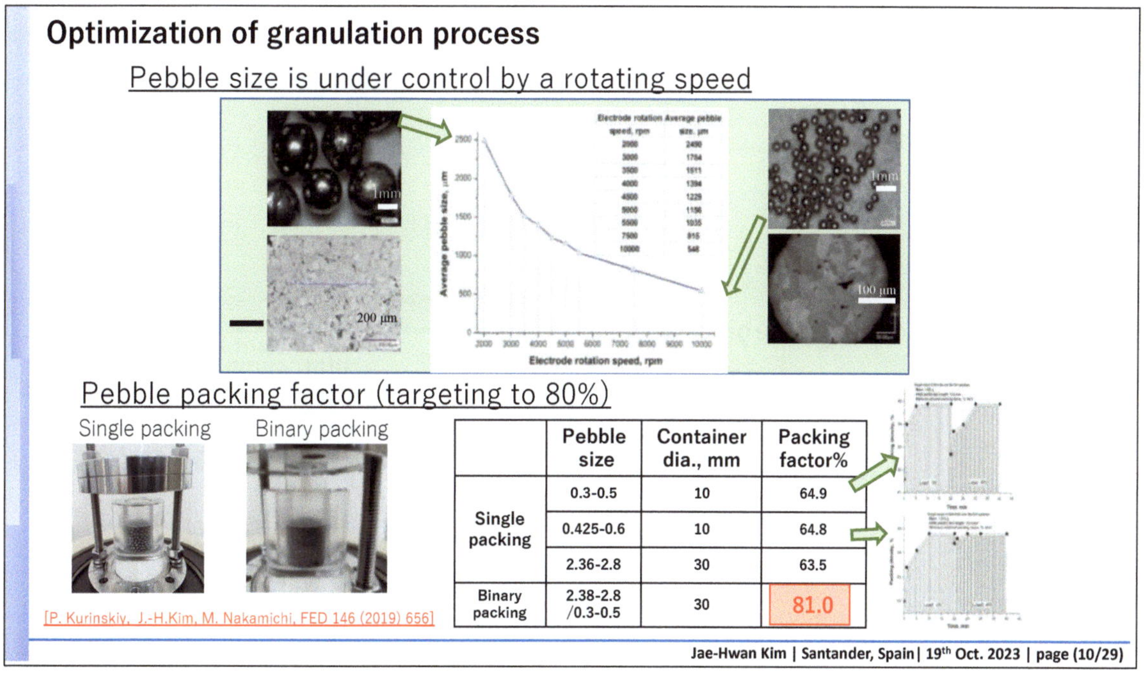

	Pebble size	Container dia., mm	Packing factor%
Single packing	0.3-0.5	10	64.9
	0.425-0.6	10	64.8
	2.36-2.8	30	63.5
Binary packing	2.38-2.8 /0.3-0.5	30	81.0

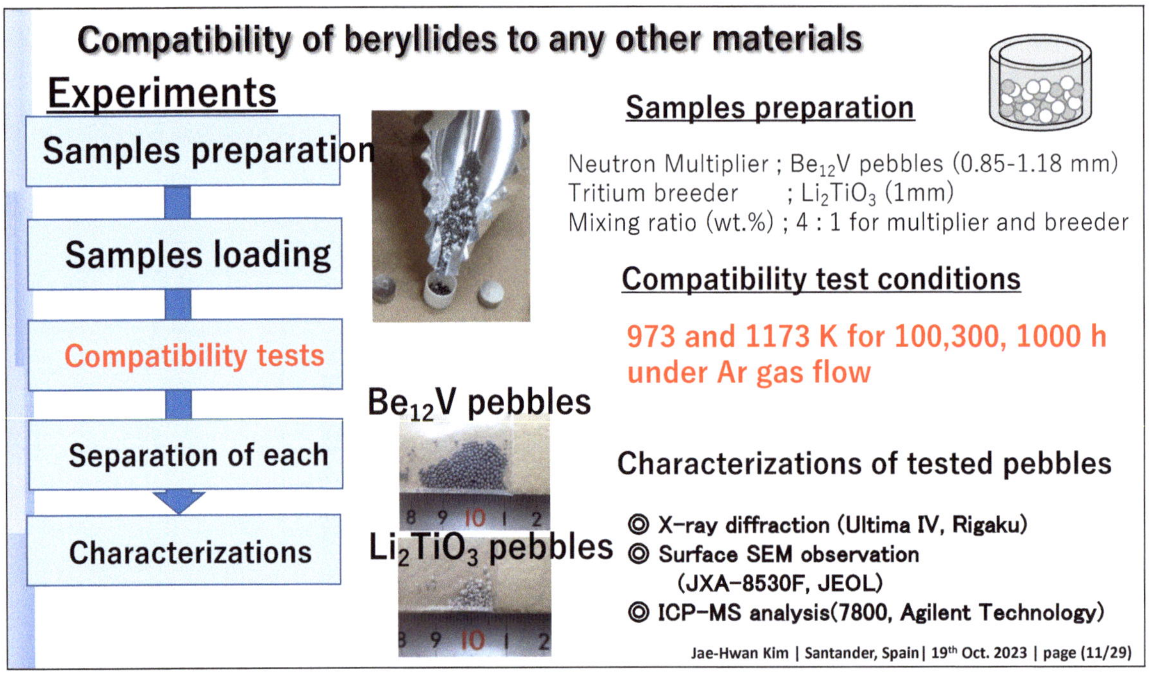
Compatibility of beryllides to any other materials
Experiments
Samples preparation
Samples loading
Compatibility tests
Separation of each
Characterizations
$Be_{12}V$ pebbles
Li_2TiO_3 pebbles
Samples preparation
Neutron Multiplier ; $Be_{12}V$ pebbles (0.85-1.18 mm)
Tritium breeder ; Li_2TiO_3 (1mm)
Mixing ratio (wt.%) ; 4 : 1 for multiplier and breeder
Compatibility test conditions
973 and 1173 K for 100,300, 1000 h
under Ar gas flow
Characterizations of tested pebbles
◎ X-ray diffraction (Ultima IV, Rigaku)
◎ Surface SEM observation
(JXA-8530F, JEOL)
◎ ICP-MS analysis(7800, Agilent Technology)
Jae-Hwan Kim | Santander, Spain| 19th Oct. 2023 | page (11/29)

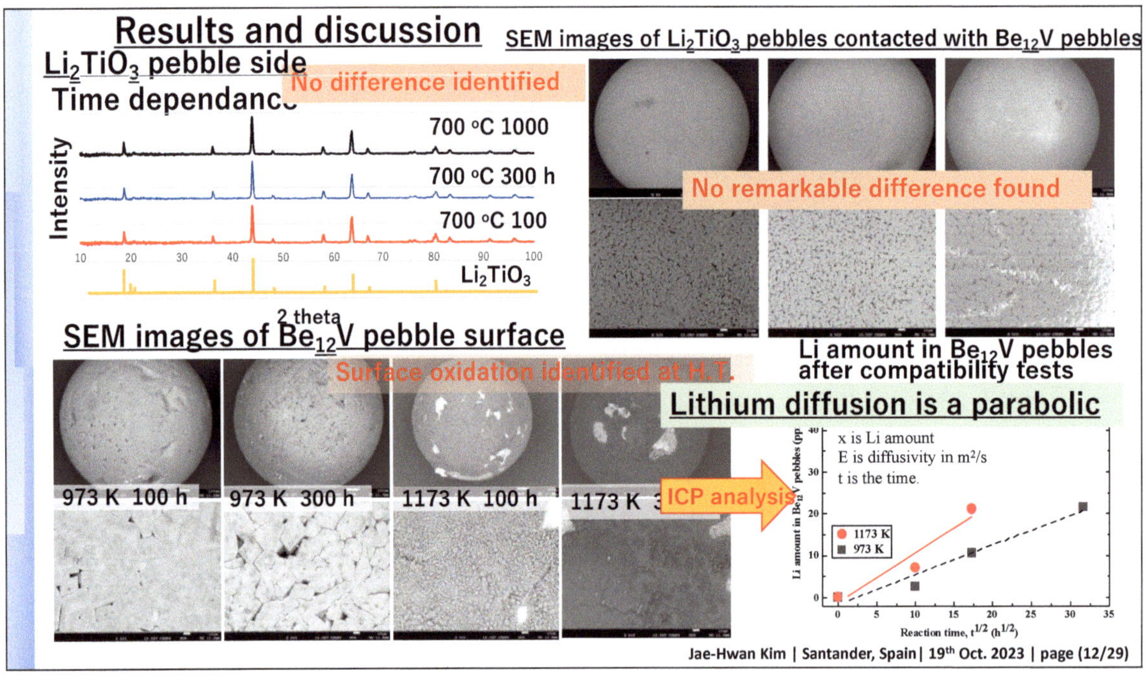
Results and discussion
Li_2TiO_3 pebble side
Time dependance
No difference identified
700 °C 1000
700 °C 300 h
700 °C 100
Li_2TiO_3
Intensity
2 theta
SEM images of Li_2TiO_3 pebbles contacted with $Be_{12}V$ pebbles
No remarkable difference found
SEM images of $Be_{12}V$ pebble surface
Surface oxidation identified at H.T.
973 K 100 h
973 K 300 h
1173 K 100 h
1173 K
ICP analysis
Li amount in $Be_{12}V$ pebbles
after compatibility tests
Lithium diffusion is a parabolic
x is Li amount
E is diffusivity in m²/s
t is the time.
1173 K
973 K
Reaction time, $t^{1/2}$ ($h^{1/2}$)
Jae-Hwan Kim | Santander, Spain| 19th Oct. 2023 | page (12/29)

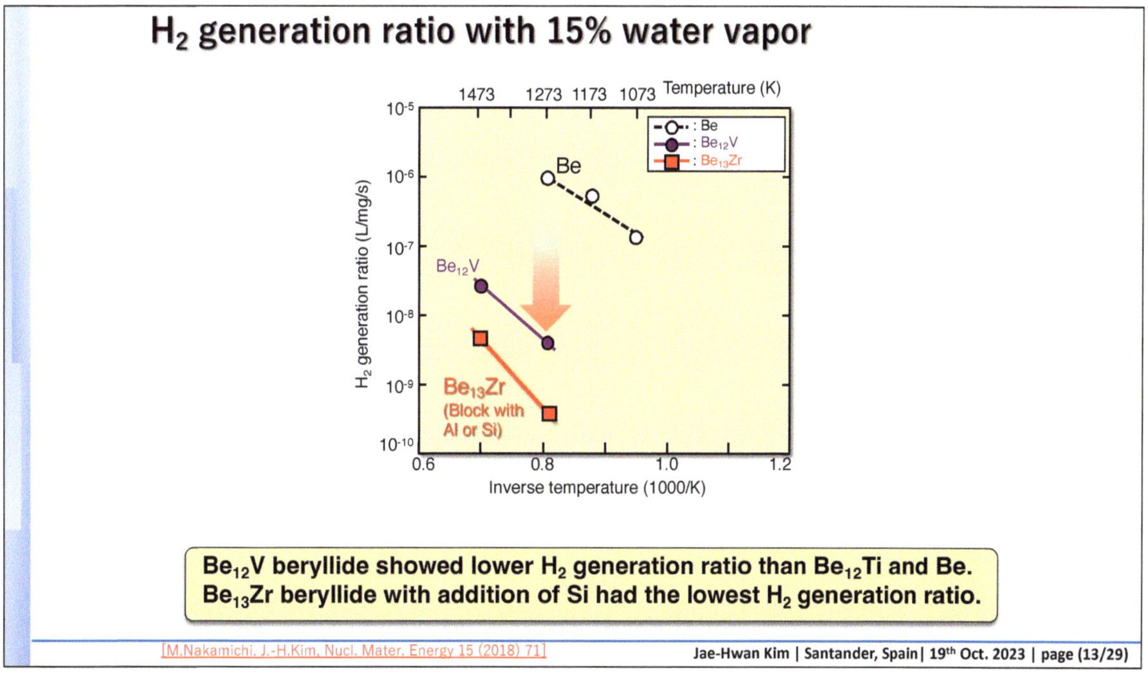
H2 generation ratio with 15% water vapor
Temperature (K)
1473 1273 1173 1073
H2 generation ratio (L/mg/s)
Inverse temperature (1000/K)
: Be
: Be12V
: Be13Zr
Be
Be12V
Be13Zr
(Block with Al or Si)
Be12V beryllide showed lower H2 generation ratio than Be12Ti and Be.
Be13Zr beryllide with addition of Si had the lowest H2 generation ratio.
[M.Nakamichi, J.-H.Kim, Nucl. Mater. Energy 15 (2018) 71]
Jae-Hwan Kim | Santander, Spain| 19th Oct. 2023 | page (13/29)

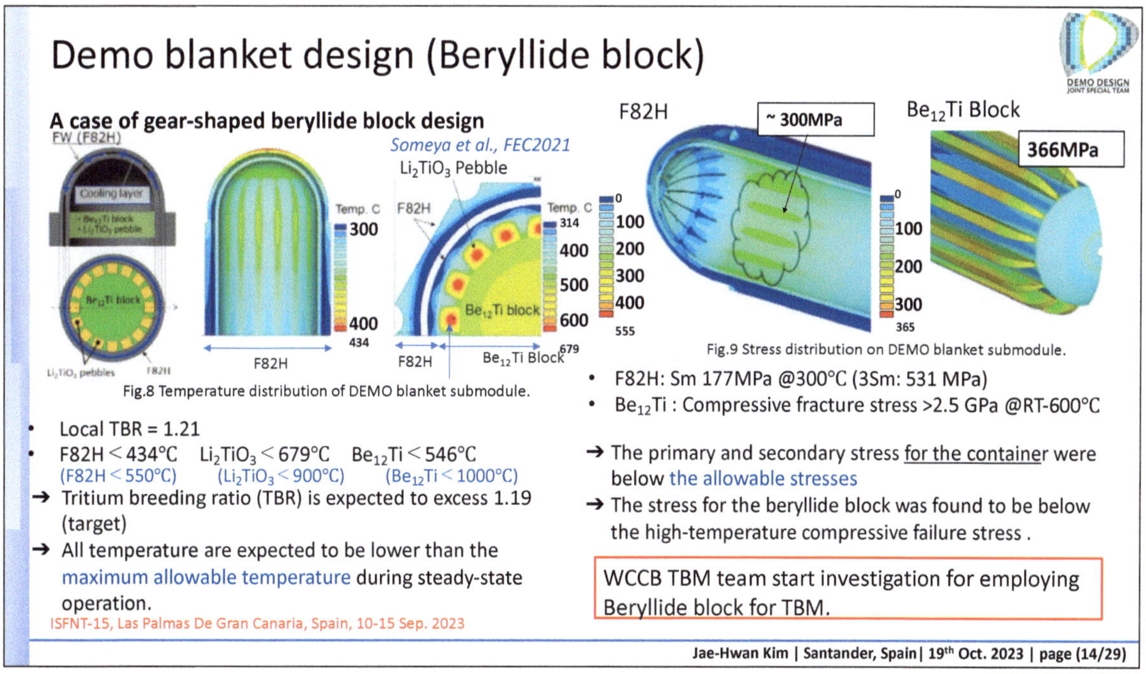
Demo blanket design (Beryllide block)
A case of gear-shaped beryllide block design
Someya et al., FEC2021
Li2TiO3 Pebble
F82H
~ 300MPa
Be12Ti Block
366MPa
Fig.8 Temperature distribution of DEMO blanket submodule.
Fig.9 Stress distribution on DEMO blanket submodule.
• F82H: Sm 177MPa @300°C (3Sm: 531 MPa)
• Be12Ti : Compressive fracture stress >2.5 GPa @RT-600°C
• Local TBR = 1.21
• F82H < 434°C Li2TiO3 < 679°C Be12Ti < 546°C
(F82H < 550°C) (Li2TiO3 < 900°C) (Be12Ti < 1000°C)
➔ Tritium breeding ratio (TBR) is expected to excess 1.19 (target)
➔ All temperature are expected to be lower than the maximum allowable temperature during steady-state operation.
ISFNT-15, Las Palmas De Gran Canaria, Spain, 10-15 Sep. 2023
➔ The primary and secondary stress for the container were below the allowable stresses
➔ The stress for the beryllide block was found to be below the high-temperature compressive failure stress .
WCCB TBM team start investigation for employing Beryllide block for TBM.
Jae-Hwan Kim | Santander, Spain| 19th Oct. 2023 | page (14/29)

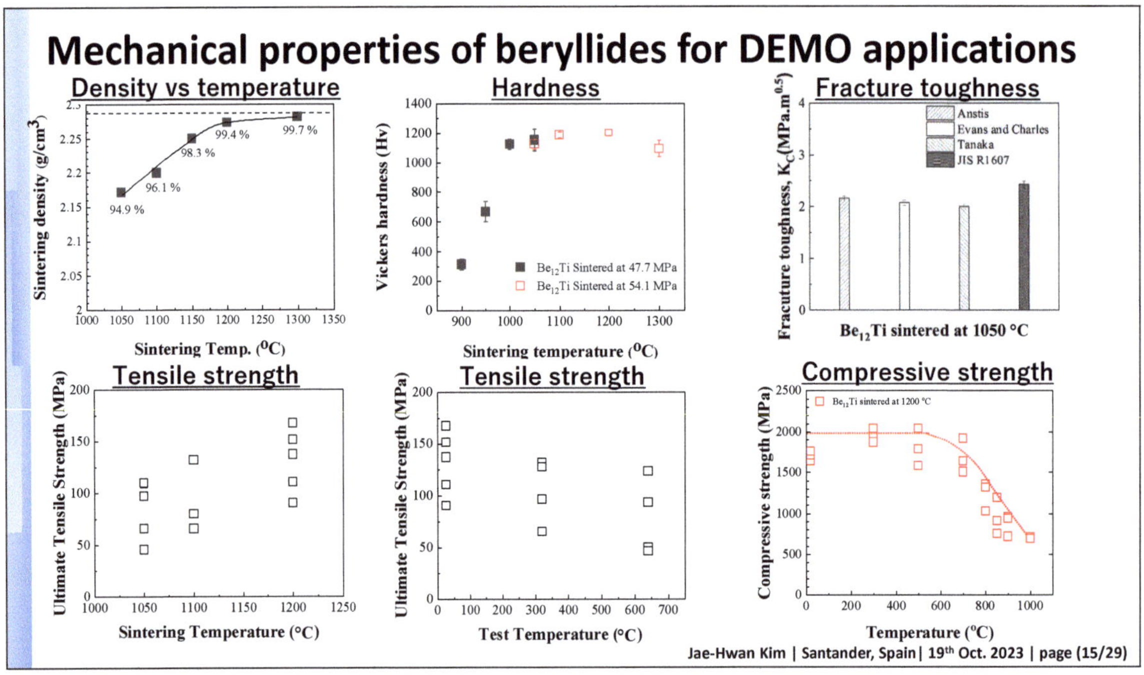
Mechanical properties of beryllides for DEMO applications
Density vs temperature
Sintering density (g/cm3)
Sintering Temp. (°C)
94.9 %
96.1 %
98.3 %
99.4 %
99.7 %
Hardness
Vickers hardness (Hv)
Be12Ti Sintered at 47.7 MPa
Be12Ti Sintered at 54.1 MPa
Sintering temperature (°C)
Fracture toughness
Fracuture toughness, KC(MPa.m0.5)
Anstis
Evans and Charles
Tanaka
JIS R1607
Be12Ti sintered at 1050 °C
Tensile strength
Ultimate Tensile Strength (MPa)
Sintering Temperature (°C)
Tensile strength
Ultimate Tensile Strength (MPa)
Test Temperature (°C)
Compressive strength
Compressive strength (MPa)
Be12Ti sintered at 1200 °C
Temperature (°C)
Jae-Hwan Kim | Santander, Spain | 19th Oct. 2023 | page (15/29)

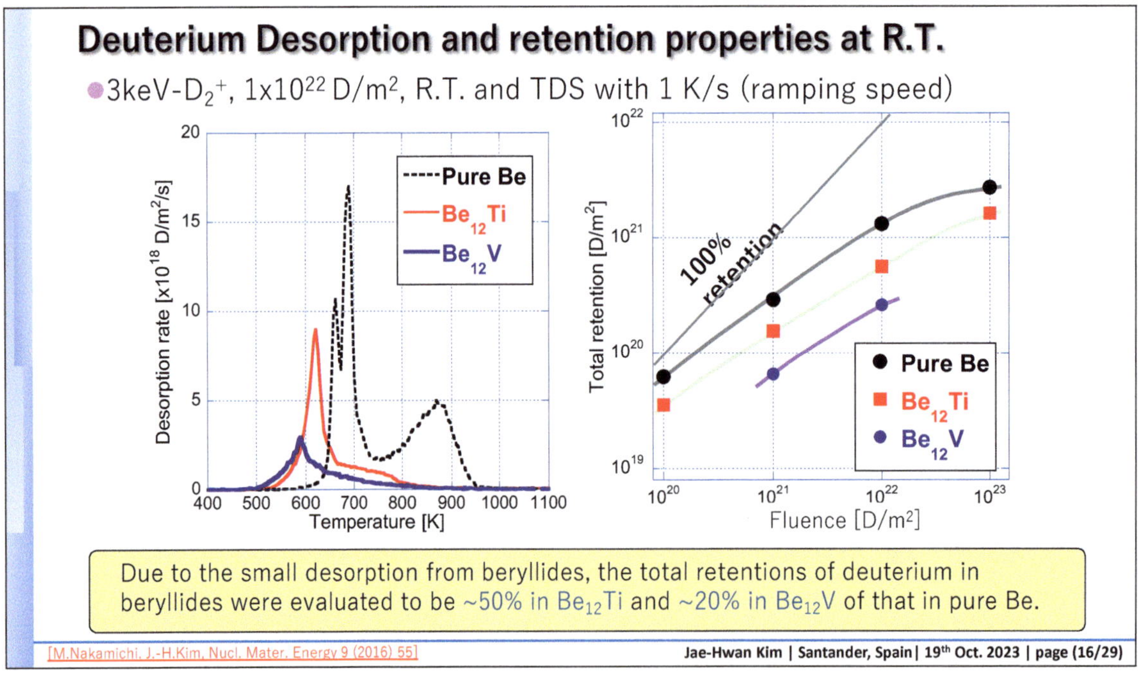
Deuterium Desorption and retention properties at R.T.
3keV-D2+, 1x1022 D/m2, R.T. and TDS with 1 K/s (ramping speed)
Desorption rate [x1018 D/m2/s]
Temperature [K]
Pure Be
Be12Ti
Be12V
Total retention [D/m2]
100% retention
Fluence [D/m2]
Pure Be
Be12Ti
Be12V
Due to the small desorption from beryllides, the total retentions of deuterium in beryllides were evaluated to be ~50% in Be12Ti and ~20% in Be12V of that in pure Be.
[M.Nakamichi, J.-H.Kim, Nucl. Mater. Energy 9 (2016) 55]
Jae-Hwan Kim | Santander, Spain | 19th Oct. 2023 | page (16/29)

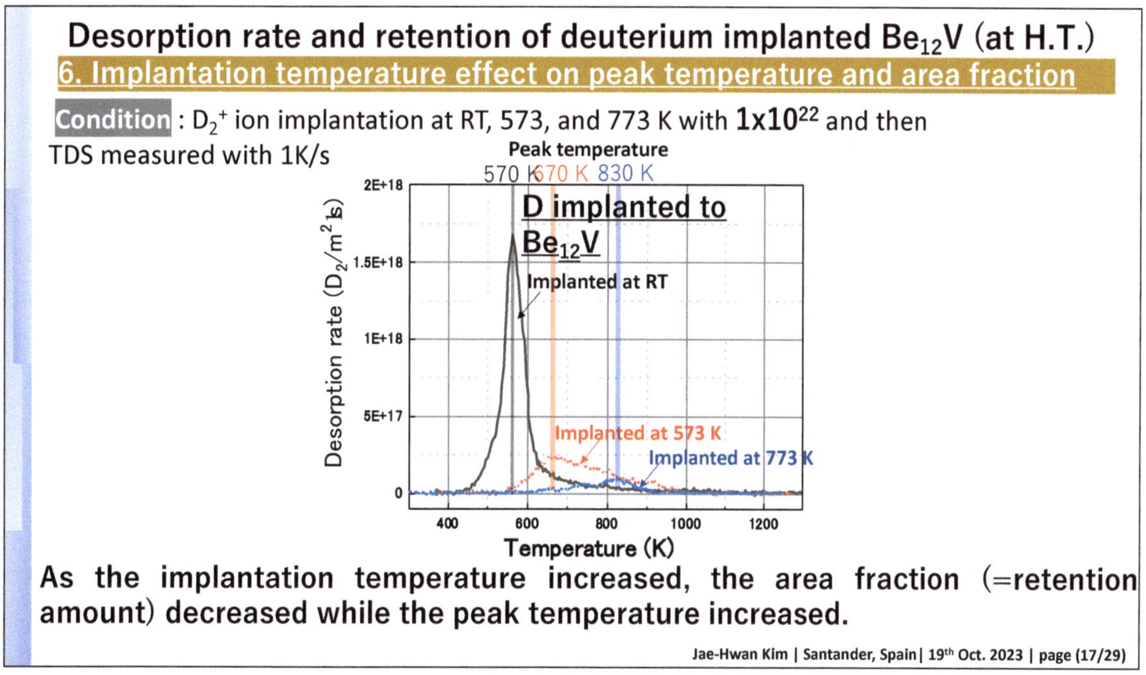
Desorption rate and retention of deuterium implanted $Be_{12}V$ (at H.T.)
6. Implantation temperature effect on peak temperature and area fraction
Condition : D_2^+ ion implantation at RT, 573, and 773 K with $1x10^{22}$ and then TDS measured with 1K/s
Peak temperature
570 K 670 K 830 K
D implanted to $Be_{12}V$
Implanted at RT
Implanted at 573 K
Implanted at 773 K
Desorption rate ($D_2/m^2 s$)
2E+18
1.5E+18
1E+18
5E+17
0
400
600
800
1000
1200
Temperature (K)
As the implantation temperature increased, the area fraction (=retention amount) decreased while the peak temperature increased.
Jae-Hwan Kim | Santander, Spain| 19th Oct. 2023 | page (17/29)

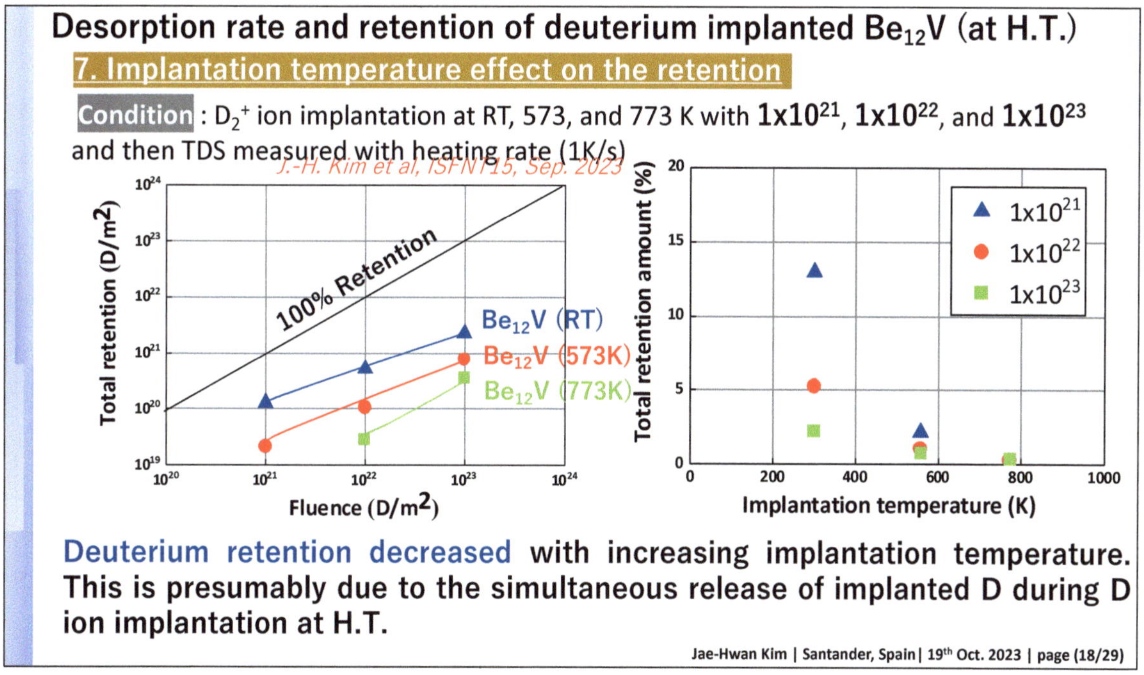
Desorption rate and retention of deuterium implanted $Be_{12}V$ (at H.T.)
7. Implantation temperature effect on the retention
Condition : D_2^+ ion implantation at RT, 573, and 773 K with $1x10^{21}$, $1x10^{22}$, and $1x10^{23}$ and then TDS measured with heating rate (1K/s)
J.-H. Kim et al, ISFNT15, Sep. 2023
Total retention (D/m^2)
10^{24}
10^{23}
10^{22}
10^{21}
10^{20}
10^{19}
100% Retention
$Be_{12}V$ (RT)
$Be_{12}V$ (573K)
$Be_{12}V$ (773K)
10^{20}
10^{21}
10^{22}
10^{23}
10^{24}
Fluence (D/m^2)
Total retention amount (%)
20
15
10
5
0
$1x10^{21}$
$1x10^{22}$
$1x10^{23}$
0
200
400
600
800
1000
Implantation temperature (K)
Deuterium retention decreased with increasing implantation temperature. This is presumably due to the simultaneous release of implanted D during D ion implantation at H.T.
Jae-Hwan Kim | Santander, Spain| 19th Oct. 2023 | page (18/29)

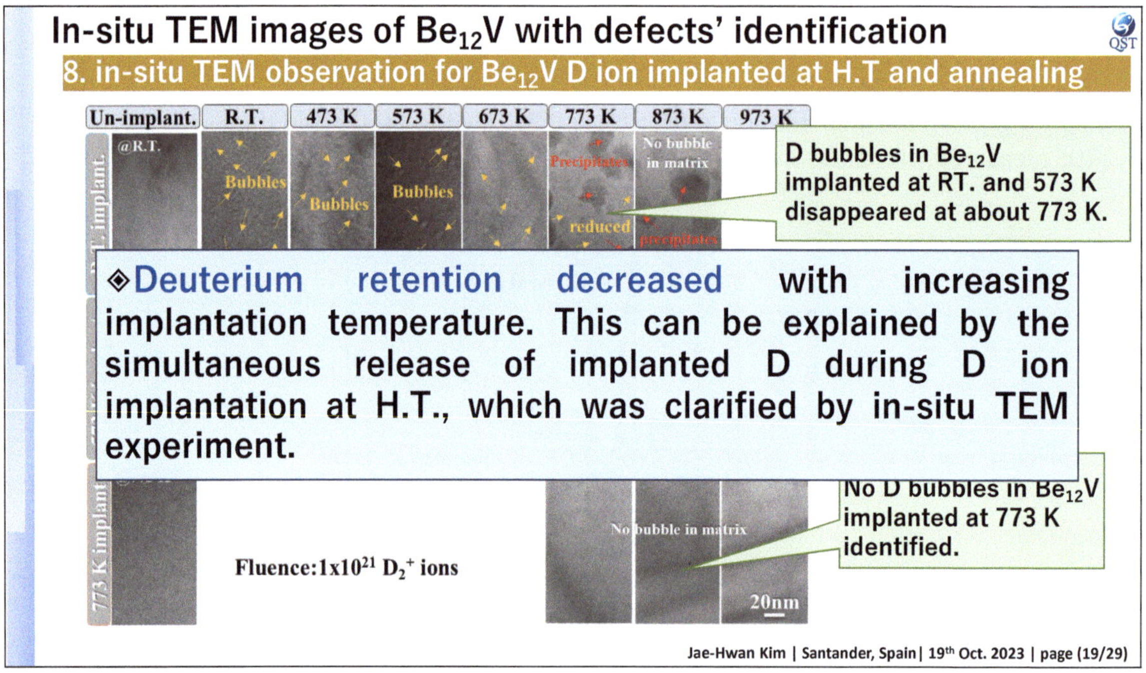
In-situ TEM images of $Be_{12}V$ with defects' identification
8. in-situ TEM observation for $Be_{12}V$ D ion implanted at H.T and annealing
Un-implant.
R.T.
473 K
573 K
673 K
773 K
873 K
973 K
@R.T.
Bubbles
Bubbles
Bubbles
Precipitates
No bubble in matrix
reduced
D bubbles in $Be_{12}V$ implanted at RT. and 573 K disappeared at about 773 K.
◈Deuterium retention decreased with increasing implantation temperature. This can be explained by the simultaneous release of implanted D during D ion implantation at H.T., which was clarified by in-situ TEM experiment.
No D bubbles in $Be_{12}V$ implanted at 773 K identified.
773 K implant.
Fluence:$1x10^{21}$ D_2^+ ions
No bubble in matrix
20nm
Jae-Hwan Kim | Santander, Spain | 19th Oct. 2023 | page (19/29)

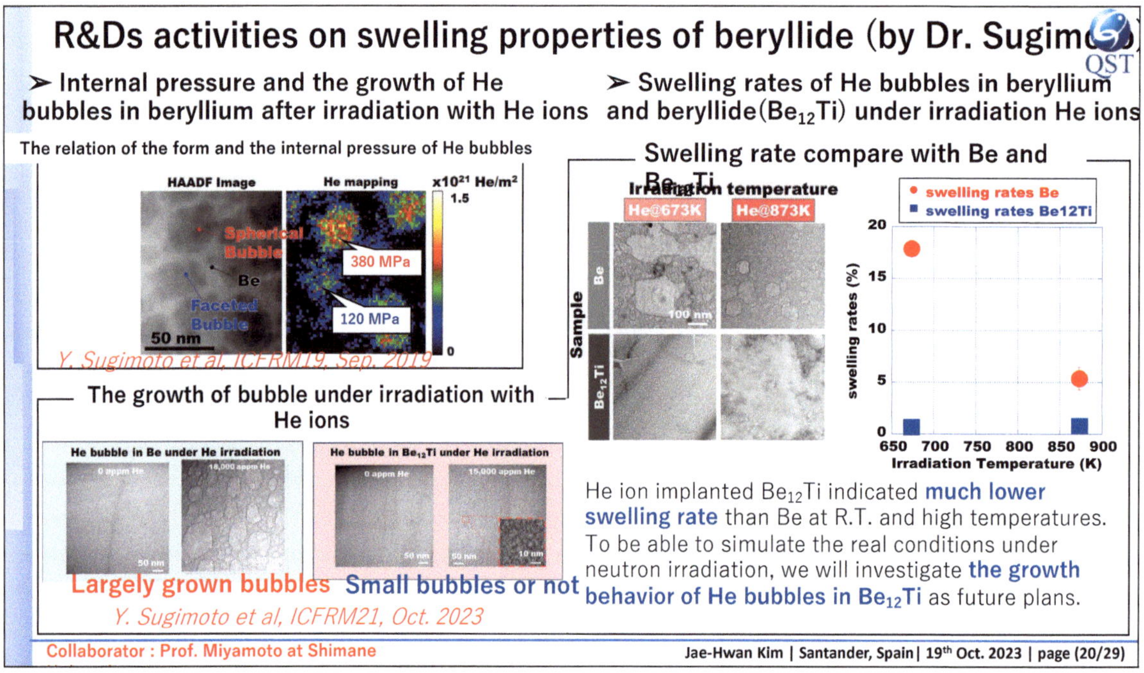
R&Ds activities on swelling properties of beryllide (by Dr. Sugimoto)
➢ Internal pressure and the growth of He bubbles in beryllium after irradiation with He ions
➢ Swelling rates of He bubbles in beryllium and beryllide($Be_{12}Ti$) under irradiation He ions
The relation of the form and the internal pressure of He bubbles
HAADF Image
He mapping
$x10^{21}$ He/m^2
1.5
Spherical Bubble
Be
Faceted Bubble
380 MPa
120 MPa
50 nm
0
Y. Sugimoto et al, ICFRM19, Sep. 2019
Swelling rate compare with Be and $Be_{12}Ti$
Irradiation temperature
He@673K
He@873K
Sample
Be
$Be_{12}Ti$
100 nm
swelling rates Be
swelling rates Be12Ti
swelling rates (%)
Irradiation Temperature (K)
The growth of bubble under irradiation with He ions
He bubble in Be under He irradiation
He bubble in $Be_{12}Ti$ under He irradiation
Largely grown bubbles
Small bubbles or not
Y. Sugimoto et al, ICFRM21, Oct. 2023
He ion implanted $Be_{12}Ti$ indicated much lower swelling rate than Be at R.T. and high temperatures. To be able to simulate the real conditions under neutron irradiation, we will investigate the growth behavior of He bubbles in $Be_{12}Ti$ as future plans.
Collaborator : Prof. Miyamoto at Shimane
Jae-Hwan Kim | Santander, Spain | 19th Oct. 2023 | page (20/29)

R&Ds activities on swelling properties of beryllide (by Dr. Sugimoto)

Swelling properties of Be&$Be_{12}Ti$
HIDOBE I&II results

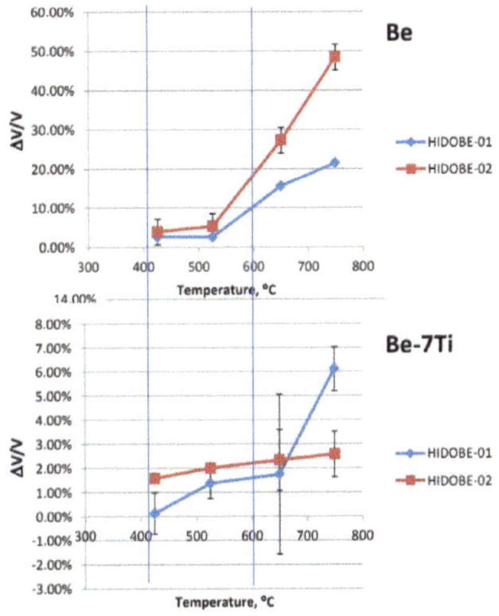

Comparison between ion implantation and neutron irradiation in Be&$Be_{12}Ti$

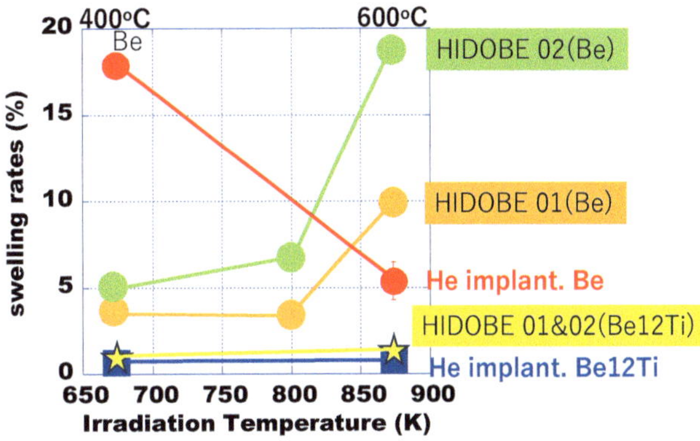

To be able to simulate the real condition under neutron environment by ion implantation, further studies will be needed.

Jae-Hwan Kim | Santander, Spain | 19th Oct. 2023 | page (21/29)

6. Refinement process of BeO and Recycling process of beryllides

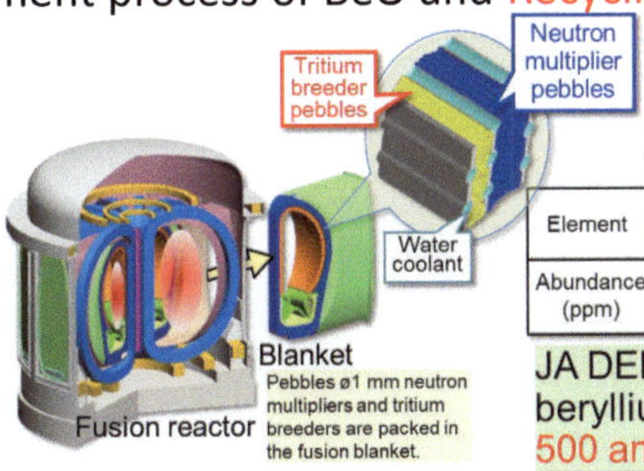

•Beryllium metal and its intermetallic compounds are candidate materials of neutron multipliers.
•Beryllium and Lithium are one of rare metals.

Element	Li	Be	Ti	In	Nd	Sm	W	Pt
Abundance (ppm)	13	1.5	5400	0.05	16	3.5	1.0	0.001

JA DEMO requires huge amount of beryllium and Lithium in the weight of 500 and 300 tons, respectively.

J.-H. Kim, SOFT2022, Croatia

J.-H. Kim, JNM 542, (2020) 152522

Be	· Total production ： about 300ton/year · Condition ： close to oligopoly · Process ： complicated, energy consuming process (incl. 2000 °C treatment)	Li	· ^{6}Li enrichment ： 90% (natural about 7.5%) · ^{6}Li burn up ： about 20 % (remained most of them) · Extremely expensive

Objective： Development of refining and recycling technologies to secure stable resources of beryllium and lithium and to optimize their costs.

Jae-Hwan Kim | Santander, Spain | 19th Oct. 2023 | page (22/29)

Novel Beryllium refinement technology

J.-H. Kim, JNM 542, (2020) 152522

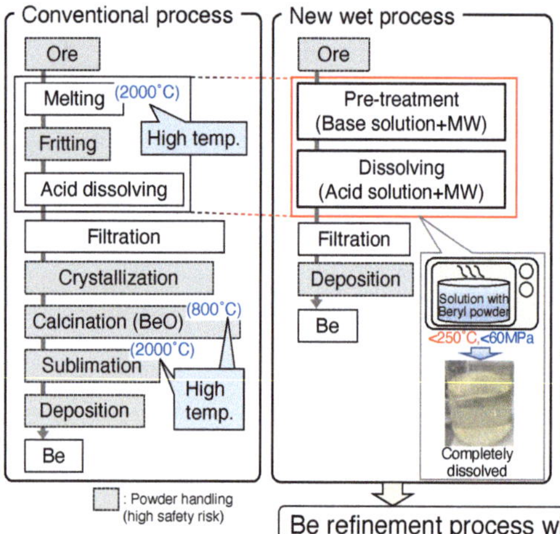

[Ore: Beyl, MW: Microwave]

Advantages

1)Wet process mainly (decrease of handling with powdery state),

2)Low processing temperature (<250˚C),

•Use of heating energy: < one hundredth

•Scale of facility: < one several hundredth

3)Capable of decreasing the number of processes (about half of conventional processes),

4)Capable of removing Uranium,

5)Applicable for reuse process.

6)Applicable for $Be(OH)_2$ / BeO refinements.

Patent: National application

Be refinement process with economical and safe was established.

S. Nakano, BeWS16, Spain

•This process will open up the prospects to stably secure beryllium resources for DEMO.

•As an energy-saving refinement technology, general metals (ores, polymetallic nodules, appliances, etc.) can be refined by this process.

Jae-Hwan Kim | Santander, Spain | 19th Oct. 2023 | page (23/29)

Extraction of impurity, Uranium

J.-H. Kim, JNM 542, (2020) 152522

Objective : Development of technology to remove uranium from beryllium (ores)

Idea : Simple and inexpensive Uranium removal technology

Issue

From the viewpoint of radioactive waste disposal
Permissible uranium concentration for shallow disposal: 0.7ppm or less

Uranium and beryllium distribution map

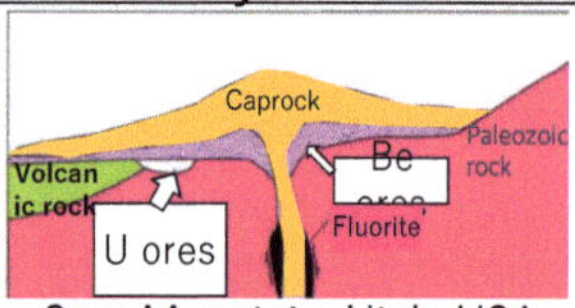

Spor Mountain, Utah, USA

· Same pegmatite deposit
· Approximately 150ppm of uranium in the U.S.

J.-H. Kim, SOFT2022, Croatia

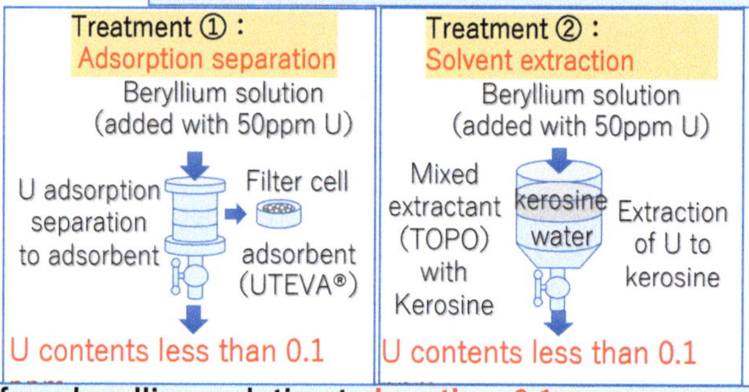

• **We succeeded in removing uranium from beryllium solution to less than 0.1ppm.**
• **Uranium removal was also achieved using a less expensive solvent extraction method (1/60th the price).**

Jae-Hwan Kim | Santander, Spain | 19th Oct. 2023 | page (24/29)

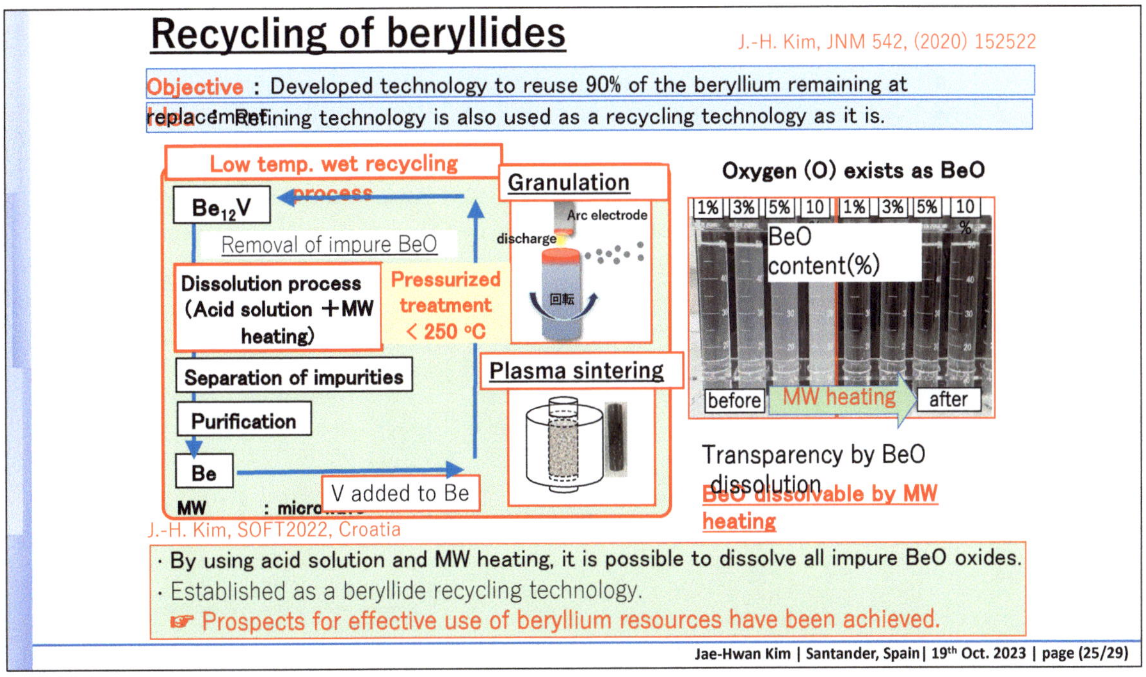

Summary

For JA DEMO blanket design, two options of pebbles and block shape for neutron multipliers are considered. Currently, great interest on the block-shaped beryllides has been paid attention. Additionally, R&Ds on the refinement and recycling process of highly cost functional materials are indispensable for early realization of fusion reactors.

1) QST has contributed to R&Ds on materials fabrication as well as characterizations for the JA DEMO blanket.
2) Mechanical properties for plasma-sintered beryllides have been investigated prior to characterizations for the neutron irradiated beryllides.
3) Not only characterizations but also refinement and recycling process has been conducted from viewpoints of cost optimization and reduction.

Jae-Hwan Kim | Santander, Spain | 19th Oct. 2023 | page (26/29)

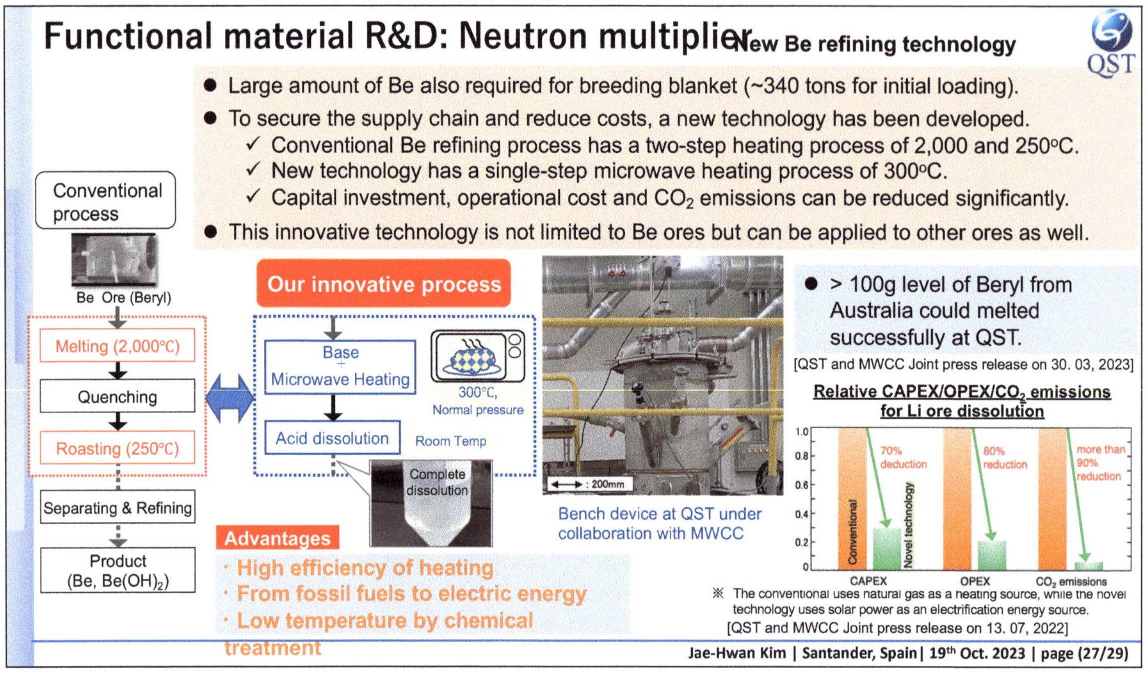

Functional material R&D: Neutron multiplier
New Be refining technology
QST
● Large amount of Be also required for breeding blanket (~340 tons for initial loading).
● To secure the supply chain and reduce costs, a new technology has been developed.
✓ Conventional Be refining process has a two-step heating process of 2,000 and 250°C.
✓ New technology has a single-step microwave heating process of 300°C.
✓ Capital investment, operational cost and CO2 emissions can be reduced significantly.
● This innovative technology is not limited to Be ores but can be applied to other ores as well.
Conventional process
Be Ore (Beryl)
Melting (2,000°C)
Quenching
Roasting (250°C)
Separating & Refining
Product (Be, Be(OH)2)
Our innovative process
Base + Microwave Heating
300°C, Normal pressure
Acid dissolution
Room Temp
Complete dissolution
Advantages
· High efficiency of heating
· From fossil fuels to electric energy
· Low temperature by chemical treatment
: 200mm
Bench device at QST under collaboration with MWCC
● > 100g level of Beryl from Australia could melted successfully at QST.
[QST and MWCC Joint press release on 30. 03, 2023]
Relative CAPEX/OPEX/CO2 emissions for Li ore dissolution
70% deduction
80% reduction
more than 90% reduction
Conventional
Novel technology
CAPEX
OPEX
CO2 emissions
※ The conventional uses natural gas as a heating source, while the novel technology uses solar power as an electrification energy source.
[QST and MWCC Joint press release on 13. 07, 2022]
Jae-Hwan Kim | Santander, Spain | 19th Oct. 2023 | page (27/29)

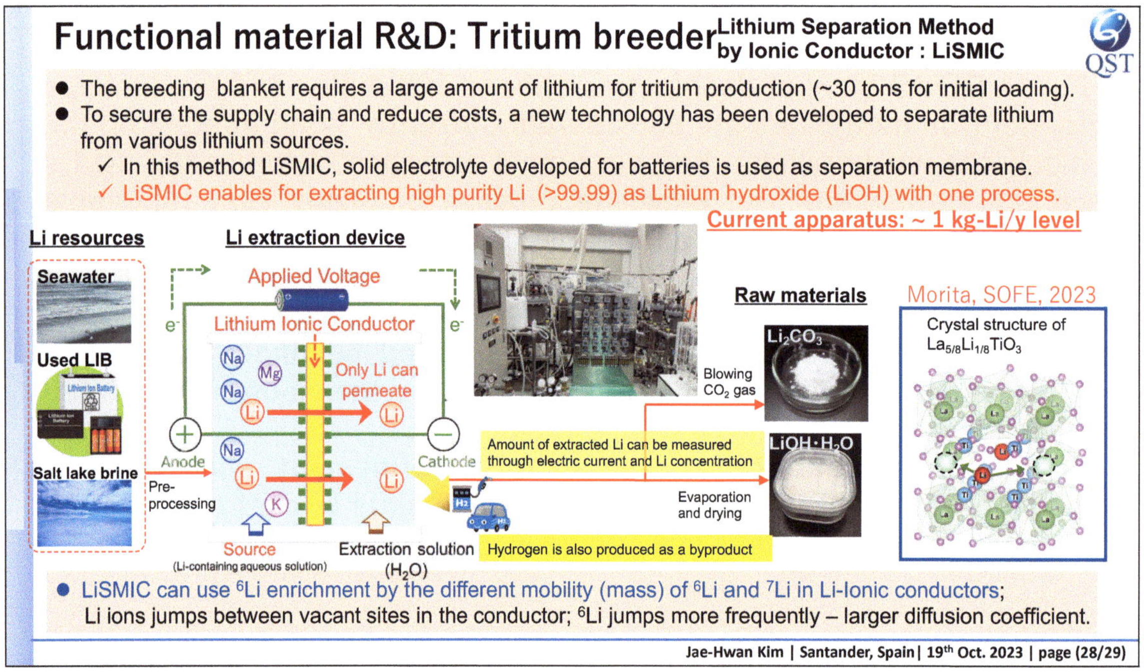

Functional material R&D: Tritium breeder
Lithium Separation Method by Ionic Conductor : LiSMIC
QST
● The breeding blanket requires a large amount of lithium for tritium production (~30 tons for initial loading).
● To secure the supply chain and reduce costs, a new technology has been developed to separate lithium from various lithium sources.
✓ In this method LiSMIC, solid electrolyte developed for batteries is used as separation membrane.
✓ LiSMIC enables for extracting high purity Li (>99.99) as Lithium hydroxide (LiOH) with one process.
Current apparatus: ~ 1 kg-Li/y level
Li resources
Seawater
Used LIB
Salt lake brine
Li extraction device
Applied Voltage
Lithium Ionic Conductor
Only Li can permeate
Anode
Cathode
Pre-processing
Source
(Li-containing aqueous solution)
Extraction solution (H2O)
Raw materials
Li2CO3
Blowing CO2 gas
LiOH·H2O
Evaporation and drying
Amount of extracted Li can be measured through electric current and Li concentration
Hydrogen is also produced as a byproduct
Morita, SOFE, 2023
Crystal structure of La5/8Li1/8TiO3
● LiSMIC can use 6Li enrichment by the different mobility (mass) of 6Li and 7Li in Li-Ionic conductors;
Li ions jumps between vacant sites in the conductor; 6Li jumps more frequently – larger diffusion coefficient.
Jae-Hwan Kim | Santander, Spain | 19th Oct. 2023 | page (28/29)

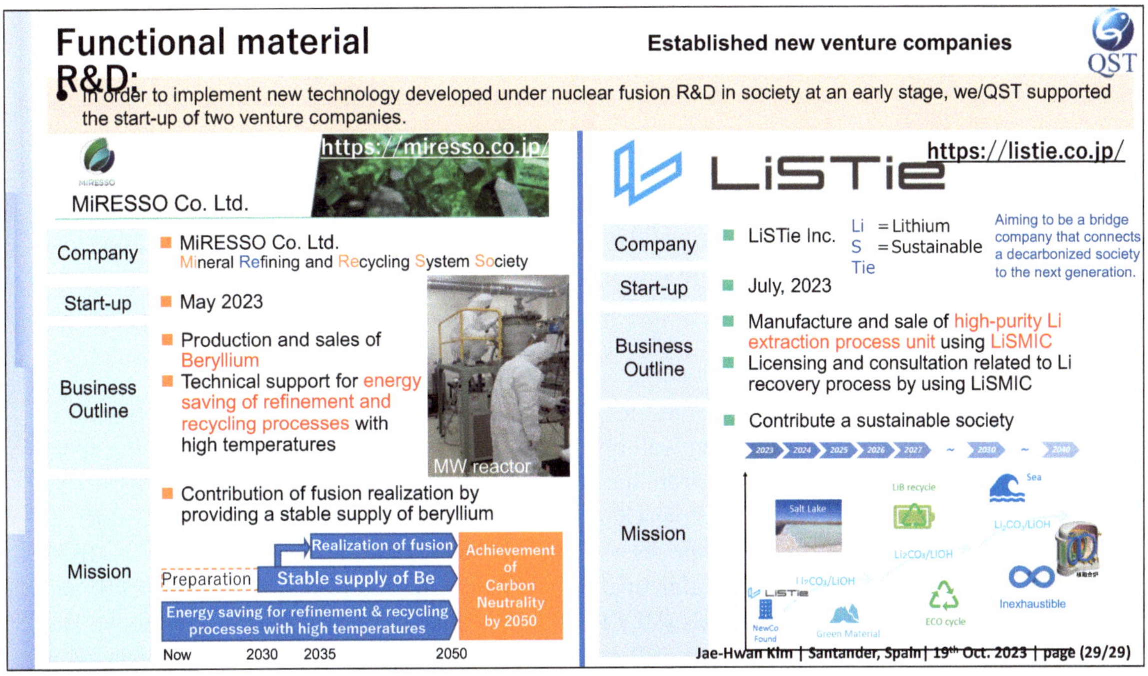
Functional material R&D:
Established new venture companies
QST
• In order to implement new technology developed under nuclear fusion R&D in society at an early stage, we/QST supported the start-up of two venture companies.
https://miresso.co.jp/
MiRESSO Co. Ltd.
Company
MiRESSO Co. Ltd.
Mineral Refining and Recycling System Society
Start-up
May 2023
Business Outline
Production and sales of Beryllium
Technical support for energy saving of refinement and recycling processes with high temperatures
MW reactor
Mission
Contribution of fusion realization by providing a stable supply of beryllium
Realization of fusion
Preparation
Stable supply of Be
Energy saving for refinement & recycling processes with high temperatures
Achievement of Carbon Neutrality by 2050
Now
2030
2035
2050
LiSTie
https://listie.co.jp/
Company
LiSTie Inc.
Li = Lithium
S = Sustainable
Tie
Aiming to be a bridge company that connects a decarbonized society to the next generation.
Start-up
July, 2023
Business Outline
Manufacture and sale of high-purity Li extraction process unit using LiSMIC
Licensing and consultation related to Li recovery process by using LiSMIC
Mission
Contribute a sustainable society
2023
2024
2025
2026
2027
2030
2040
Salt Lake
LiB recycle
Sea
Li2CO3/LiOH
Inexhaustible
ECO cycle
Green Material
NewCo Found
Jae-Hwan Kim | Santander, Spain | 19th Oct. 2023 | page (29/29)

2.2 Overview of analysis methods for beryllium from controlled fusion devices

Marek Rubel, P. Petersson, D. Primetzhofer, A. Widdowson, L. Dittrich, E. Fortuna-Zaleśna, I. Jepu, T.T. Tran, Y. Zayachuk

Beryllium has been used as plasma-facing material (PFM) in controlled fusion devices since eighties of the 20th century [1]. On a small scale it was applied in Unitor and ISX-B machines, and on large scale in the JET tokamak where - over the years – Be coatings and tons of bulk metal have been used. In parallel to experience in plasma operation, it became necessary to develop handling procedures and analysis methods of Be exploited in the nuclear environment of a fusion reactor [2,3]. The morphology of PFM is modified by plasma-wall interactions (PWI) comprising a wide range of processes including co-implantation and co-deposition of hydrogen isotopes (fuel) and plasma impurity atoms thus leading to the presence of radioactive tritium in materials retrieved from the torus for ex-situ studies.

Analyses are carried out for bulk Be wall tiles, evaporated Be coatings and for tiny objects such as dust particles and splashes resulting from wall erosion. Analysis techniques based on probing specimens with either electrons (Auger electron spectroscopy AES, energy and wavelength dispersive X-ray spectroscopy EDS and WDS), photons (X-ray photoelectron spectroscopy XPS, X-ray diffraction XRD) or ions. The latter group comprises a range of accelerator based ion beam analysis (IBA) methods such as 3He-based nuclear reaction analysis (NRA), enhanced proton scattering (EPS) and heavy ion elastic recoil detection analysis (HI-ERDA). They play a prominent role in sensitive and selective quantification and depth profiling of Be-containing materials. In addition, micro-beam NRA and EPS are essential in studies of Be dust. In connection with IBA, recent advances in measurements of reaction cross-sections and stopping powers in Be will be presented in detail.

A critical overview of methods will be presented with particular emphasis of their advantages and limitations, both in studies of PFM and in the characterization of breeding and neutron multiplying materials, i.e. beryllium compounds. The need and benefits of combination with microscopy methods will be addressed. Concrete examples will be discussed, also from basic laboratory experiments.

REFERENCES

[1] G. De Temmermann et al., Nucl. Mater. Energy 27 (2021) 100994.
[2] A. Widdowson et al., Phys. Scr. T167 (2016) 014057.
[3] M. Rubel et al., Eur. Phys. J. Techniques and Instrumentation 10:3 (2023).

Presenting Author: Marek Rubel
Presenting Author's Affiliation: KTH Royal Institute of Technology, Sweden
Email: rubel@kth.se

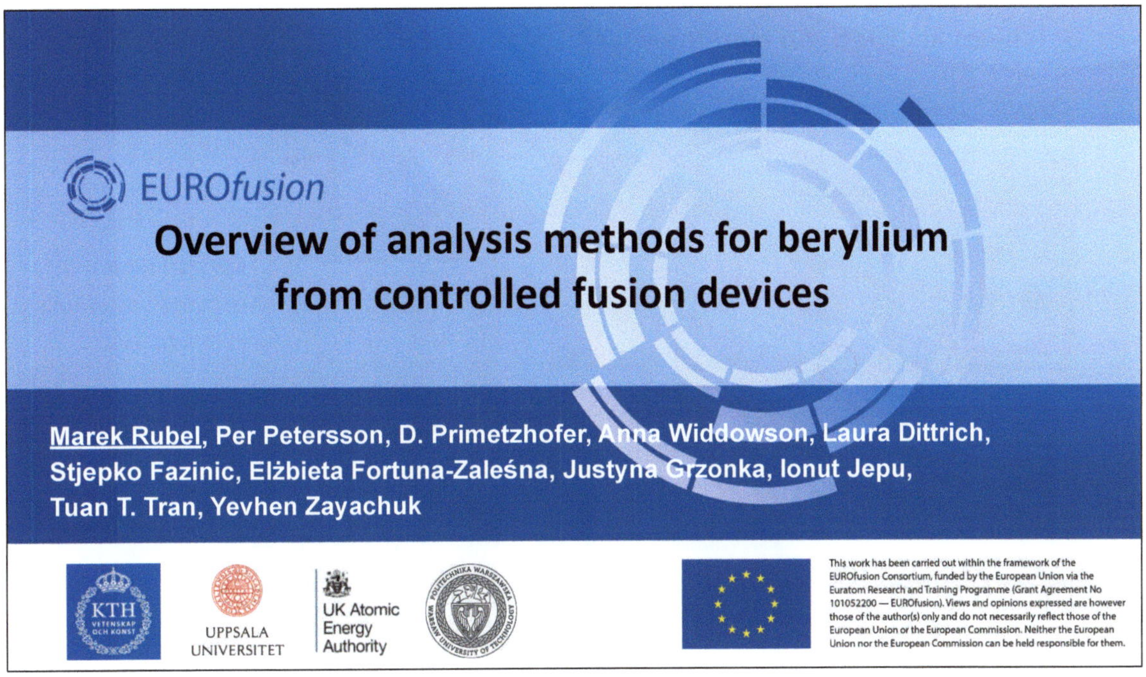
EUROfusion
Overview of analysis methods for beryllium from controlled fusion devices
Marek Rubel, Per Petersson, D. Primetzhofer, Anna Widdowson, Laura Dittrich, Stjepko Fazinic, Elżbieta Fortuna-Zaleśna, Justyna Grzonka, Ionut Jepu, Tuan T. Tran, Yevhen Zayachuk
KTH VETENSKAP OCH KONST
UPPSALA UNIVERSITET
UK Atomic Energy Authority
This work has been carried out within the framework of the EUROfusion Consortium, funded by the European Union via the Euratom Research and Training Programme (Grant Agreement No 101052200 — EUROfusion). Views and opinions expressed are however those of the author(s) only and do not necessarily reflect those of the European Union or the European Commission. Neither the European Union nor the European Commission can be held responsible for them.

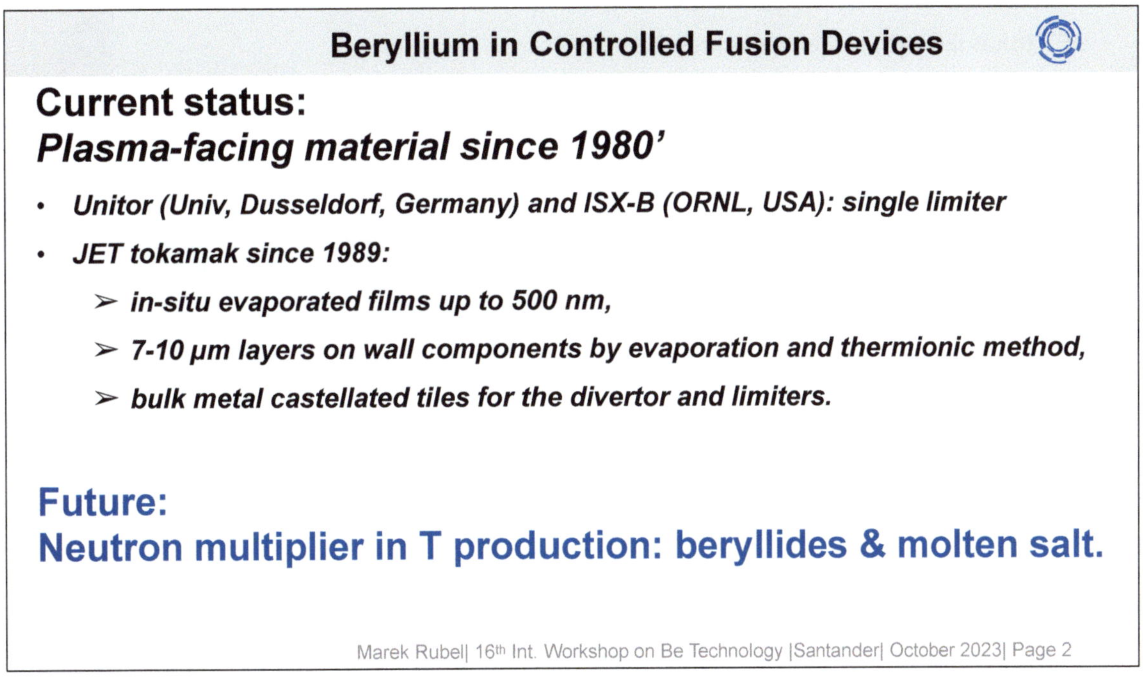
Beryllium in Controlled Fusion Devices
Current status:
Plasma-facing material since 1980'
• Unitor (Univ, Dusseldorf, Germany) and ISX-B (ORNL, USA): single limiter
• JET tokamak since 1989:
➢ in-situ evaporated films up to 500 nm,
➢ 7-10 µm layers on wall components by evaporation and thermionic method,
➢ bulk metal castellated tiles for the divertor and limiters.
Future:
Neutron multiplier in T production: beryllides & molten salt.
Marek Rubel| 16th Int. Workshop on Be Technology |Santander| October 2023| Page 2

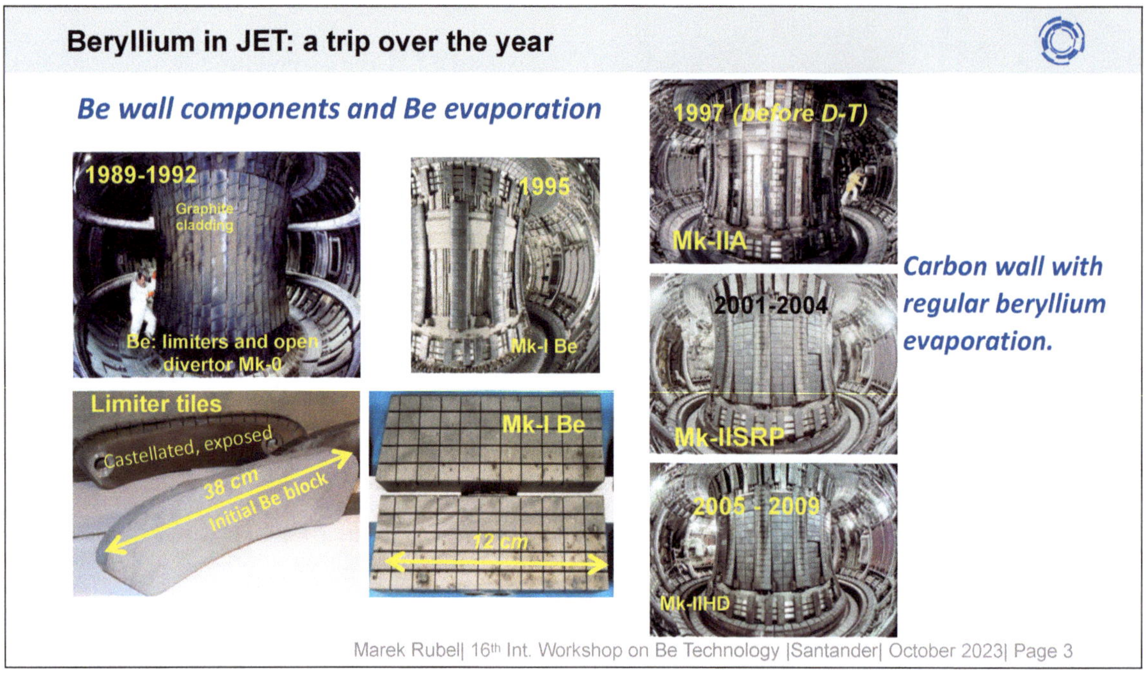
Beryllium in JET: a trip over the year
Be wall components and Be evaporation
1989-1992
Graphite cladding
Be: limiters and open divertor Mk-0
1995
Mk-I Be
1997 (before D-T)
Mk-IIA
2001-2004
Mk-IISRP
2005 - 2009
Mk-IIHD
Carbon wall with regular beryllium evaporation.
Limiter tiles
Castellated, exposed
38 cm
Initial Be block
Mk-I Be
12 cm
Marek Rubel| 16th Int. Workshop on Be Technology |Santander| October 2023| Page 3

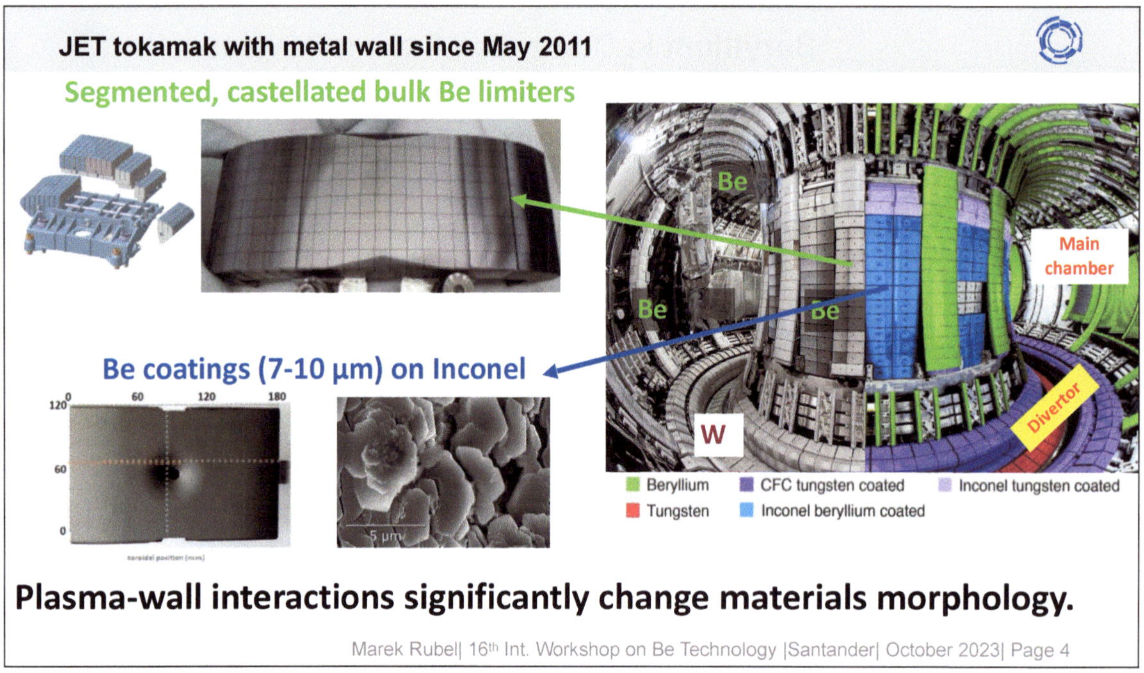
JET tokamak with metal wall since May 2011
Segmented, castellated bulk Be limiters
Be coatings (7-10 µm) on Inconel
Be
Be
Be
Main chamber
Divertor
W
Beryllium
Tungsten
CFC tungsten coated
Inconel beryllium coated
Inconel tungsten coated
5 µm
Plasma-wall interactions significantly change materials morphology.
Marek Rubel| 16th Int. Workshop on Be Technology |Santander| October 2023| Page 4

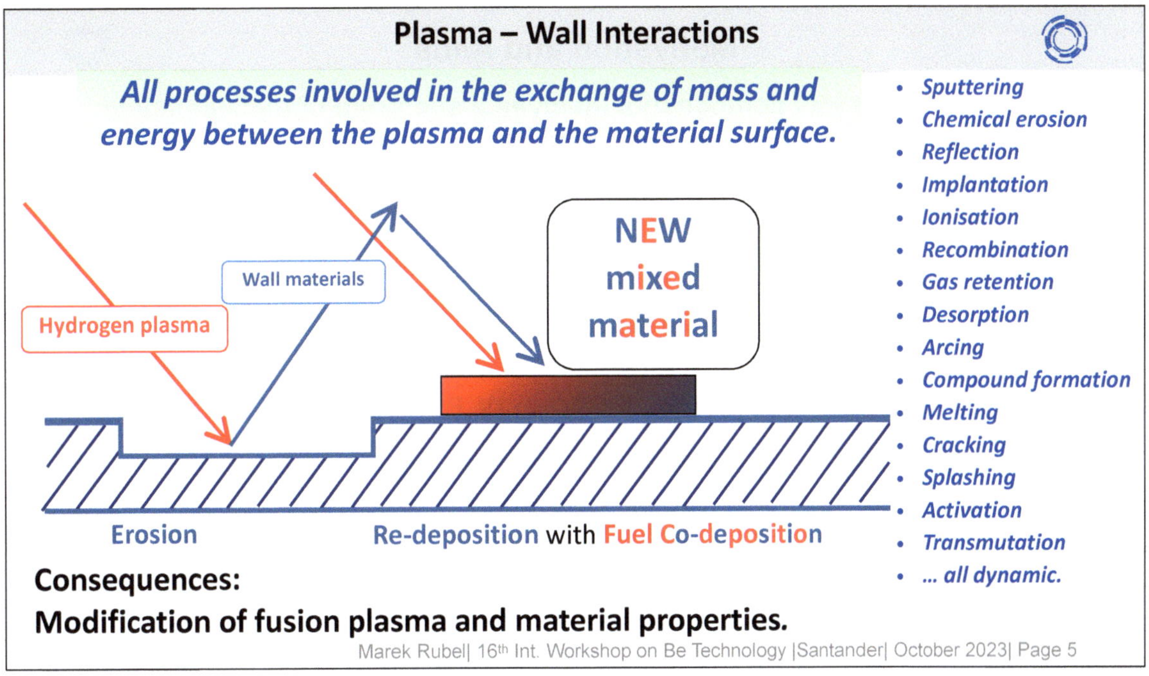
Plasma – Wall Interactions
All processes involved in the exchange of mass and energy between the plasma and the material surface.
Hydrogen plasma
Wall materials
NEW mixed material
Erosion
Re-deposition with Fuel Co-deposition
• Sputtering
• Chemical erosion
• Reflection
• Implantation
• Ionisation
• Recombination
• Gas retention
• Desorption
• Arcing
• Compound formation
• Melting
• Cracking
• Splashing
• Activation
• Transmutation
• ... all dynamic.
Consequences:
Modification of fusion plasma and material properties.
Marek Rubel| 16th Int. Workshop on Be Technology |Santander| October 2023| Page 5

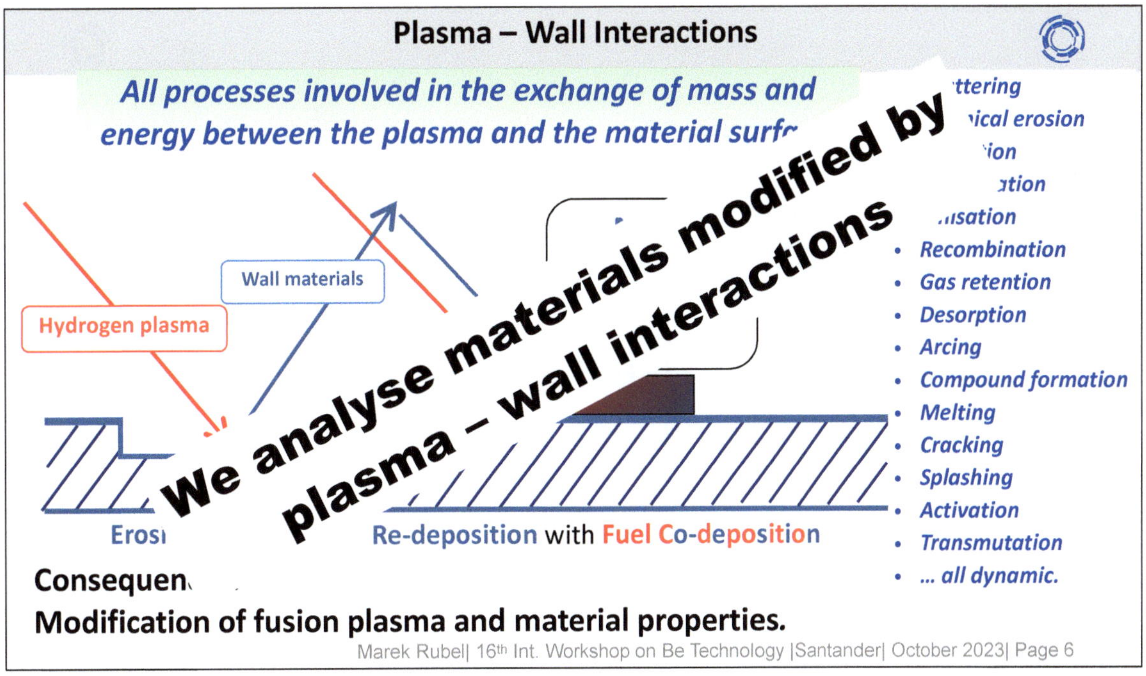
Plasma – Wall Interactions
All processes involved in the exchange of mass and
Hydrogen plasma
Wall materials
We analyse materials modified by plasma – wall interactions
Re-deposition with Fuel Co-deposition
• Recombination
• Gas retention
• Desorption
• Arcing
• Compound formation
• Melting
• Cracking
• Splashing
• Activation
• Transmutation
• ... all dynamic.
Modification of fusion plasma and material properties.
Marek Rubel| 16th Int. Workshop on Be Technology |Santander| October 2023| Page 6

Motivation and Aims

Operation with Be wall components involves a spectrum of nuclear safety issues arising from plasma-wall interactions:

- *Erosion, migration re-deposition (co-deposition),*
- *Fuel retention (including tritium),*
- *Melt damage, melt layer propagation & metal splashing,*
- *Dust generation and mobilisation by various processes,*
- *Erosion products on in-vessel diagnostics,*

All above aspects are to be properly assessed and quantified.

→ Analysis of all materials retrieved from the vessel :

- ***Tiles of plasma-facing components (divertor, limiters)***
- ***Diagnostic components***
- ***Dust***
- ***Smears of Be from remote handling robotic arms used for in-torus work.***

Broad application of Be in JET has boosted and forced development of material handling procedures and analysis methods.

Analysis is NOT an isolated activity, but an crucial step in material migration studies.

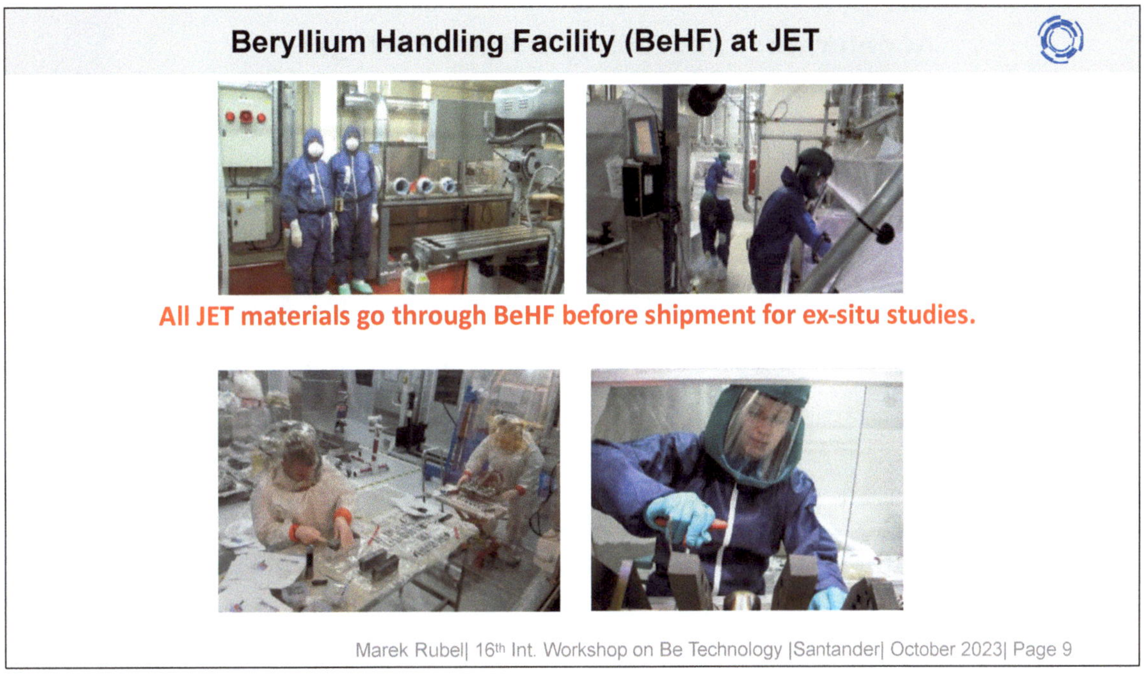
Beryllium Handling Facility (BeHF) at JET
All JET materials go through BeHF before shipment for ex-situ studies.
Marek Rubel| 16th Int. Workshop on Be Technology |Santander| October 2023| Page 9

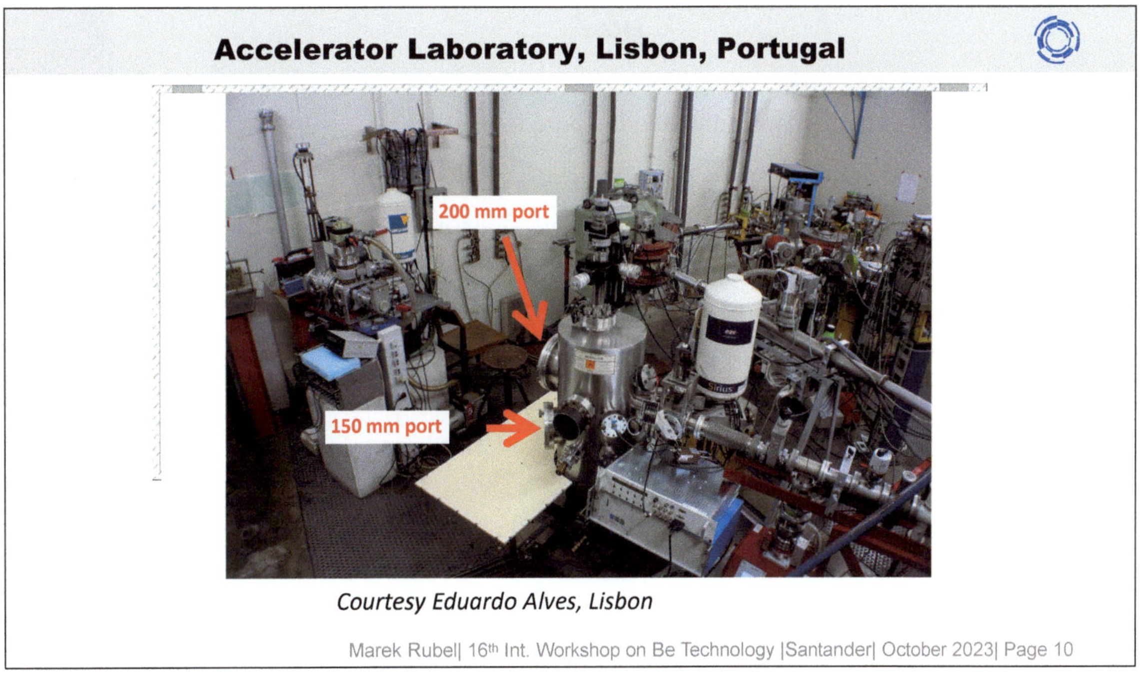
Accelerator Laboratory, Lisbon, Portugal
200 mm port
150 mm port
Courtesy Eduardo Alves, Lisbon
Marek Rubel| 16th Int. Workshop on Be Technology |Santander| October 2023| Page 10

Accelerator Laboratory, Lisbon, Portugal

Courtesy Eduardo Alves, Lisbon

Marek Rubel| 16th Int. Workshop on Be Technology |Santander| October 2023| Page 10

How?

A mosaic (alternatively: ”ZOO”) **of over 50 different methods has been used in our studies of wall components from many controlled fusion devices.**

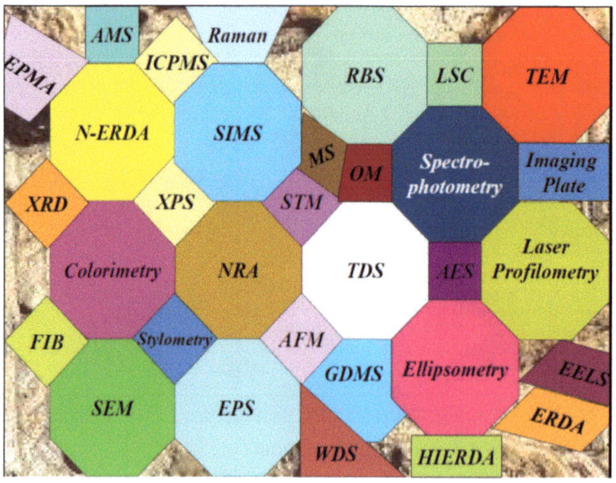

Marek Rubel| 16th Int. Workshop on Be Technology |Santander| October 2023| Page 11

Electron-based methods

Photoelectrons and Aurger electrons

Be $1s^2 2s^2$

X-ray Photoelectron Spectr. (XPS)

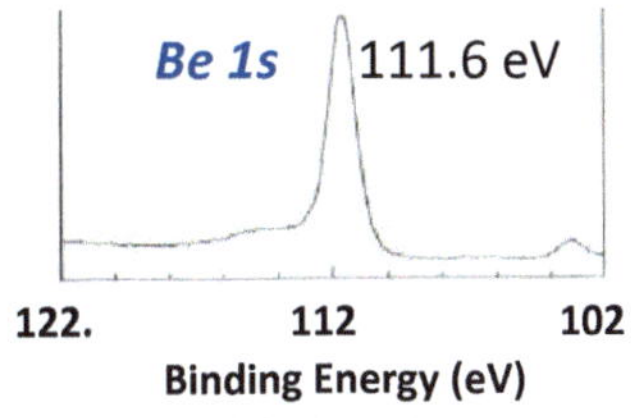

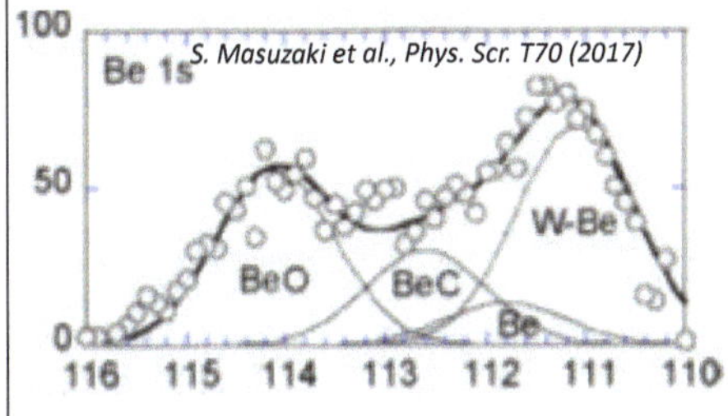

Auger Electron Spectr. (AES)

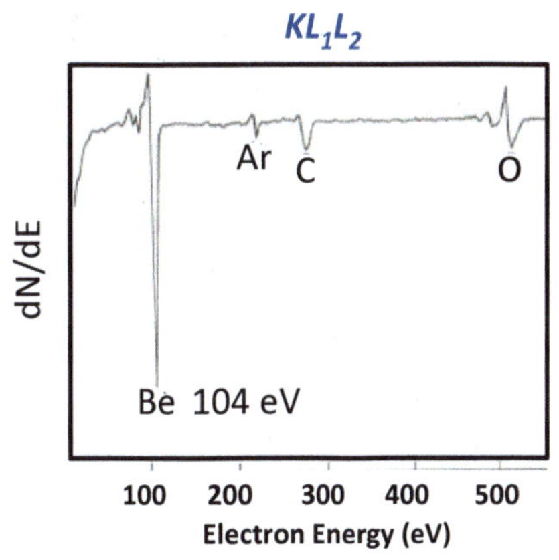

Very shallow information depth.

Depth profiling requires sputtering.

Signal attenuation by impurity species.

13

Energy Disperisve X-ray Spectroscopy: Dust from JET

EDX: X-ray emission: Kα_1 108.5 eV → strong signal attenuation by other species on surfaces.

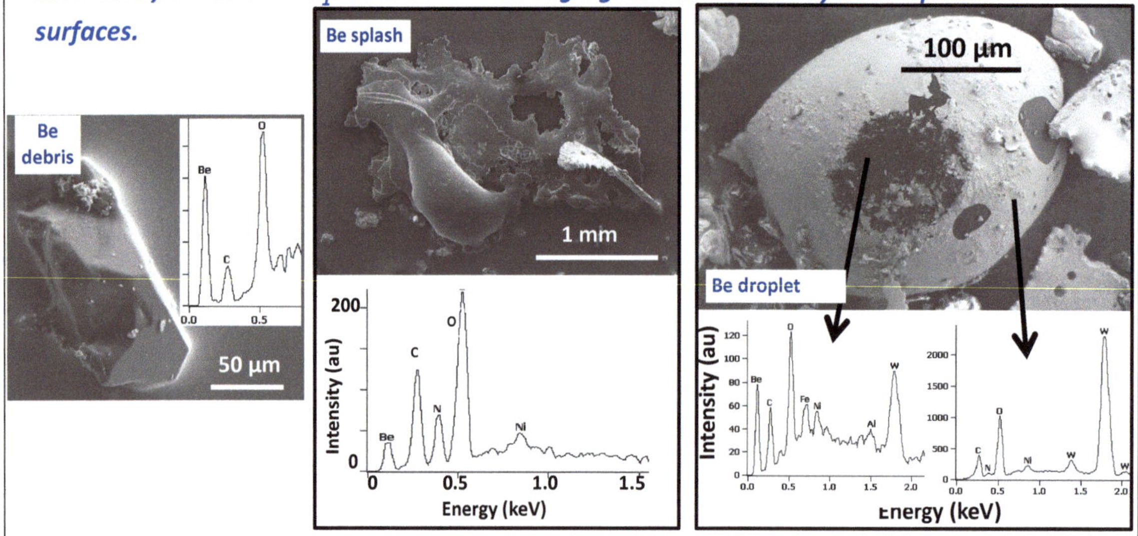

14

Be dust: co-deposits and their internal structure

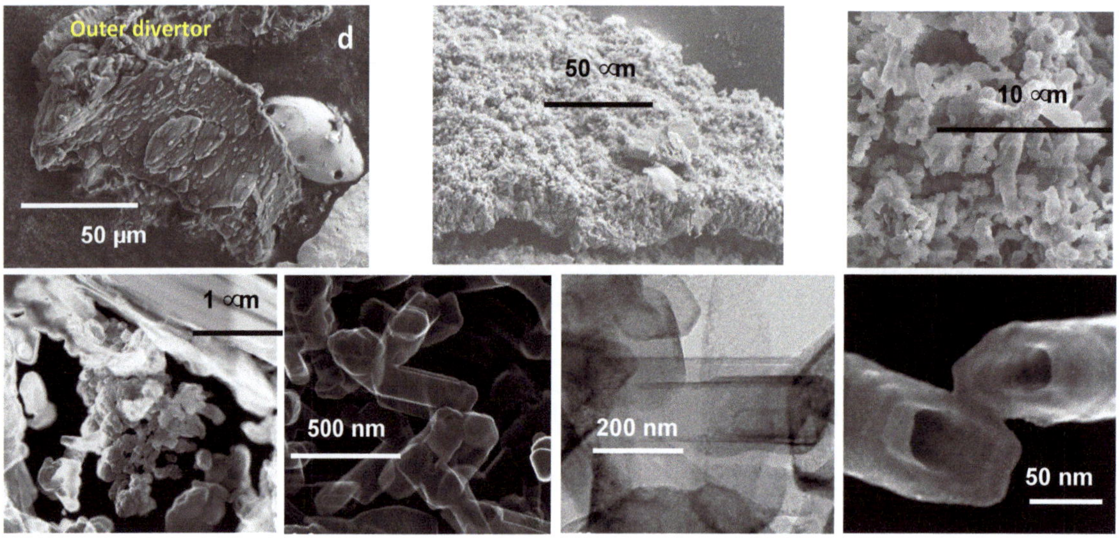

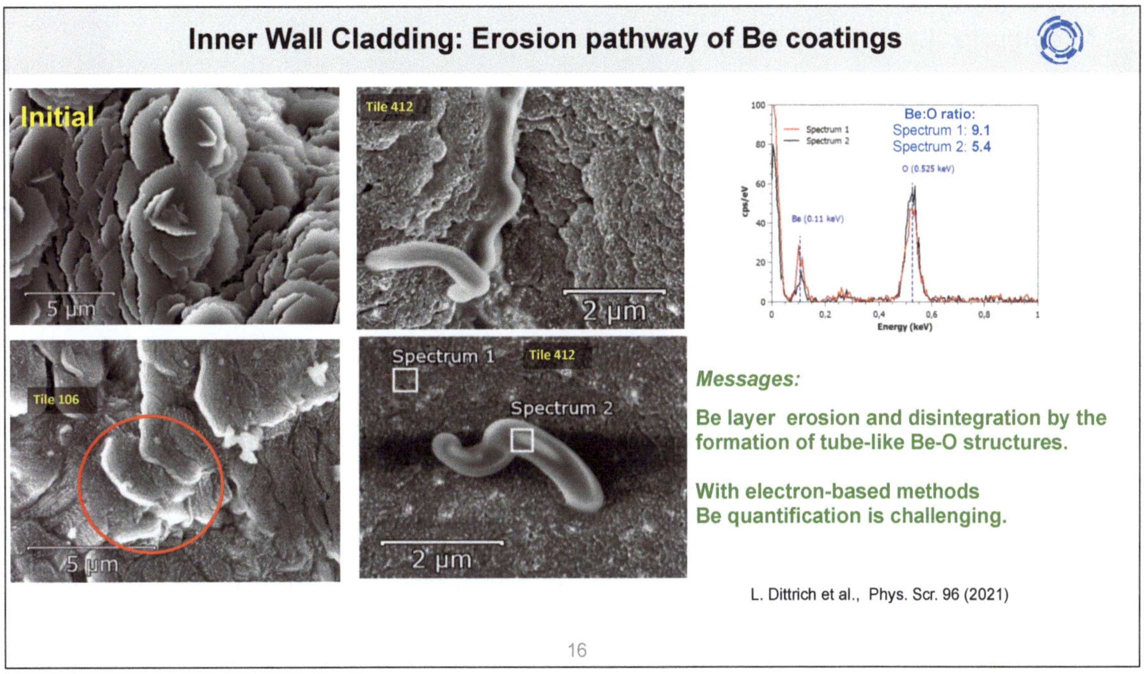
Inner Wall Cladding: Erosion pathway of Be coatings
Initial
5 µm
Tile 412
2 µm
Tile 106
5 µm
Spectrum 1
Tile 412
Spectrum 2
2 µm
Spectrum 1
Spectrum 2
Be:O ratio:
Spectrum 1: 9.1
Spectrum 2: 5.4
O (0.525 keV)
Be (0.11 keV)
cps/eV
Energy (keV)
Messages:
Be layer erosion and disintegration by the formation of tube-like Be-O structures.
With electron-based methods
Be quantification is challenging.
L. Dittrich et al., Phys. Scr. 96 (2021)
16

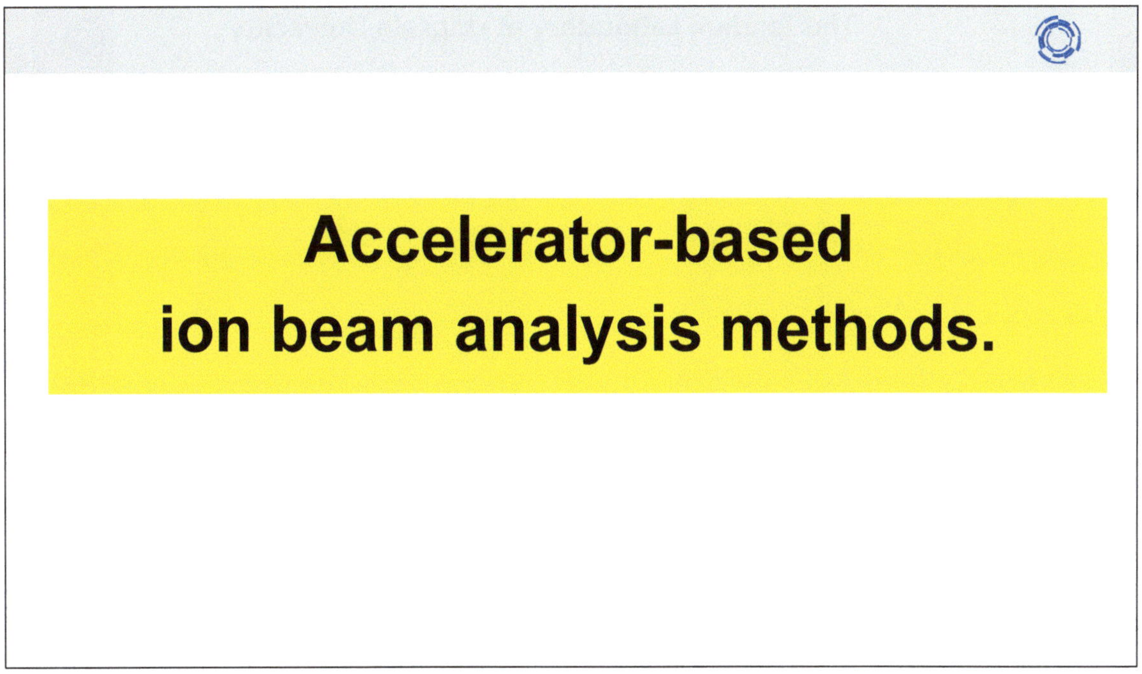
Accelerator-based
ion beam analysis methods.

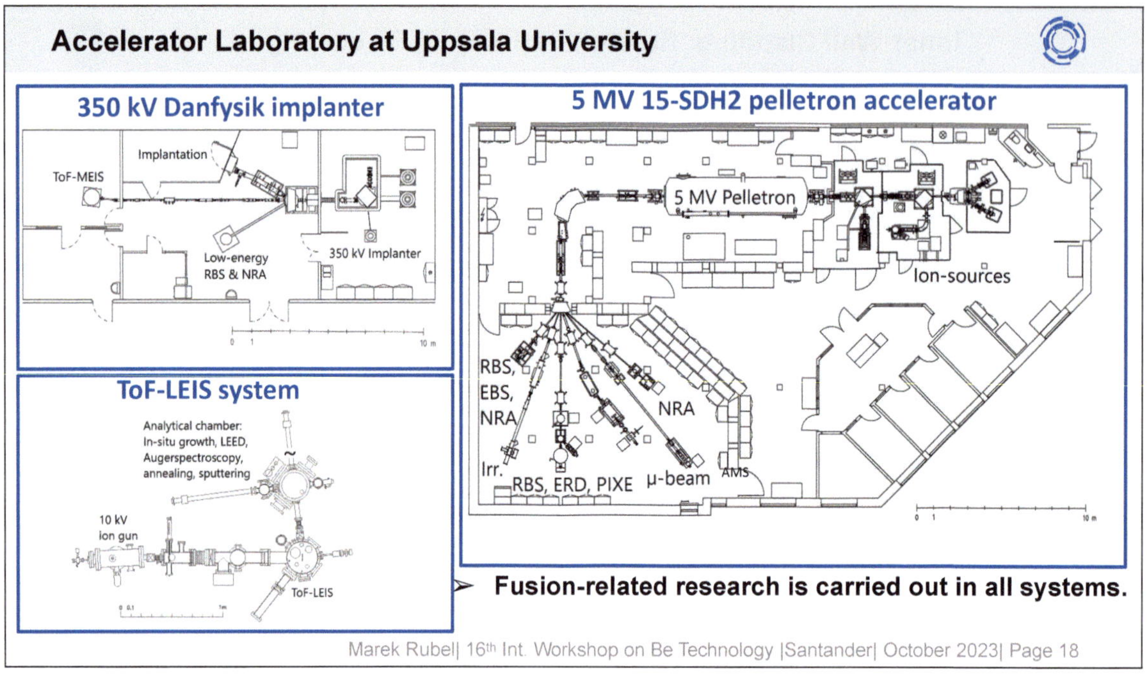
Accelerator Laboratory at Uppsala University
350 kV Danfysik implanter
Implantation
ToF-MEIS
Low-energy
RBS & NRA
350 kV Implanter
0 1
10 m
5 MV 15-SDH2 pelletron accelerator
5 MV Pelletron
Ion-sources
RBS,
EBS,
NRA
NRA
Irr.
RBS, ERD, PIXE
μ-beam
AMS
ToF-LEIS system
Analytical chamber:
In-situ growth, LEED,
Augerspectroscopy,
annealing, sputtering
10 kV
ion gun
ToF-LEIS
Fusion-related research is carried out in all systems.
Marek Rubel| 16th Int. Workshop on Be Technology |Santander| October 2023| Page 18

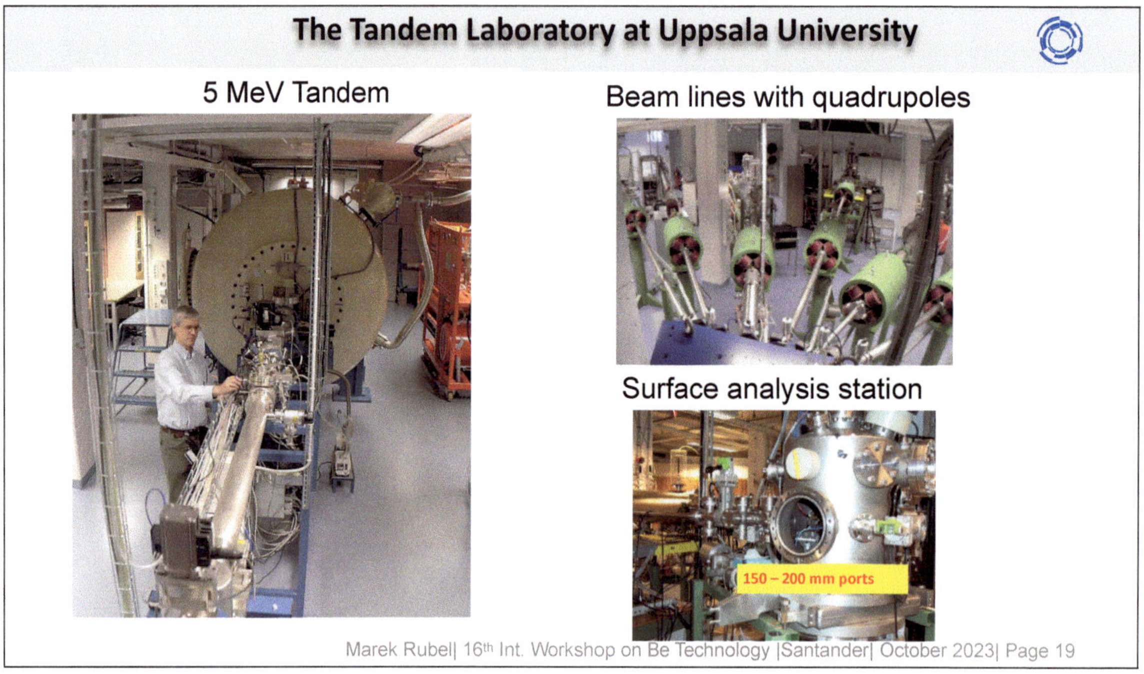
The Tandem Laboratory at Uppsala University
5 MeV Tandem
Beam lines with quadrupoles
Surface analysis station
150 – 200 mm ports
Marek Rubel| 16th Int. Workshop on Be Technology |Santander| October 2023| Page 19

Why accelerator-based IBA techniques?

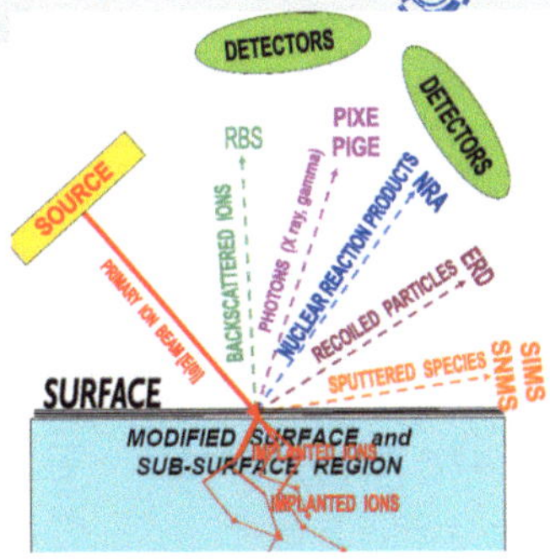

- **Efficiency:**
  - *Combination of various techniques in one system.*
  - *Analysis of many elements and isotopes in the same system.*
  - *Relatively quick analysis over large areas.*
- **Sensitivity & Selectivity & Quantification** *(no standards).*
- **Neither special sampling nor sample preparation needed** *(in many cases).*
- **Depth profiling** *(limited in some cases).*
- **Chemical state of atoms is often of secondary importance.** *(materials retrieved from devices are transported in air).*

Marek Rubel| 16th Int. Workshop on Be Technology |Santander| October 2023| Page 20

Nuclear Processes for ion beam analysis of Be

Scattering

$^{9}Be(p,p)^{9}Be$

$^{9}Be(d,d)^{9}Be$

$^{9}Be(\alpha,\alpha)^{9}Be$

Reaction	Q value (MeV)
$^{9}Be(p,d)^{8}Be$	0.56
$^{9}Be(p,a)^{6}Li$	2.13
$^{9}Be(d,p)^{10}Be$	4.59
$^{9}Be(p,t)^{8}Be$	4.59
$^{9}Be(d,a)^{7}Li$	7.15
$^{9}Be(^{3}He,p)^{11}B$	**10.32**
$^{9}Be(^{6}Li,p)^{14}C$	15.13

Marek Rubel| 16th Int. Workshop on Be Technology |Santander| October 2023| Page 21

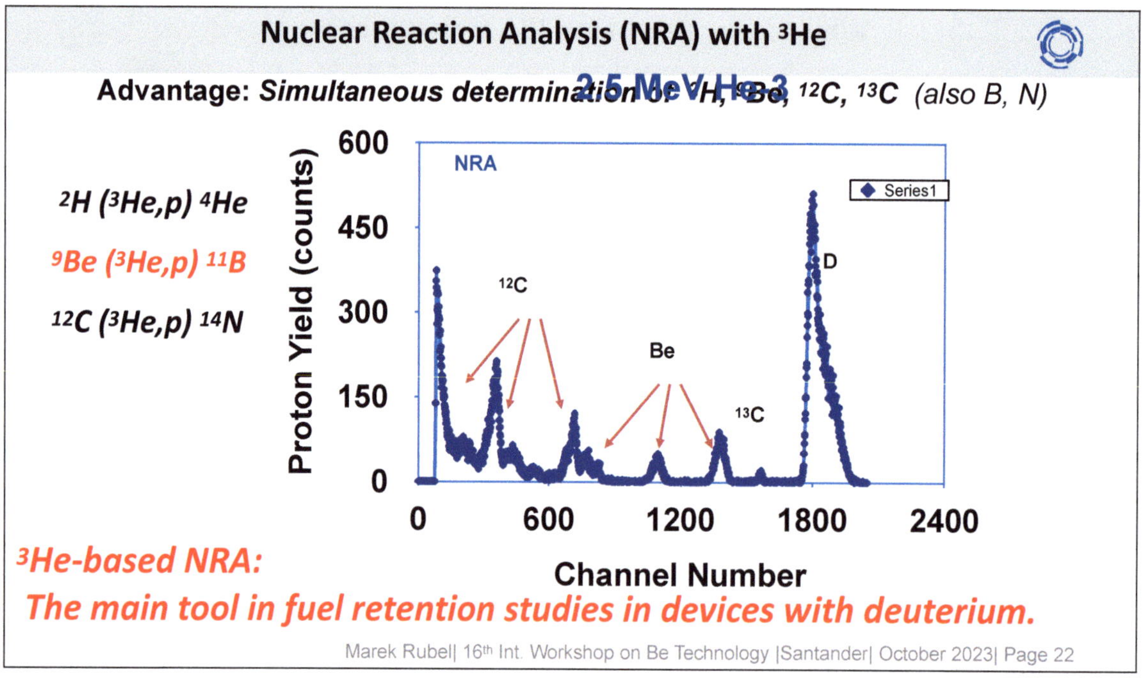
Nuclear Reaction Analysis (NRA) with 3He
Advantage: Simultaneous determination of 2H, 9Be, 12C, 13C (also B, N)
2.5 MeV He-3
2H (3He,p) 4He
9Be (3He,p) 11B
12C (3He,p) 14N
Proton Yield (counts)
600
450
300
150
0
NRA
Series1
12C
Be
13C
D
0
600
1200
1800
2400
Channel Number
3He-based NRA:
The main tool in fuel retention studies in devices with deuterium.
Marek Rubel| 16th Int. Workshop on Be Technology |Santander| October 2023| Page 22

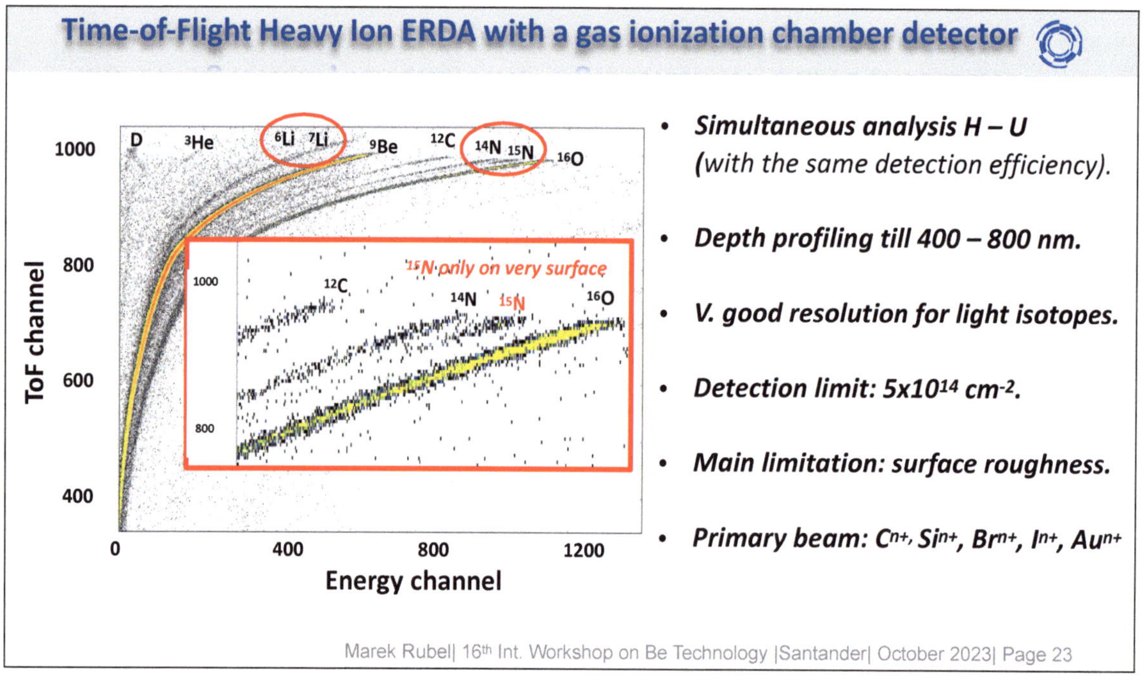
Time-of-Flight Heavy Ion ERDA with a gas ionization chamber detector
ToF channel
1000
800
600
400
D
3He
6Li 7Li
9Be
12C
14N 15N
16O
15N only on very surface
12C
14N 15N
16O
1000
800
0
400
800
1200
Energy channel
Simultaneous analysis H – U (with the same detection efficiency).
Depth profiling till 400 – 800 nm.
V. good resolution for light isotopes.
Detection limit: 5x10^14 cm^-2.
Main limitation: surface roughness.
Primary beam: Cn+, Sin+, Brn+, In+, Aun+
Marek Rubel| 16th Int. Workshop on Be Technology |Santander| October 2023| Page 23

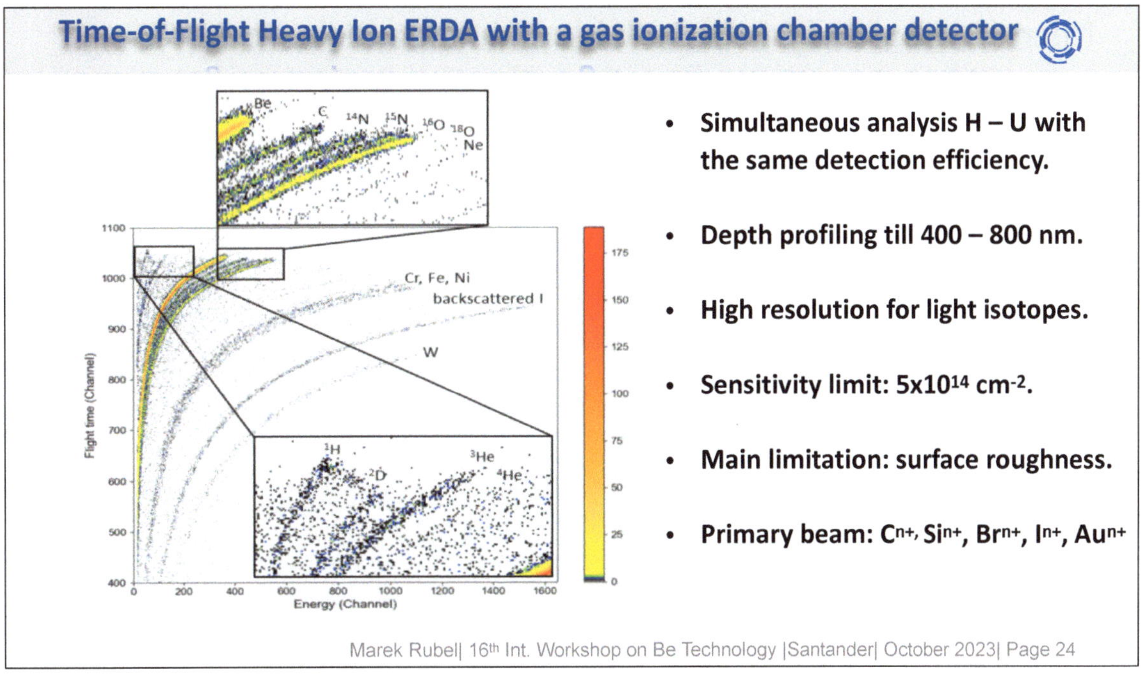
Time-of-Flight Heavy Ion ERDA with a gas ionization chamber detector
Be
C
^{14}N
^{15}N
^{16}O
^{18}O
Ne
Cr, Fe, Ni
backscattered I
W
^{1}H
^{2}D
^{3}He
^{4}He
Flight time (Channel)
Energy (Channel)
• Simultaneous analysis H – U with the same detection efficiency.
• Depth profiling till 400 – 800 nm.
• High resolution for light isotopes.
• Sensitivity limit: 5x10^{14} cm^{-2}.
• Main limitation: surface roughness.
• Primary beam: C^{n+}, Si^{n+}, Br^{n+}, I^{n+}, Au^{n+}
Marek Rubel| 16th Int. Workshop on Be Technology |Santander| October 2023| Page 24

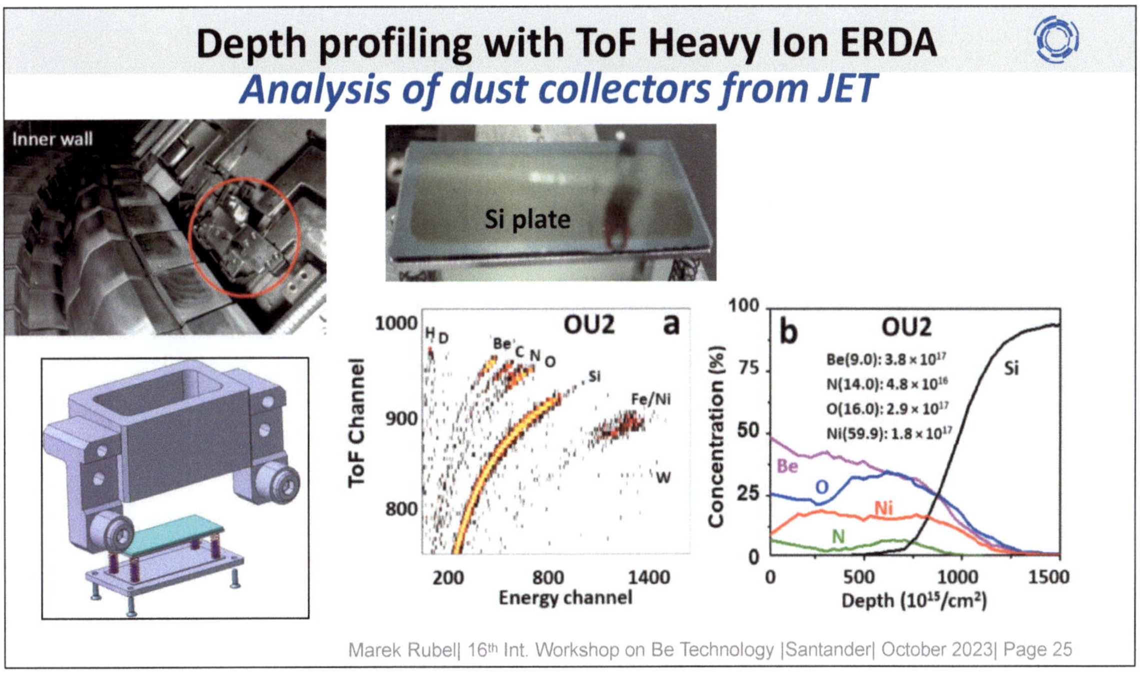
Depth profiling with ToF Heavy Ion ERDA
Analysis of dust collectors from JET
Inner wall
Si plate
OU2 a
ToF Channel
Energy channel
H D
Be C N O
Si
Fe/Ni
W
b OU2
Be(9.0): 3.8 × 10^{17}
N(14.0): 4.8 × 10^{16}
O(16.0): 2.9 × 10^{17}
Ni(59.9): 1.8 × 10^{17}
Si
Be
O
Ni
N
Concentration (%)
Depth (10^{15}/cm^{2})
Marek Rubel| 16th Int. Workshop on Be Technology |Santander| October 2023| Page 25

Microbeam: Analysis of individual dust particles from JET

Combination of microscopy, EDS μ-NRA and μ-PIXE

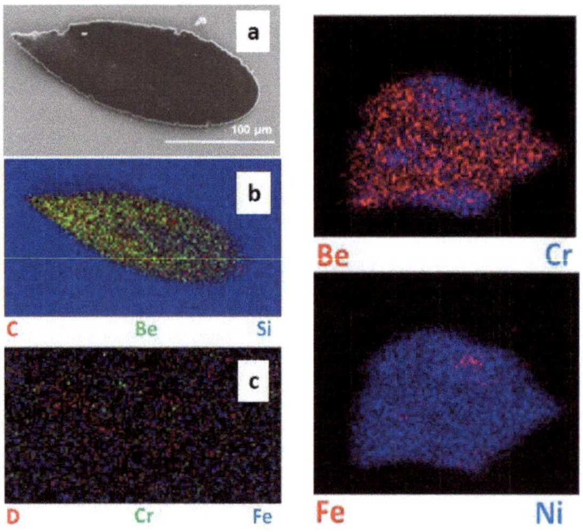

Can such analytical procedure be applied in studies of small beryllides pellets?

Marek Rubel| 16th Int. Workshop on Be Technology |Santander| October 2023| Page 26

Cross-sections

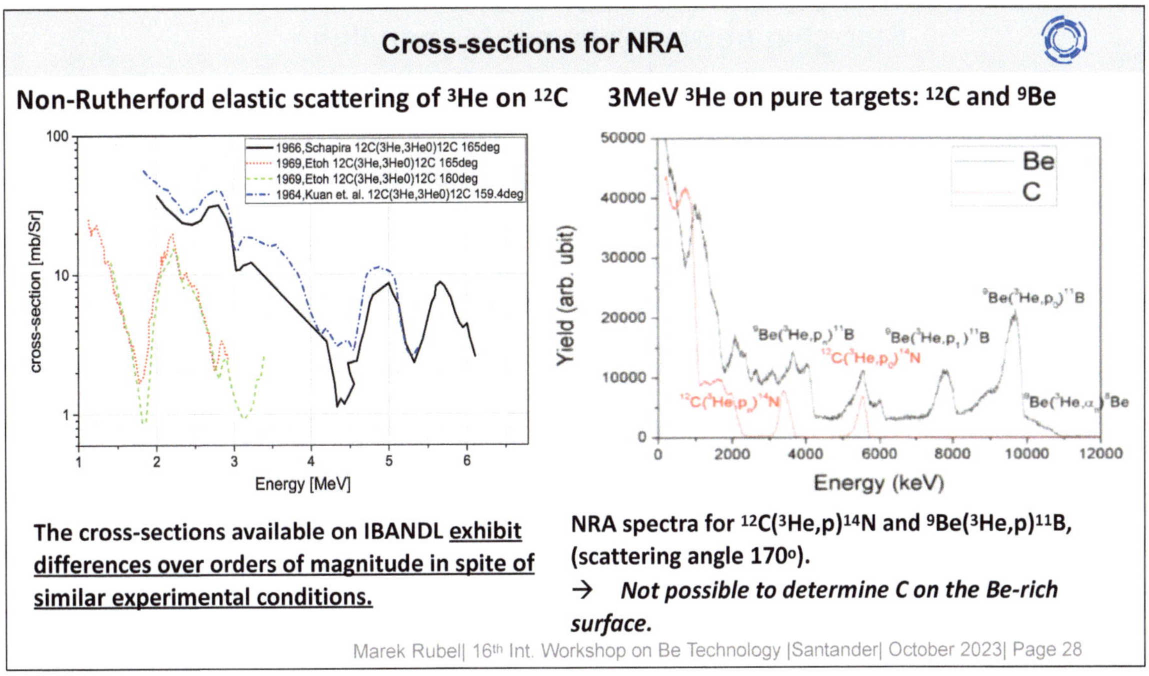
Cross-sections for NRA
Non-Rutherford elastic scattering of ³He on ¹²C
3MeV ³He on pure targets: ¹²C and ⁹Be
1966,Schapira 12C(3He,3He0)12C 165deg
1969,Etoh 12C(3He,3He0)12C 165deg
1969,Etoh 12C(3He,3He0)12C 160deg
1964,Kuan et. al. 12C(3He,3He0)12C 159.4deg
cross-section [mb/Sr]
Energy [MeV]
Yield (arb. ubit)
Energy (keV)
Be
C
⁹Be(³He,p₀)¹¹B
⁹Be(³He,p₂)¹¹B
⁹Be(³He,p₁)¹¹B
¹²C(³He,p₀)¹⁴N
¹²C(³He,p₁)¹⁴N
⁹Be(³He,α)⁸Be
The cross-sections available on IBANDL exhibit differences over orders of magnitude in spite of similar experimental conditions.
NRA spectra for ¹²C(³He,p)¹⁴N and ⁹Be(³He,p)¹¹B, (scattering angle 170°).
→ Not possible to determine C on the Be-rich surface.
Marek Rubel| 16th Int. Workshop on Be Technology |Santander| October 2023| Page 28

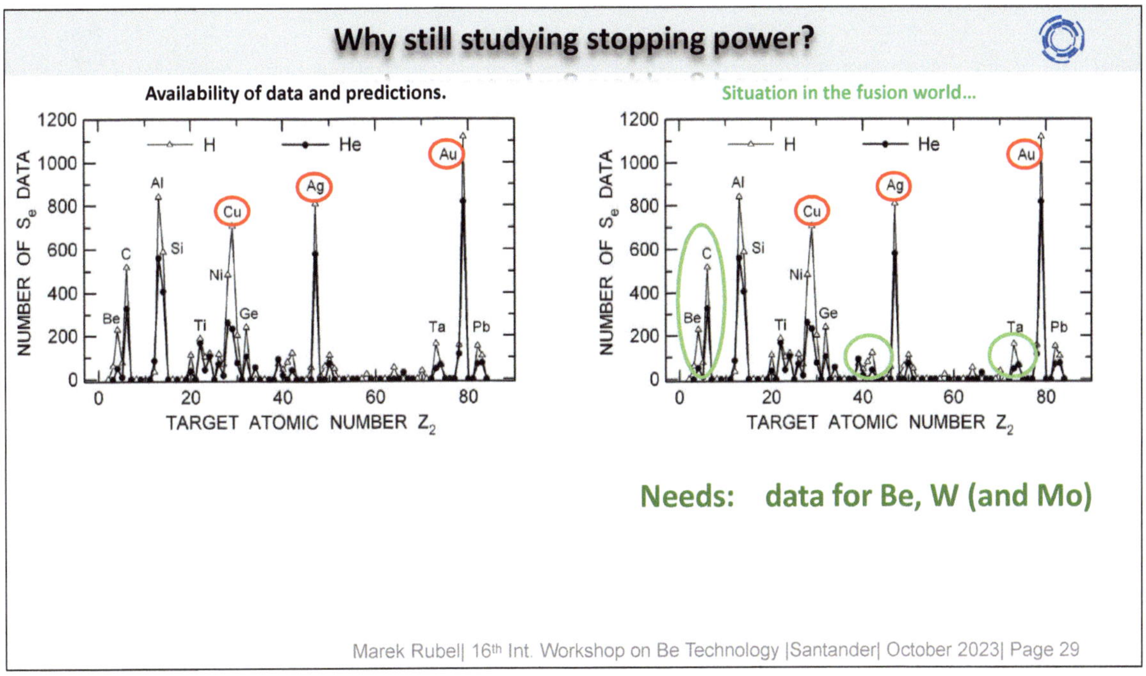
Why still studying stopping power?
Availability of data and predictions.
Situation in the fusion world...
NUMBER OF Sₑ DATA
TARGET ATOMIC NUMBER Z₂
H
He
Be
C
Al
Si
Ti
Ni
Cu
Ge
Ag
Ta
Au
Pb
Needs: data for Be, W (and Mo)
Marek Rubel| 16th Int. Workshop on Be Technology |Santander| October 2023| Page 29

Stopping powers relevant for beryllium

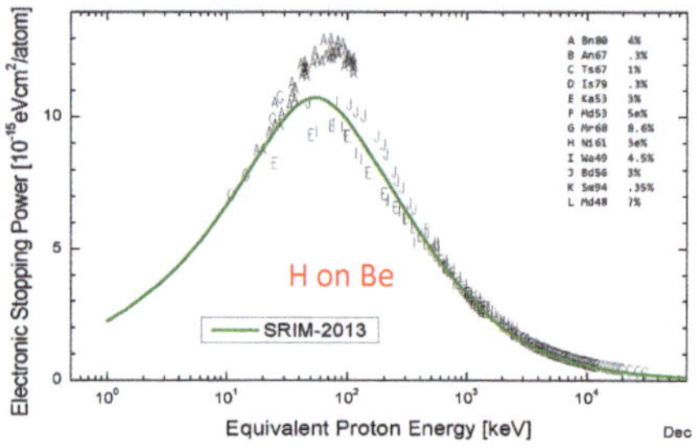

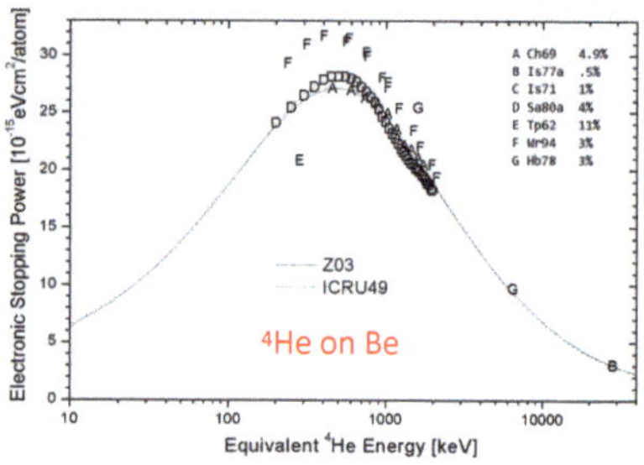

- ***Limited data at low energies for both H and He ions.***
- ***Discrepancy at higher energies.***
- ***For classical IBA-energies two distinct datasets – SRIM represents an average.***
- ***Significant impact on analysis and simulation of erosion – deposition.***
- ***Coordinated Research Project ongoing at IAEA to build a data base.***

Electronic Stopping Power of Matter for Ions: available data from International Atomic Energy Agency (IAEA): [https://www-nds.iaea.org/stopping/stopping_hydr.html.

Marek Rubel| 16th Int. Workshop on Be Technology |Santander| October 2023| Page 30

The Toolbox

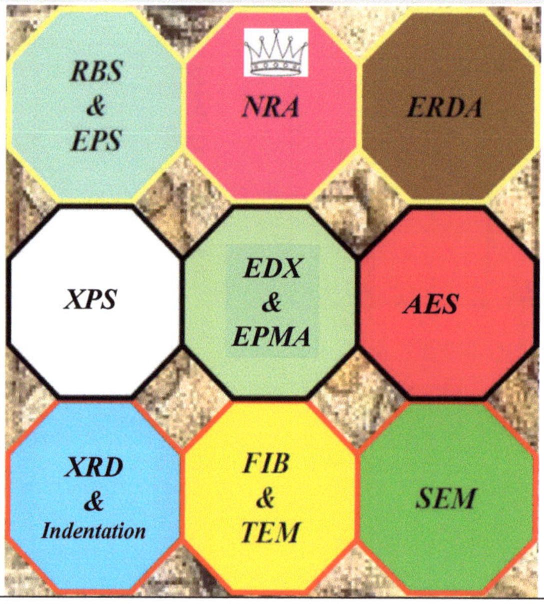

SESSION 3

Fabrication & Recycling

3.1 Reaction kinetics in the leaching process of beryllium from beryl

Suguru Nakano, Jae-Hwan Kim, Yoshiaki Akatsu, Taehyun Hwang, Shota Yokohama, Yutaka Sugimoto, Ryuta Kasada, Masaru Nakamichi

Beryllium is refined from ores including beryllium. Beryl ($Be_3Al_2Si_6O_18$), one of the beryllium ores, has a very stable crystal structure and is insoluble in acids. In the conventional beryllium refining process from beryl, beryl is melted at a high temperature of nearly 2,000 °C and then quenched to destroy its crystal structure. As a result, beryl is transformed into a form that is soluble in acid. This energy-intensive process not only raises the cost of beryllium, but also has a significant environmental impact, thus requiring technological innovation. On the other hand, the blanket of a fusion reactor is loaded with a large amount of beryllium as a neutron multiplier. It is our extremely important mission to secure stable beryllium resources for the realization of fusion reactors. Therefore, we have developed a new innovative beryllium refining technology.

We have succeeded in completely dissolving beryl at low temperatures below 300 °C by combining chemical treatment and microwave heating. The chemical reaction with beryl that occurs under basic aqueous solution plays an important role in this new technology. In this study, we investigated the reaction process between beryl and aqueous sodium hydroxide solution under microwave heating from a kinetic point of view. As a result, we found that a mixed kinetic model based on process that proceeds by instantaneous surface nucleation and diffusion-controlled processes can well explain the dissolution reaction of beryllium in beryl. The apparent activation energies of beryllium, aluminum, and silicon in beryl were estimated to be 67-72 kJ/mol.

Presenting Author: Suguru Nakano
Presenting Author's Affiliation: National Institutes for Quantum and Radiological Science and Technology, Japan
Email: nakano.suguru@qst.go.jp

1

The 16th International Workshop on Beryllium Technology (BeWS-16)
Santander, Spain
19-20.10.2023

Reaction kinetics in the leaching process of beryllium from beryl

○Suguru Nakano[a], Jae-Hwan Kim[a], Yoshiaki Akatsu[a], Taehyun Hwang[a], Yutaka Sugimoto[a], Ryuta Kasada[b], Masaru Nakamichi[a]

[a]National Institutes for Quantum Science and Technology
[b]Institute for Materials Research, Tohoku University

QST

2

Introduction ; Beryllium in Fusion Reactors

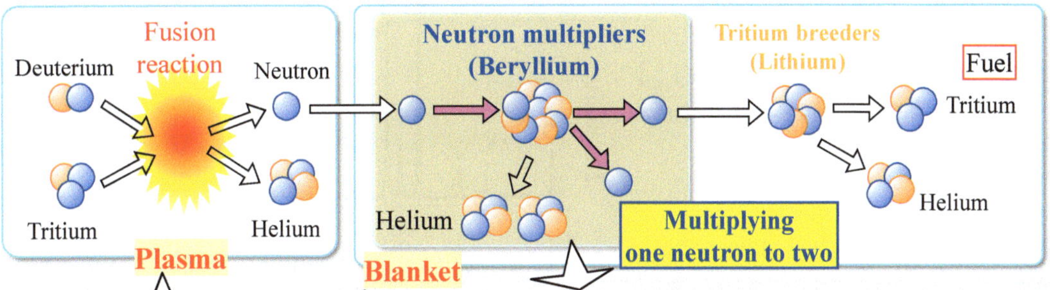

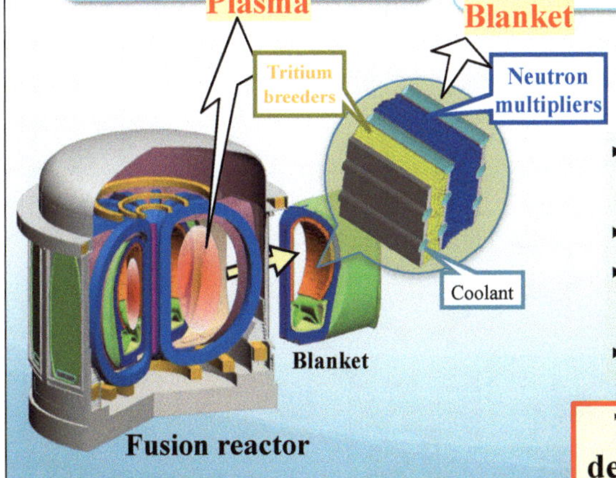

- As neutron multipliers, **Beryllium metal or its alloys** are candidates.
  - About 350 tons of Be is required for a fusion DEMO reactor.
  - Blanket must be replaced about every 4 years.
  - Total production of Be in the world is about 300 tons/year.
  - Beryllium production is not enough for fusion.

To secure stable beryllium resources, we are developing new beryllium refining technologies.

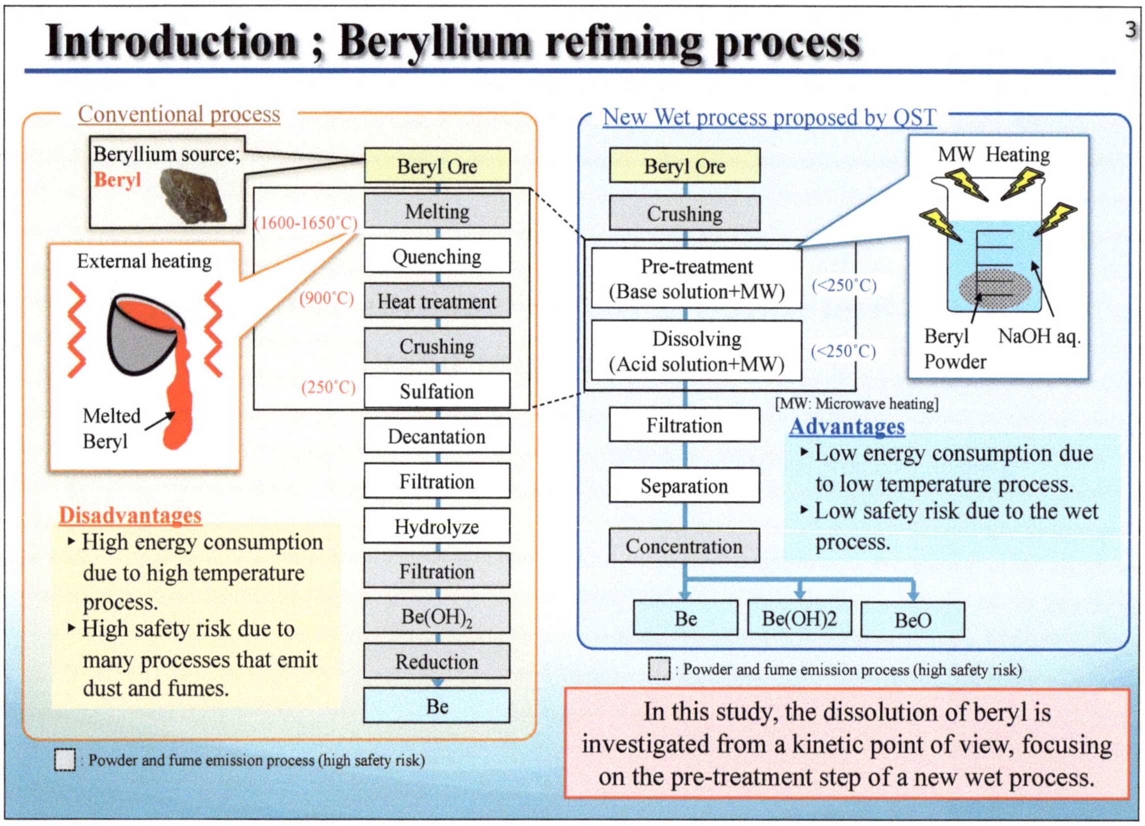

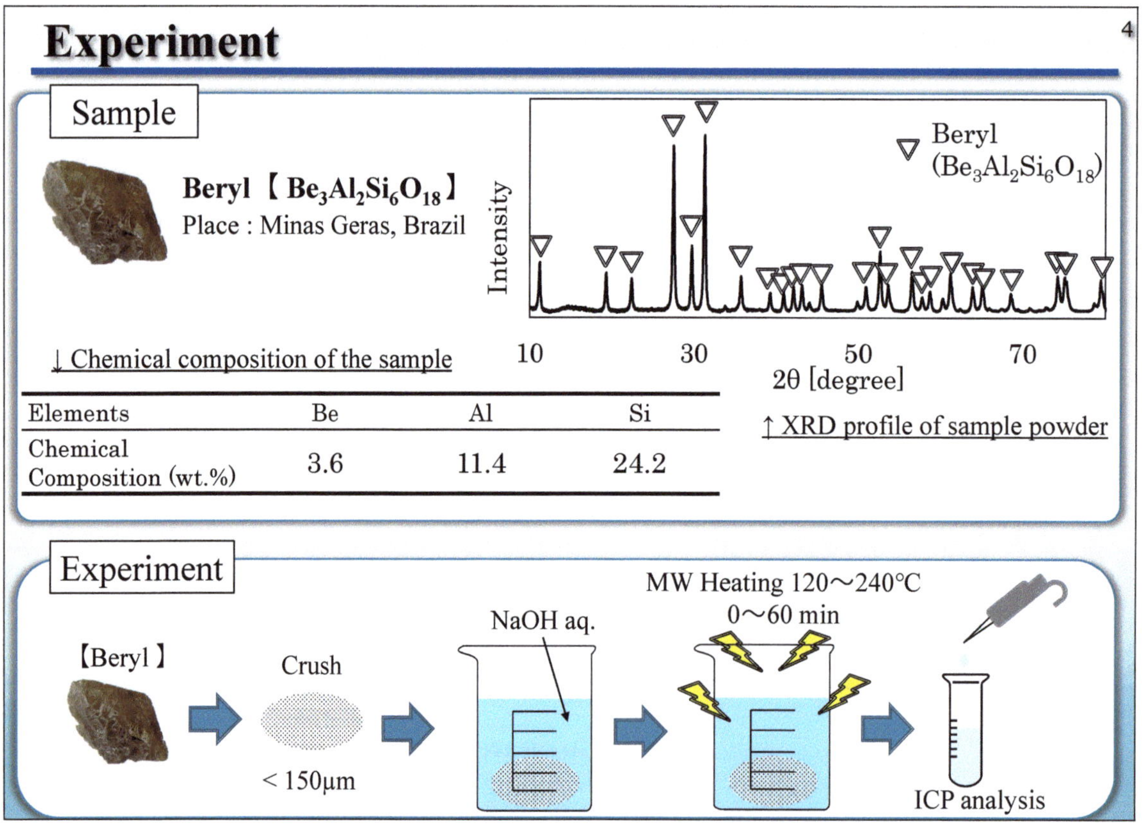

Elements	Be	Al	Si
Chemical Composition (wt.%)	3.6	11.4	24.2

5

Results and Discussion ; Shrinking Core Model

Leaching Rate

$$f = \frac{(\text{Mass of element dissolved in solution [mg]})}{(\text{Mass of element in beryl [mg]})}$$

- Apply four shrinking core models to determine the reaction process between beryl and NaOH solution in the pre-treatment process.

Schematic diagram of the shrinking core model

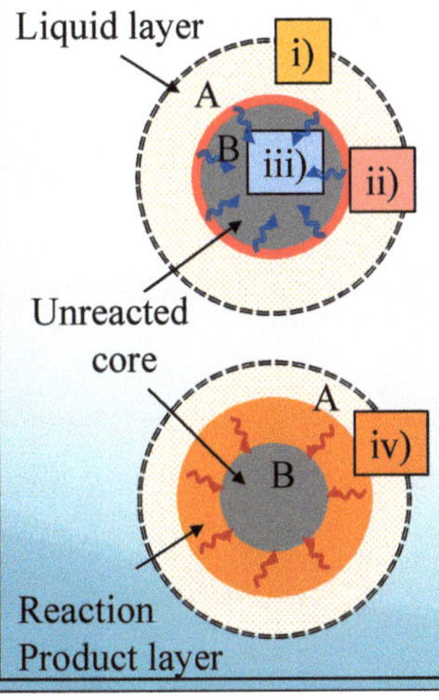

i) $f = k_f t$

→In cases where the leaching rate is controlled by diffusion through the liquid layer.

ii) $1 - (1 - f)^{1/3} = k_r t$

→ In cases where the reaction is controlled by the chemistry of the solid surface.

iii) $1 - 3(1 - f)^{2/3} + 2(1 - f) = k_d t$

→ In cases where the reaction is controlled by diffusion through the reaction product layer.

iv) $\left[1 - (1 - f)^{1/3}\right]^2 = k_m t$

→Mixed kinetic model based on processes by instantaneous surface nucleation and diffusion-controlled processes.

6

Results and Discussion ; Model Fittings

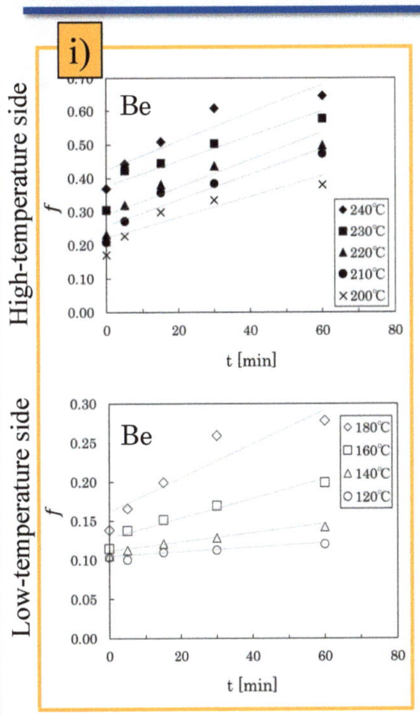

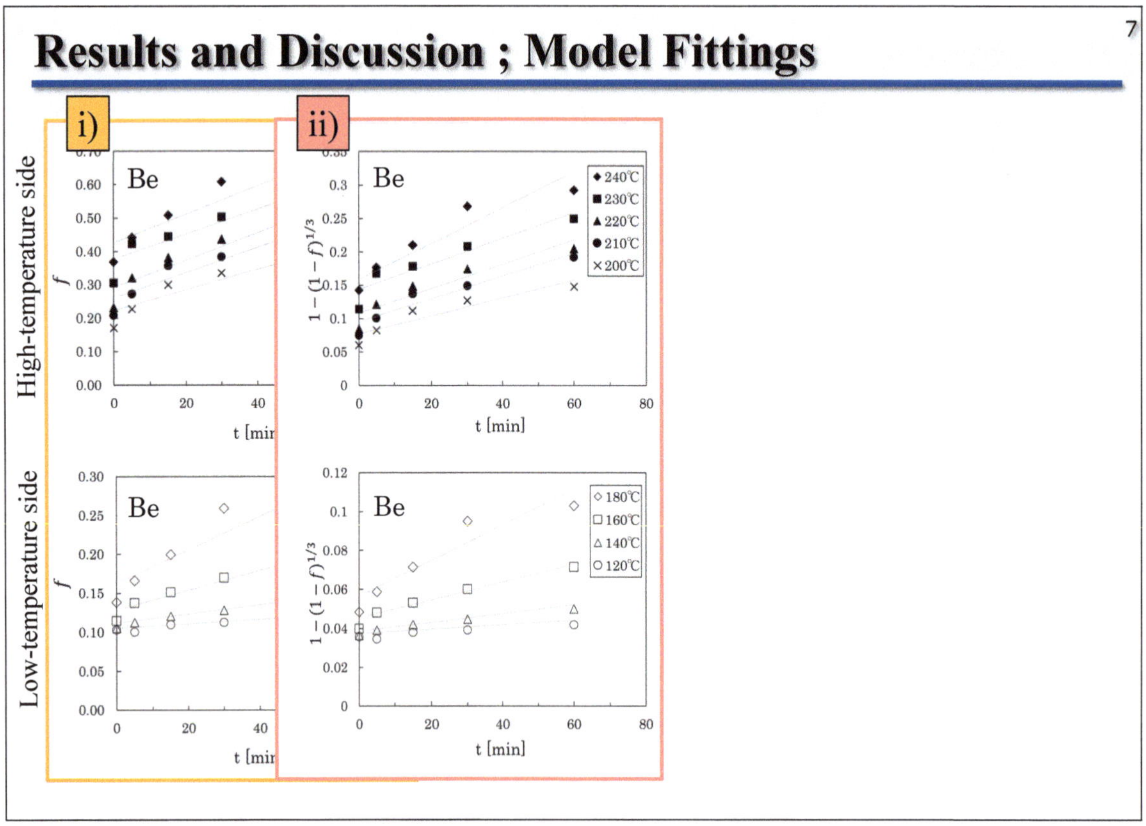
Results and Discussion ; Model Fittings
7
i)
ii)
High-temperature side
Low-temperature side
Be
f
1 − (1 − f)^1/3
t [min]
◆ 240℃
■ 230℃
▲ 220℃
● 210℃
× 200℃
◇ 180℃
□ 160℃
△ 140℃
○ 120℃

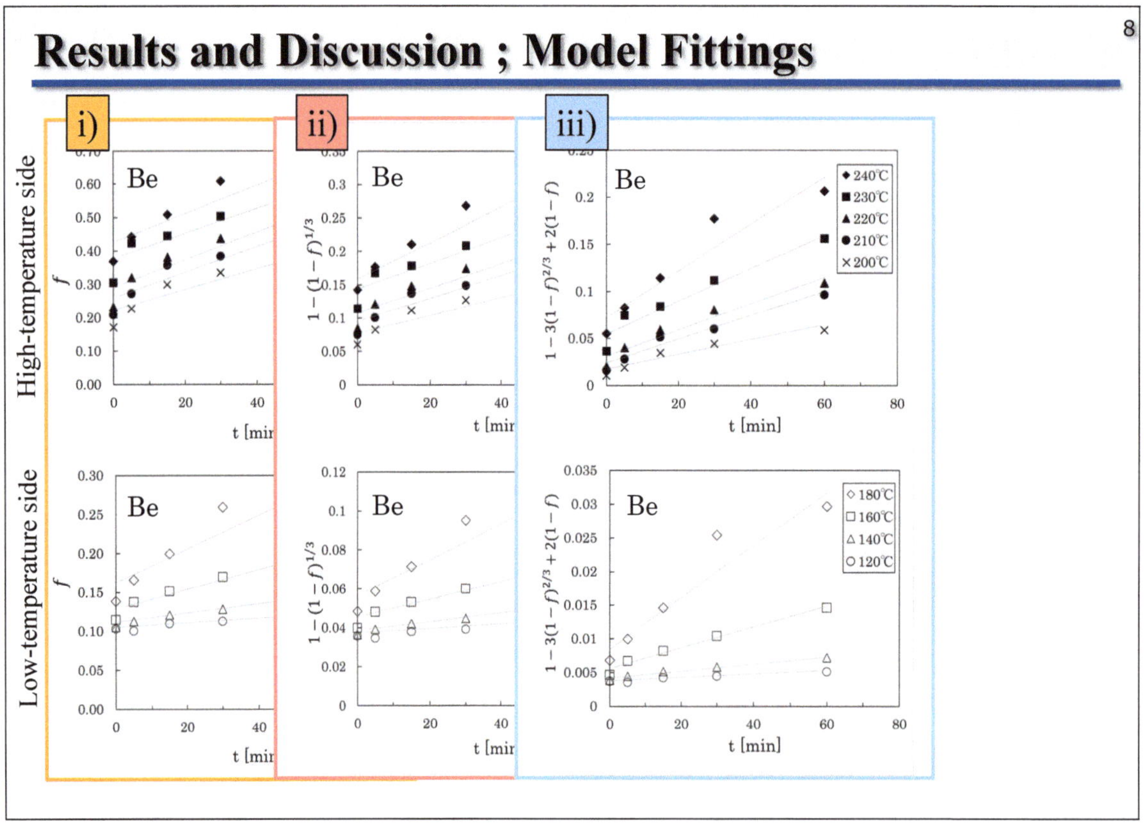
Results and Discussion ; Model Fittings
8
i)
ii)
iii)
High-temperature side
Low-temperature side
Be
f
1 − (1 − f)^1/3
1 − 3(1 − f)^2/3 + 2(1 − f)
t [min]
◆ 240℃
■ 230℃
▲ 220℃
● 210℃
× 200℃
◇ 180℃
□ 160℃
△ 140℃
○ 120℃

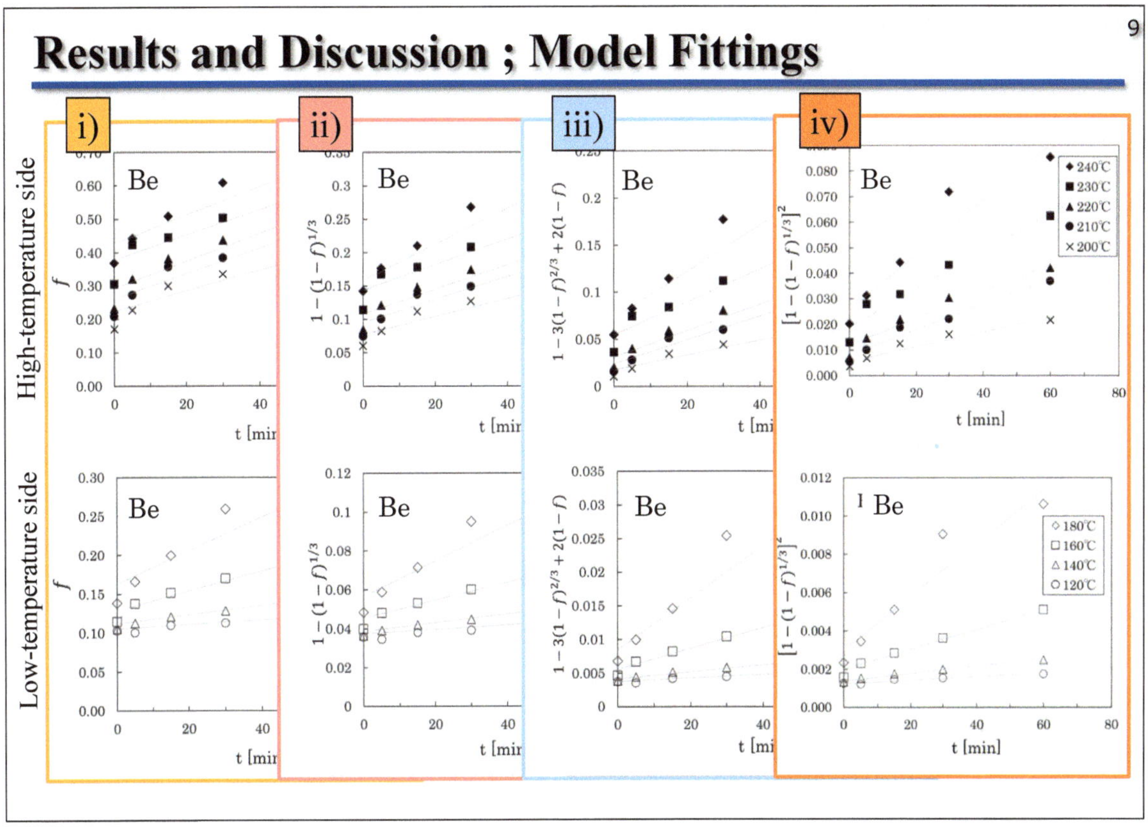
9
Results and Discussion ; Model Fittings
i)
ii)
iii)
iv)
High-temperature side
Low-temperature side
Be
f
1 − (1 − f)^1/3
1 − 3(1 − f)^2/3 + 2(1 − f)
[1 − (1 − f)^1/3]^2
t [min]
◆240℃
■230℃
▲220℃
●210℃
×200℃
◇180℃
□160℃
△140℃
○120℃

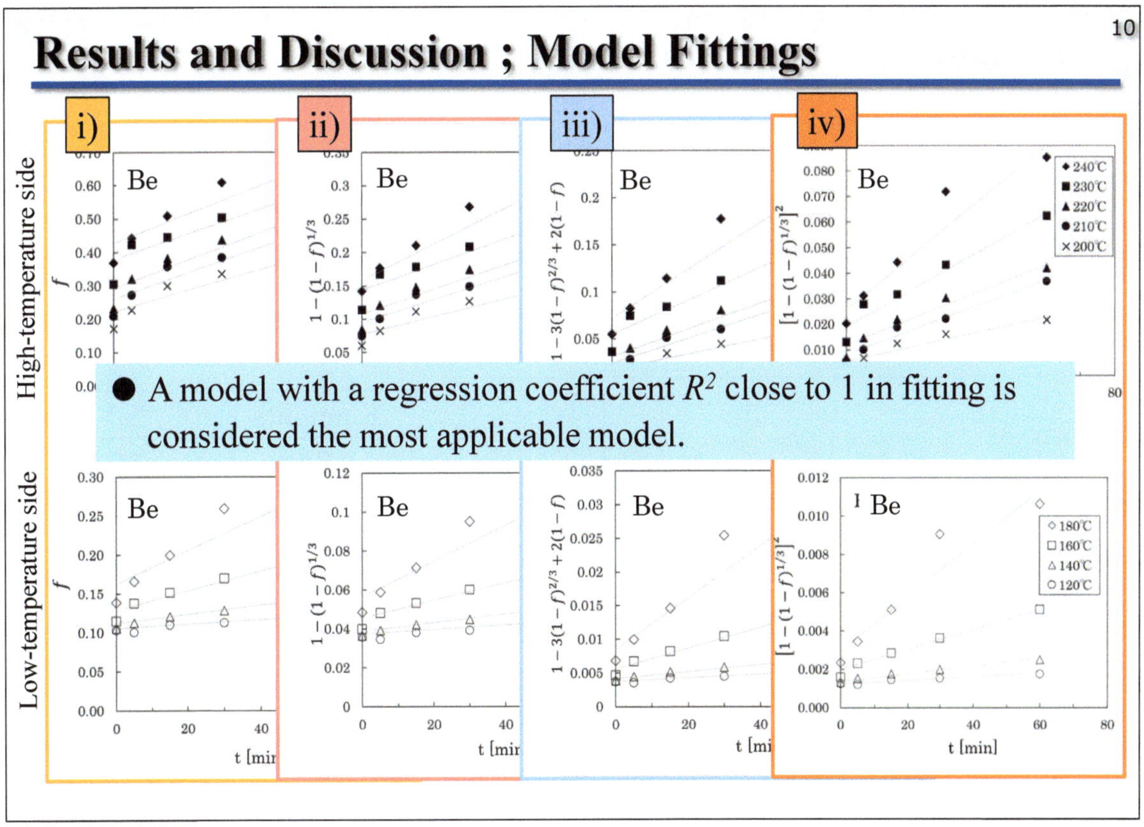
10
Results and Discussion ; Model Fittings
i)
ii)
iii)
iv)
High-temperature side
Low-temperature side
Be
● A model with a regression coefficient R² close to 1 in fitting is considered the most applicable model.
◆240℃
■230℃
▲220℃
●210℃
×200℃
◇180℃
□160℃
△140℃
○120℃
t [min]

Regression coefficients for each model fittings

11

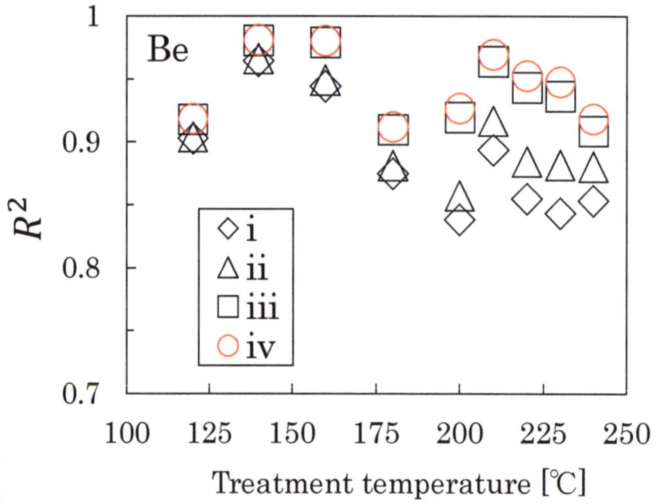

◆The value of the regression coefficient R^2 for each fitting was relatively high for that of the model in equation iv).

◆During the pre-treatment process, beryl was partly dissolved by the NaOH solution, suggesting that the reaction process is dominated by a mixed kinetic model.

Arrhenius plots in the pretreatment process

12

$$k = Ae^{-E_a/RT}$$

◆An Arrhenius plot was drawn from the obtained reaction rate constants and the apparent activation energy E_a was determined.

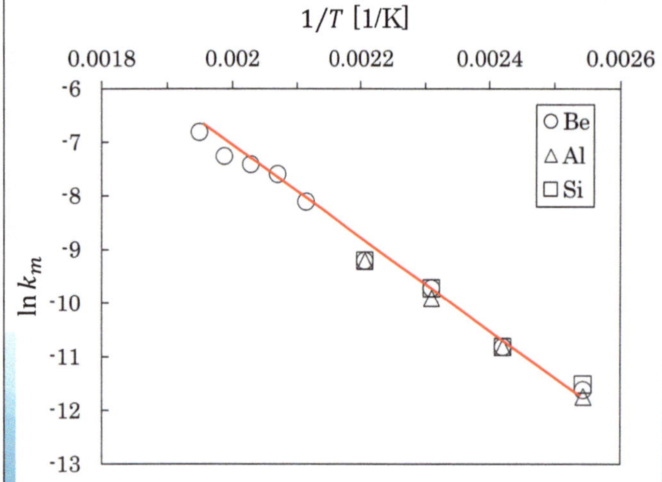

Elements	E_a [kJ/mol]
Be	67.9
Al	71.2
Si	67.5

◆For all Be, Al and Si elements, the slopes of the Arrhenius plots were almost the same.

◆Beryl was found to be uniformly dissolved by NaOH solution.

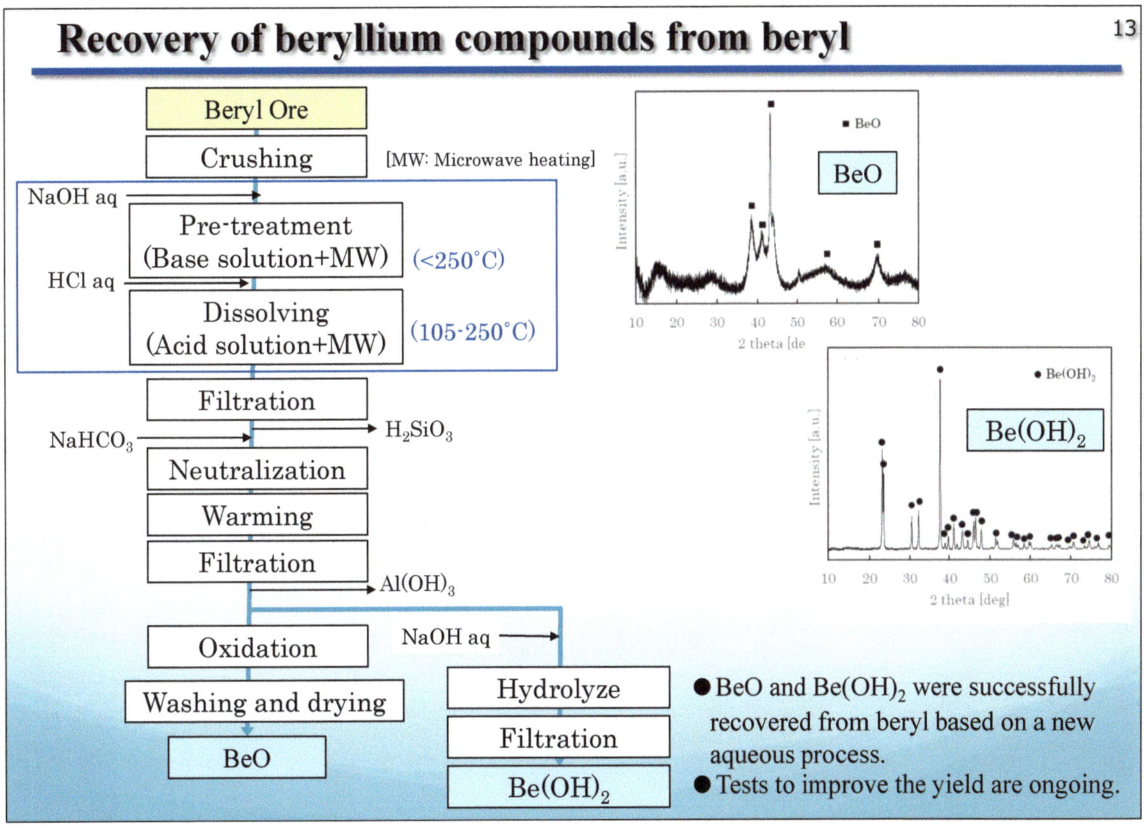
Recovery of beryllium compounds from beryl
13
Beryl Ore
Crushing
[MW: Microwave heating]
NaOH aq
Pre-treatment
(Base solution+MW)
(<250°C)
HCl aq
Dissolving
(Acid solution+MW)
(105-250°C)
Filtration
NaHCO3
H2SiO3
Neutralization
Warming
Filtration
Al(OH)3
Oxidation
NaOH aq
Washing and drying
BeO
Hydrolyze
Filtration
Be(OH)2
BeO
Be(OH)2
2 theta [deg]
Intensity [a.u.]
● BeO and Be(OH)2 were successfully recovered from beryl based on a new aqueous process.
● Tests to improve the yield are ongoing.

MiRESSO
MiRESSO Co. Ltd.
Corporate Profile
October 2023
Copyright © 2023 MiRESSO Inc. All Rights Reserved

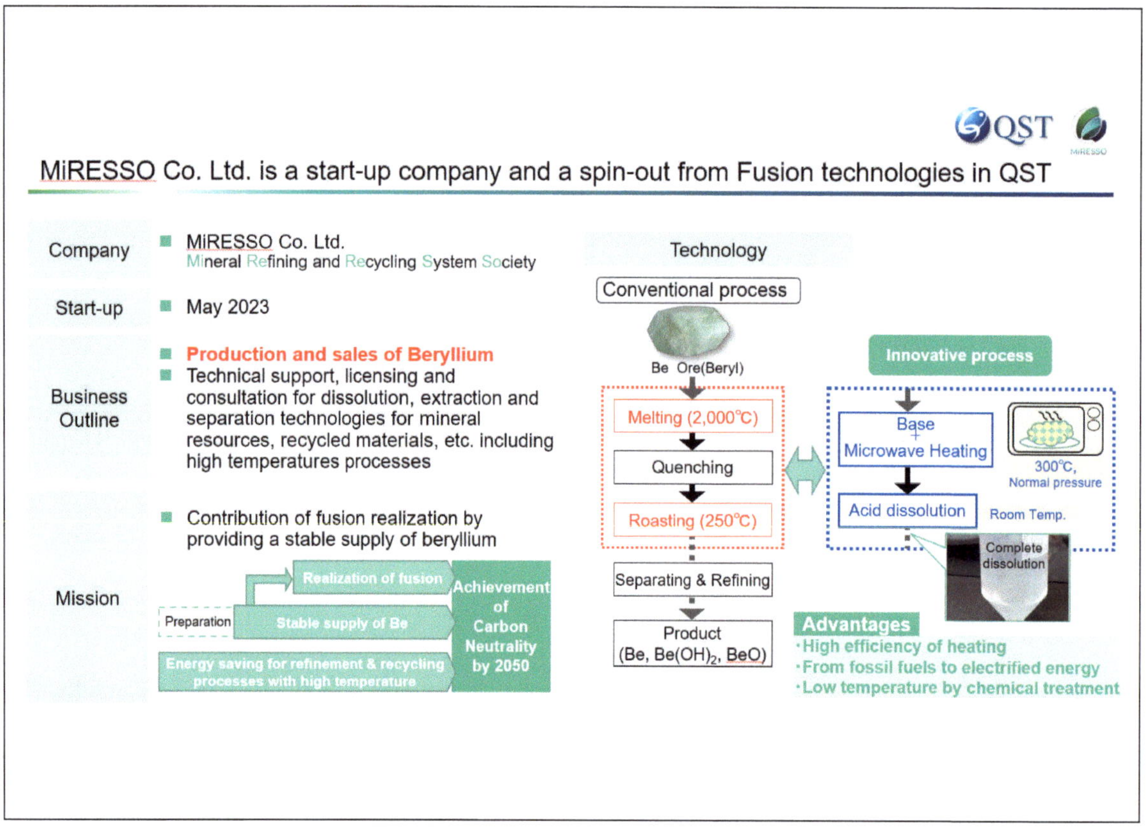

Summary

16

- The new wet process developed at QST is a new beryllium refining technology that can be processed safely at low temperatures.
- During the pre-treatment process, beryl was partly dissolved by the NaOH solution, suggesting that the reaction process is dominated by a mixed kinetic model.
- For all Be, Al and Si elements, the slopes of the Arrhenius plots were almost the same. Beryl was found to be uniformly dissolved by NaOH solution.
- Also, based on a new wet process, beryllium compounds have been successfully recovered from beryl ores.
- The new wet process can refine beryllium in an energy-saving method, making it easy to increase production, and is expected to make a significant contribution to the demand for large amounts of beryllium resources in DEMO reactors.

3.2 Recycling of neutron-irradiated beryllium and titanium beryllide

Vladimir Chakin, Ramil Gaisin, Pavel Pereslavtsev

Beryllides are promising materials for advanced neutron multipliers of the DEMO solid breeding blanket. The key point is a large amount of beryllium that needs to be used even for only one DEMO reactor. Therefore, development of effective recycling processes of neutron-irradiated beryllium is essential to implement the fusion reactor project.

This study proposes an approach to the problem of decontamination of irradiated beryllium, which consist in the removal of transmuted tritium from irradiated beryllium and titanium beryllide by high-temperature annealing and a removal of metal radioactive isotopes (^{60}Co, ^{57}Co, ^{55}Fe, ^{63}Ni, ^{45}Ca, ^{46}Sc, etc.). Tritium can be completely removed from the irradiated beryllium-based materials at 600-1000 °C. The annealing temperature and the exposure time depend on the irradiation temperature. The removal of the radioactive impurities occurs by cyclic dissolving in chloric acid, addition of $La(NO_3)_3$ and sodium hydroxide, filtration and calcination at 900 °C. After first tests, a level of decontamination of irradiated beryllium from radioactive metallic impurities was 96% but after improving the method, it can reach 100% efficiency.

Regarding the recycling process, general approaches are proposed and analyzed to solve a recovery problem for irradiated titanium beryllide blocks after operation in the DEMO reactor.

Presenting Author: Vladimir Chakin
Presenting Author's Affiliation: Karlsruhe Institute of Technology, Germany
Email: vladimir.chakin@kit.edu

The 16th International Workshop on Beryllium Technology (BeWS-16)
Santander, Spain, 19-20.10.2023

Recycling of neutron-irradiated beryllium and titanium beryllide

Vladimir Chakin[1], Ramil Gaisin[1], Pavel Pereslavtsev[2]

[1]*Karlsruhe Institute of Technology, Institute for Applied Materials, Hermann-von-Helmholtz-Platz 1,*
76344 Eggenstein-Leopoldshafen, Germany
[2]*Karlsruhe Institute of Technology, Institute of Neutron Physics and Reactor Technology, Hermann-von-Helmholtz-Platz 1,*

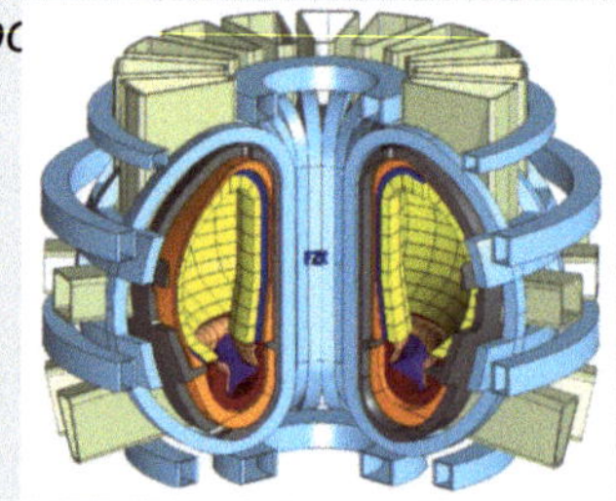

Outline

Irradiated beryllium materials must be recycled after use in fusion reactor blankets

Results of decontamination of irradiated beryllium obtained within ISTC project No. 3381: 2007-2009, RIAR-SCK-INL

➢Surface decontamination

➢Tritium removal by high temperature annealing

➢Removal of solid radioactive impurities

Conclusions

Irradiated beryllium materials must be recycled

KIT Karlsruhe Institute of Technology

"Dissolution and recovery of beryllium from beryl using a novel wet process with microwave heating" by Suguro Nakano, Jae-Hwan Kim, Taehyun Hwang, Ryuta Kasada, Masaru Nakamichi, Nuclear Materials and Energy 30 (2022) 101113

Citation:

"For the DEMO fusion reactors and beyond to achieve selfsufficiency of tritium fuel, an initial loading of beryllium is required at approximately 500 tons per fusion reactor. In addition, beryllium blankets are replaced every four years, requiring new neutron-multiplier materials. However, the current worldwide production of beryllium is only approximately 300 tons per year.

Hence, securing stable beryllium resources is an urgent issue for the operation of DEMO fusion reactors".

2 Institute for Applied Materials – Applied Materials Physics

ISTC project No. 3381 „Elaboration of safe methods for radioactive beryllium waste decontamination after operation in a nuclear reactor"

KIT Karlsruhe Institute of Technology

Research Institute of Atomic Reactors, Dimitrovgrad, RF
The work performed from August 1, 2007 to July 31, 2009

Collaborators:
SCK·CEN, Mol, Belgium, Dr. Frank Druits
INL, USA, Dr. Glen Longhurst

Tasks of the project:

1. THEORETICAL DEFINITION OF THE FINAL ELEMENTARY COMPOSITION AND CONTRIBUTION OF RADIOACTIVE ELEMENTS INTO THE RESULTING RADIOACTIVITY OF IRRADIATED BERYLLIUM BLOCKS
2. STUDY ON THE EFFECT OF HIGH-DOSE NEUTRON IRRADIATION ON PHYSICAL-CHEMICAL PROPERTIES AND MICROSTRUCTURE OF IRRADIATED BERYLLIUM
3. EXPERIMENTAL STUDY ON THE FINAL ELEMENTARY COMPOSITION AND THE RADIOACTIVE ELEMENTS CONTRIBUTION INTO THE TOTAL RADIOACTIVITY OF IRRADIATED BERYLLIUM BLOCKS
4. EXPERIMENTAL EVALUATION OF THE SURFACE DECONTAMINATION EFFECTIVENESS OF IRRADIATED BERYLLIUM BLOCKS
5. DEVELOPMENT OF DECONTAMINATION METHODS FOR REMOVAL OF RADIOACTIVE METAL TRANSMUTANT-IMPURITIES FROM IRRADIATED BERYLLIUM
6. EXAMINATION OF THE HIGH-TEMPERATURE ANNEALING EFFECTS ON VOLUMETRIC DECONTAMINATION OF IRRADIATED BERYLLIUM
7. ESTIMATION OF THE RADIOACTIVITY LEVEL AND PHYSICAL–CHEMICAL STATE OF IRRADIATED BERYLLIUM BLOCK AFTER ALL TYPES OF ITS TREATMENT

CONCLUSION

3 Institute for Applied Materials – Applied Materials Physics

Irradiated Be blocks from SM-3 reactor for decontamination

Core of SM-3 reactor

Moderator Be block from neutron trap

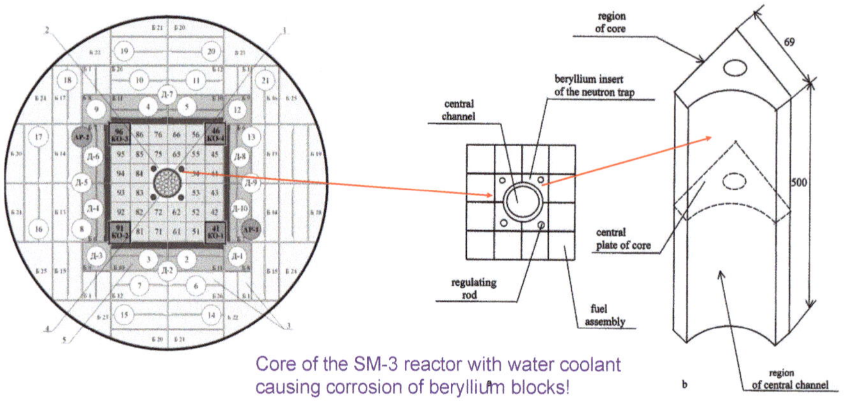

Core of the SM-3 reactor with water coolant causing corrosion of beryllium blocks!

View of irradiated Be blocks after operation until the resource neutron fluence of $6 \cdot 10^{22}$ n/cm² (E>0.1 MeV)

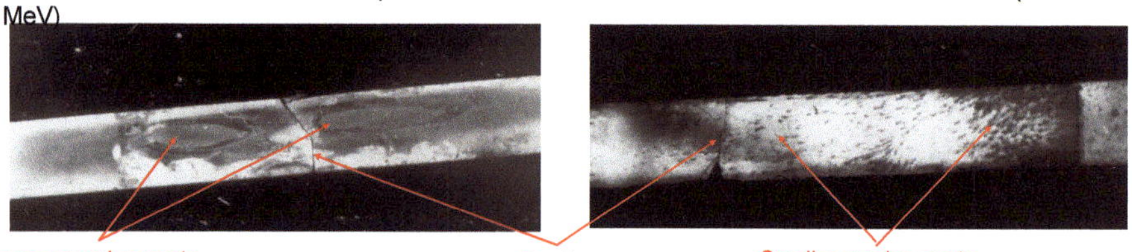

Chemical composition of block material (TE-56 Be grade by UMP production)

Element	Content, wt.%	Element	Content, wt.%	Element	Content, wt.%
Al	0.0200	Eu	0.0002	Ru	0.0010
Au	0.0001	Fe	0.1500	Sc	0.0010
B	0.0030	Hf	0.0010	Si	0.0130
Be	98.6310	K	0.0010	Sr	0.0001
Bi	0.0010	Mg	0.0005	Ta	0.0001
C	0.0800	Mn	0.0100	Ti	0.0170
Ca	0.0070	Mo	0.0020	Tl	0.0002
Cd	0.0001	Nb	0.0001	U	0.0003
Co	0.0002	Ni	0.0170	V	0.0010
Cr	0.0410	P	0.0005	W	0.0010
Cu	0.0040	Pt	0.0002	Zr	0.0150
Dy	0.0002	Rb	0.0001		

- Main impurities are as follows: Fe, C, Cr, Al, Ni, Ti, Zr, Si, Mn, Ca, Cu, B.
- U content is very low.

Contribution of radioactive α, β, γ - isotopes into resulting radioactivity of irradiated Be block

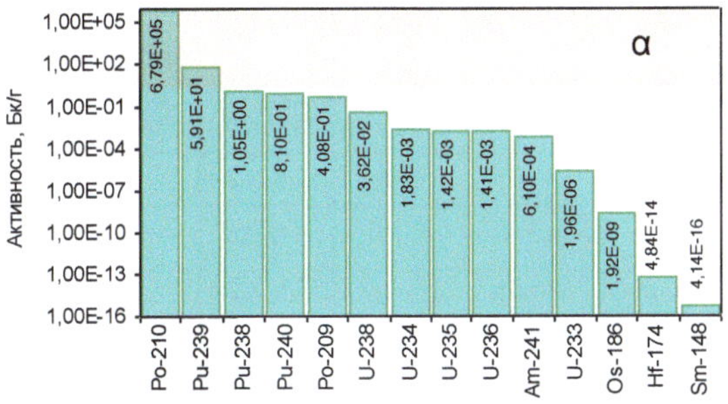

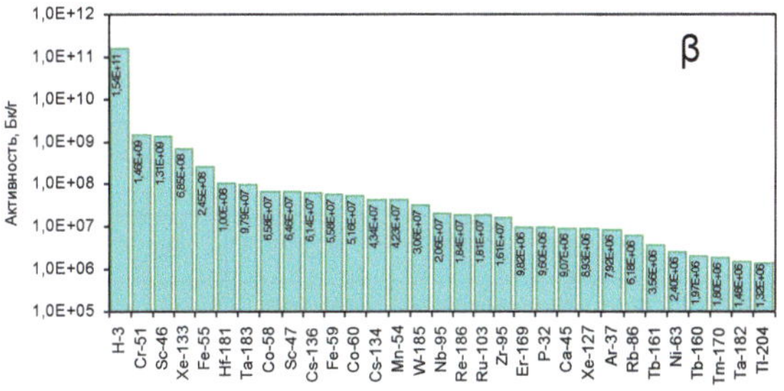

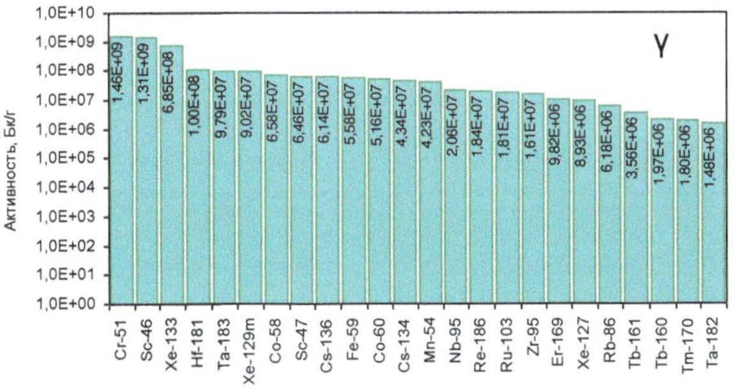

According to calculations the following isotopes make the largest contribution to total Be block activity:

H-3, Cr-51, Sc-46, Fe-55, Hf-181, Ta-183, Co-58, Sc-47, Fe-59, Co-60, Mn-54, W-185, Nb-95, Re-186, Ru-103, Zr-95, Er-169, P-32, Ca-45, Ar-37, Rb-86, Tb-161, Ni-63, Tb-160, Tm-170, Ta-18, Tl-204.

6 Institute for Applied Materials – Applied Materials Physics

Decrease of activity of main radioactive isotopes in irradiated Be on time due to decay

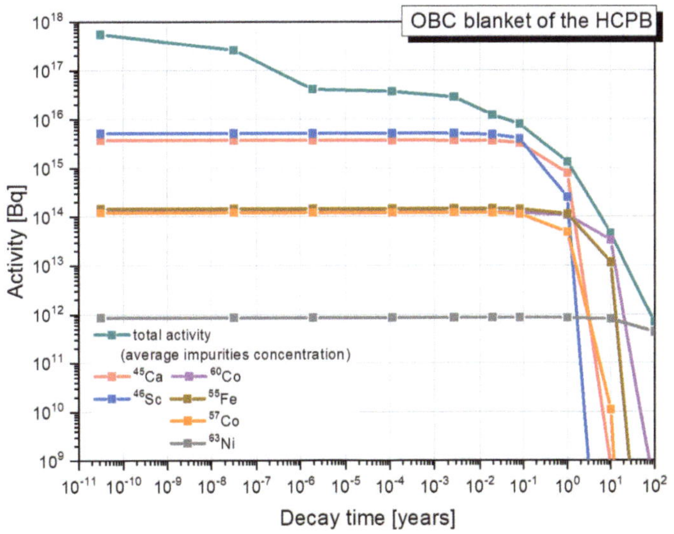

The total activity of irradiated Be waste can be significantly decreased (to 10^{12} Bq in total) due to decay of main radioactive isotopes in 100 years only.

7 Institute for Applied Materials – Applied Materials Physics

Surface decontamination of irradiated Be block

PosNo.	Medium	Specimen mass,g	Alpha-activity, Bq	Content of radionuclides in solutions, Bq					
				^{60}Co	^{54}Mn	^{137}Cs	^{134}Cs	^{144}Ce	^{154}Eu
1	Water	0,2235	10	240	40	$3\cdot10^3$	240	830	80
2	1M HNO_3	0,2229	58	460	70	$2\cdot10^3$	160	$3,5\cdot10^3$	400
3	2M HNO_3	-	40	$2\cdot10^3$	300	740	370	$2,5\cdot10^3$	100
4	1M HCl (0,5 часа)	0,2223	30	$9\cdot10^3$	560	$2\cdot10^3$	$2\cdot10^3$	$3,5\cdot10^3$	140
5	1M HCl	0,2148	-	$8\cdot10^4$	$5\cdot10^3$	$1,7\cdot10^3$	$2,3\cdot10^3$	$1,3\cdot10^3$	no
6	1M HCl	0,1954	-	$3,3\cdot10^4$	$1,8\cdot10^3$	no	no	no	no
7	2M HCl	0,1574	-	$7\cdot10^4$	$4,2\cdot10^3$	no	no	no	no
8	2M HCl	0,0941	-	$9\cdot10^4$	$5,5\cdot10^3$	no	no	no	no
9	4M HCl	complete dissolution	-	$7\cdot10^4$	$5\cdot10^3$	no	no	no	no
Balance per pos. 1-9		0,2248	138 (pos. 1-4)	$3,5\cdot10^5$	$2,2\cdot10^4$	$9\cdot10^3$ (pos. 1-5)	$5\cdot10^3$ (pos. 1-5)	$1,1\cdot10^4$ (pos. 1-5)	$7\cdot10^2$ (pos. 1-5)

- ➢ The best surface solvent for beryllium is HCl acid.
- ➢ In position 6, the decontamination solution contains Be, Co, Mn but they are still remaining in the undissolved sample.
- ➢ Cs, Ce и Eu came completely to solution.
- ➢ The results show that Cs, Ce и Eu are surface contaminators. They can be effectively removed by etching the beryllium block to depth of about 80-100 µm.

8 Institute for Applied Materials – Applied Materials Physics

Tritium removing from irradiated Be block

Specimen	Mass of specimen, mg	Average rate of specimen heating, °C/min	Start of tritium release, °C	Molar fraction of tritium in a specimen, appm
Be-2	28,0	11,0	860	372 ± 25
Be-6	17,2	12,2	880	338 ± 22

Specimen	Mass of specimen, mg	Molar fraction of tritium in a specimen, appm	Mol of tritium in a specimen, mol	Mass of tritium in a specimen, g	Activity of tritium in a specimen, Bq
Be-2	28.0	372 ± 25	$3.64\cdot10^{-7}$	$7.53\cdot10^{-11}$	$2.67\cdot10^4$
Be-6	17.2	338 ± 22	$7.81\cdot10^{-7}$	$5.09\cdot10^{-11}$	$1.80\cdot10^4$

The performed experiments allow us to recommend the following modes of irradiated Be treatment to remove tritium:

- ➢Annealing temperature – 900÷950 °C;
- ➢Heating rate – 10÷15°C/min;
- ➢Holding at annealing temperature – 0.5÷1 h;
- ➢Vacuum – no worse than $1\cdot10^{-4}$ Pa.
- ➢These annealing parameters allow practically all tritium to be removed from irradiated Be.

9 Institute for Applied Materials – Applied Materials Physics

Tritium removing from fragments of irradiated Be matrix of BR2 reactor

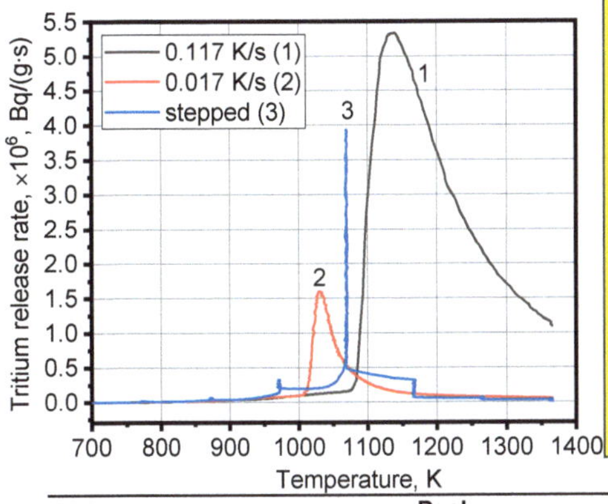

S-200E beryllium grade by Materion, USA production via the vacuum hot-pressing method and irradiated in BR2 reactor between 1980 and 1995 as Be matrix. Chemical composition as>98 wt. % Be, <2.0 wt. % BeO, <0.18 wt. % Fe, <0.15 wt. % C, <0.15 wt. % Al, <0.08 wt. % Mg, <0.08 wt. % Si and <0.04 wt. % of other elements.
The beryllium matrix was subjected to irradiation for 15 years at 323 K up to a fission neutron fluence of 5.32·× 10^{26} m^{-2}, which can be converted to a fast neutron fluence of 4.67·× 1026 m^{-2} (E > 1 MeV).
The estimated production levels of helium and tritium in the irradiated beryllium matrix were 22,000 appm of ^{4}He and 2000 appm of ^{3}H. Due to the radioactive decay of tritium from 1995 to 2022, the content of tritium in beryllium to the time of the thermal desorption tests decreased to 440 appm.

Heating Mode	Peak Temperature T_m, K	Start of Tritium Release, K	Height of Peak, ×10^6, Bq/(g·s)	Peak Width at Half Height $W_{1/2}$, K
Permanent at rate 0.117 K/s	1136	800–870	5.34	141
Permanent at rate 0.017 K/s	1031	800–870	1.58	36
Stepped at average rate	1069	800–870	3.93	4

The results obtained on irradiated Be from BR2 reactor are similar to data from the ISTC project (Be blocks from SM-3 reactor): the same location of tritium release peaks.

10 Institute for Applied Materials – Applied Materials Physics

Procedure of irradiated Be decontamination from radioactive impurities

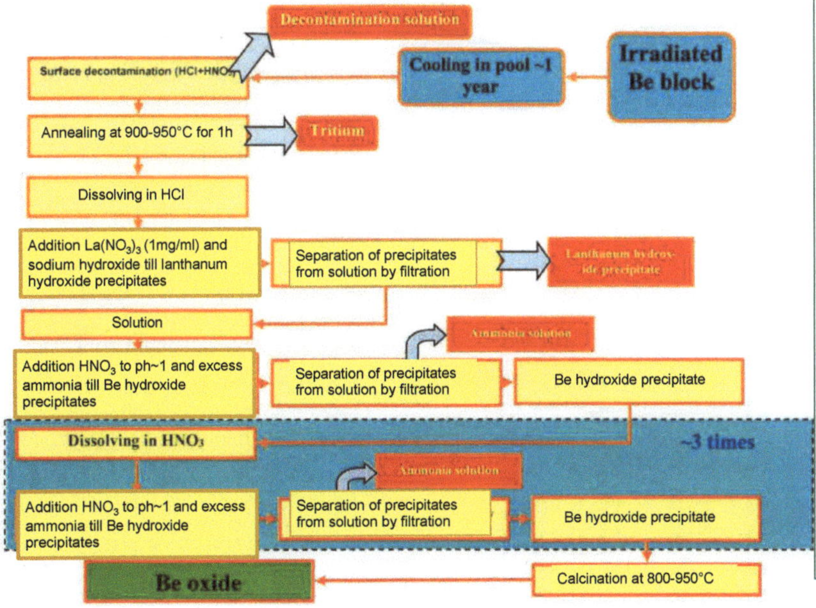

Stages of decontamination:
1.Surface decontamination by ($HCl+HNO_3$)
2.Tritium removing by annealing at 900-950°C for 1h
3. Addition $La(NO_3)_3$ and filtration.
4.Addition in solution HNO_3 and filtration to have Be hydroxide precipitate
5.Then, again addition in solution HNO_3 and filtration to have Be hydroxide precipitate
This cycle should be repeated 3 or more times to decrease the rest activity of the Be hydroxide precipitate
6. Calcination at 800-950°C to have BeO as a finale product

11 Institute for Applied Materials – Applied Materials Physics

Activity of radionuclides after each precipitation

Radionuclide	Content of radionuclides, Bq/g (%)						
	After irrad. ($F=6\cdot10^{22}$ cm^{-2})	Surface decont.	Annealing	La hydroxide precip.	1st precip. $Be(OH)_2$	2nd precip. $Be(OH)_2$	3rd precip. $Be(OH)_2$
ΣAlpha-active	~$5\cdot10^3$ (100)	~$4,6\cdot10^3$ (92)	~$4,6\cdot10^3$ (92)	(<1%)	-	-	-
H-3	$1,0\cdot10^6$ (100)	$1,0\cdot10^6$ (100)	-	-	-	-	-
Co-60	$3,2\cdot10^7$ (100)	$3,14\cdot10^7$ (98)	$3,14\cdot10^7$ (98)	$1,28\cdot10^7$(40)	$4,48\cdot10^6$ (14)	$2,24\cdot10^6$ (7)	$1,28\cdot10^6$ (4)
Mn-54	$7,2\cdot10^5$ (100)	$7,06\cdot10^5$ (98)	$7,06\cdot10^7$ (98)	$3,6\cdot10^5$ (50)	$1,3\cdot10^5$ (18)	$5,76\cdot10^4$ (8)	$4,32\cdot10^4$ (6)
Cs-137	$1,3\cdot10^6$ (100)	$1,27\cdot10^6$ (97)	$1,27\cdot10^6$ (97)	$5,2\cdot10^5$ (40)	$2,34\cdot10^5$ (18)	$9,1\cdot10^4$ (7)	$6,5\cdot10^4$ (5)
Ce-144	$3,5\cdot10^4$	(0)	-	-	-	-	-

Each precipitation of $Be(OH)_2$ leads to the significant activity decrease of each radionuclide and total activity as well.

12

Institute for Applied Materials – Applied Materials Physics

Strong degradation of properties of Be block material irradiated at 70°C in SM-3 reactor

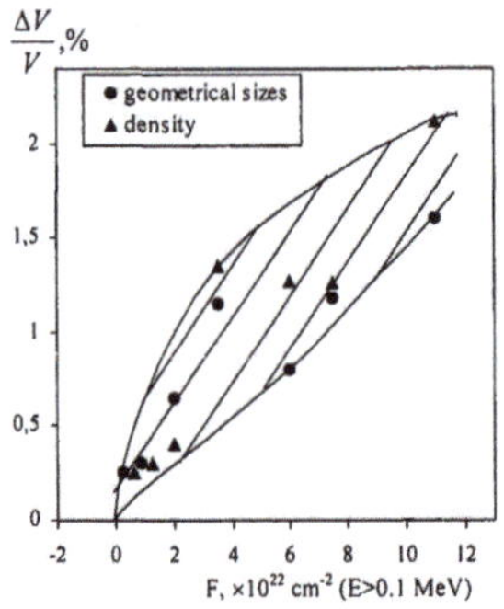

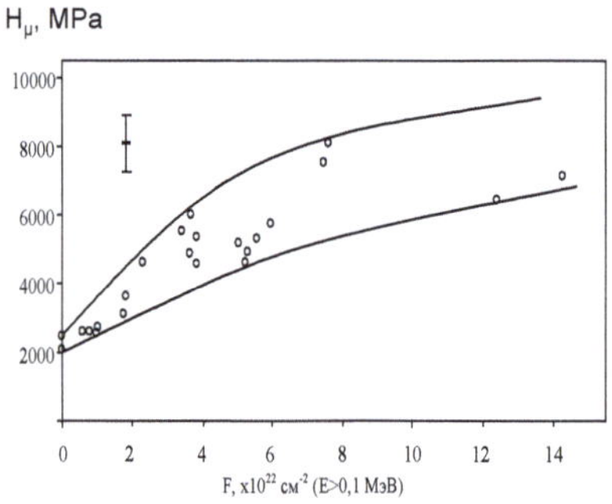

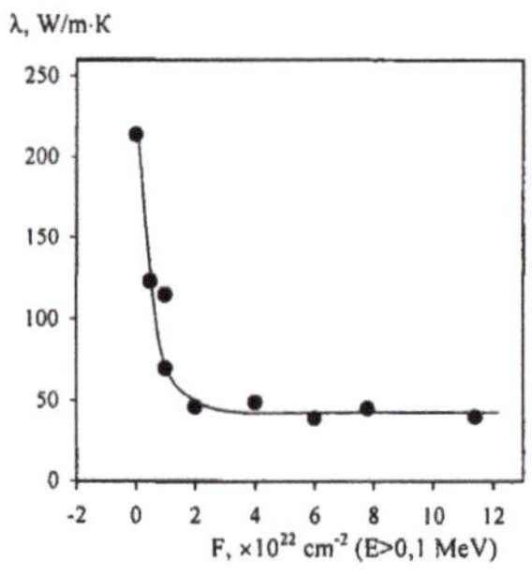

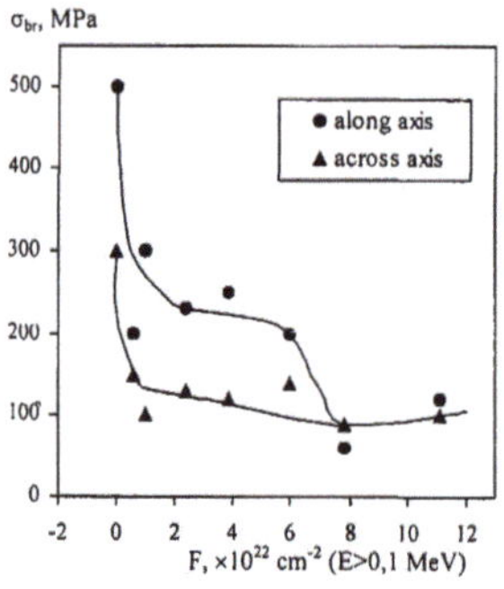

Be swelling at 70°C can achieve 1.5-2% at dose that exceeds by two times the resource one.

High increase of microhardness occurs to high neutron doses.

Strong radiation embrittlement and decrease of brittle strength at tensile testing of irradiated samples occurs already in low dose ranges to $(1\text{-}2)\cdot10^{26}$ m^{-2} (E>0.1 MeV). The rest strength is only 100 MPa at highest doses.

Thermal conductivity decreases by several times under irradiation that leads to high inner stresses into the Be block.

13

Institute for Applied Materials – Applied Materials Physics

Be microstructure evolution under neutron irradiation at 70°C in SM-3 reactor and high-temperature annealing

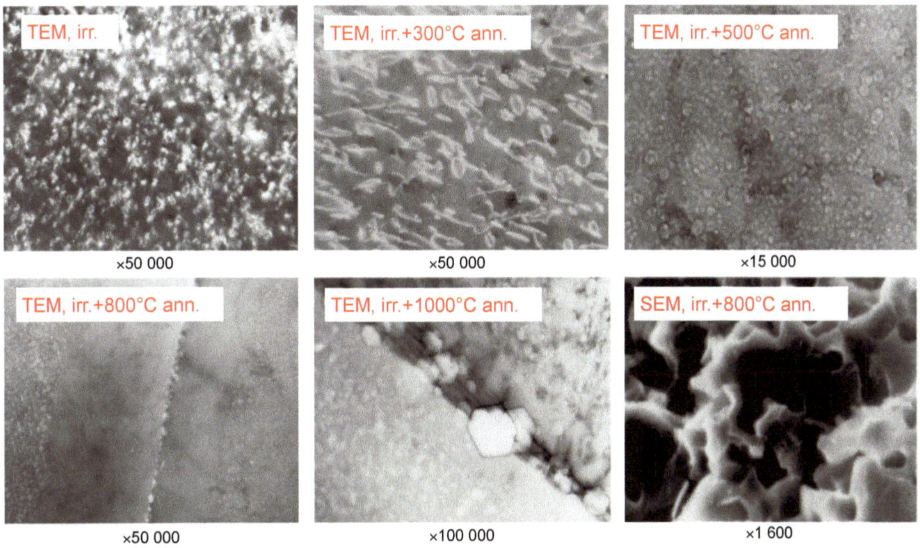

TEM study of irradiated Be annealed at high temperatures shows that the dislocation loops are formed but they are changed greatly after annealing for 1 h at 300°C. The dislocation loops increase in size and density decrease. At 500°C for 1h, the dislocation loops evolved into dislocation grid. First gas bubbles can be seen at 500°C for 1h. At higher temperatures, the bubbles increase in size.
At high annealing temperatures, so-called «sponge» structure in irradiated Be is formed. It is material with high open porosity.

14 Institute for Applied Materials – Applied Materials Physics

Conclusions

The results of ISTC Project allow development the procedure for decontamination of irradiated Be irradiated in SM-3 reactor from radioactive impurities. The procedure contains the following basic operations:

1.Cooling of irradiated Be block in pool for 1 year to decrease activity.

2.Defragmentation of irradiated Be block.

2.Surface decontamination with hydrochloric and nitric acid solutions.

3.Annealing at 950°C for an 1h to remove tritium.

4.Dissolving in chloric acid, addition of $La(NO_3)_3$ and sodium hydroxide, filtration and calcination at 900°C.

5.The level of decontamination of irradiated Be from radioactive impurities is 96% that makes irradiated beryllium waste an intermediate-level waste.

6.To completely remove radioactive impurities from irradiated Be, it is necessary to increase the number of cycles of dissolving Be hydroxide precipitate in nitric acid from 3 to 5-10 times.

7. The main disadvantage of this method of processing and recycling irradiated Be is appearance of large volume of liquid radioactive waste after Be is dissolved in acids.

15 Institute for Applied Materials – Applied Materials Physics

3.3 Recycling of Beryllides: Novel Approach for Cost-Effective and Efficient Fabrication

R. Gaisin, R. Rolli, V. Chakin, S. Udartsev, A. Goraieb, P. Vladimirov

The use of the Helium Cooled Pebble Bed (HCPB) breeding blanket in the EU DEMO fusion reactor anticipates an extensive demand for titanium beryllide blocks, estimated between 430 to 600 tons, to serve as neutron multipliers. This necessitates a significant supply of beryllium, yet global production barely reaches 200 to 300 tons annually. Additionally, the manufacturing process of beryllides via vacuum hot pressing (VHP) typically generates 25–40% of residual waste material. Compounding this situation, after approximately 5.2 years of utilization (with a continuous runtime of 1.56 years), the HCPB blanket will require a sustainable and cost-effective way of the beryllides recycling due to the worldwide scarcity and elevated costs associated with beryllium.

This study explores three distinct materials manufactured through varying techniques using recycled titanium beryllide residual material. In each instance, powder metallurgy methods were employed, yielding materials with a density exceeding 98%. In the first material, beryllide residual material was crushed into 3–10 mm fragments, blended with new titanium beryllide powder (<56 μm) in a 1:1 ratio, and subjected to VHP. Notably, this process resulted in identifiable boundaries between processed beryllide particles and fresh titanium beryllide powder as a chain of pores and variations in grain size. The second method involved grinding of beryllide residual material into <56 μm powder, which was then mixed in a 1:1 ratio with the same new titanium beryllide powder, followed by VHP. This material exhibited homogeneity, with imperceptible powder boundaries. The third material was derived entirely from 100% titanium beryllide scraps, ground into a <100 μm powder. This powder was loaded into a capsule for hot isothermal pressing, producing a material with non-uniform grain size distribution, a presence of more beryllium oxide particles along the boundaries, but minimal porosity.

Mechanical compression tests on the first and second materials demonstrated high strength at room temperature, similar to that of the previously tested two-phase material, $TiBe_{12}$+7%Be. Notably, high-temperature tests showed higher strength levels reaching that of single-phase $TiBe_{12}$. Thus, despite the different structure, the overall result was high-quality materials with strength close to the original titanium beryllide. These findings present the potential for an efficient and cost-effective recycling process for beryllides and lay the theoretical groundwork for the recycling of beryllides after their utilization in the HCPB blanket.

Presenting Author: Rami Gaisin
Presenting Author's Affiliation: Karlsruhe Institute of Technology, Germany
Email: ramil.gaisin@kit.edu

KIT
Karlsruhe Institute of Technology
BeWS-16,
October 19-20, Santander, Spain
Recycling of Beryllides: Novel Approach for Cost-Effective and Efficient Fabrication
R. Gaisin1, R. Rolli1, V. Chakin1, S. Udartsev2, A. Goraieb3, P. Vladimirov1
1 Karlsruhe Institute of Technology, Eggenstein-Leopoldshafen, Germany
2 Ulba Metallurgical Plant JSC, Ust-Kamenogorsk, Kazakhstan
3 Karlsruhe Beryllium Handling Facility, Eggenstein-Leopoldshafen, Germany
Ti
Be
KIT – The Research University in the Helmholtz Association
www.kit.edu

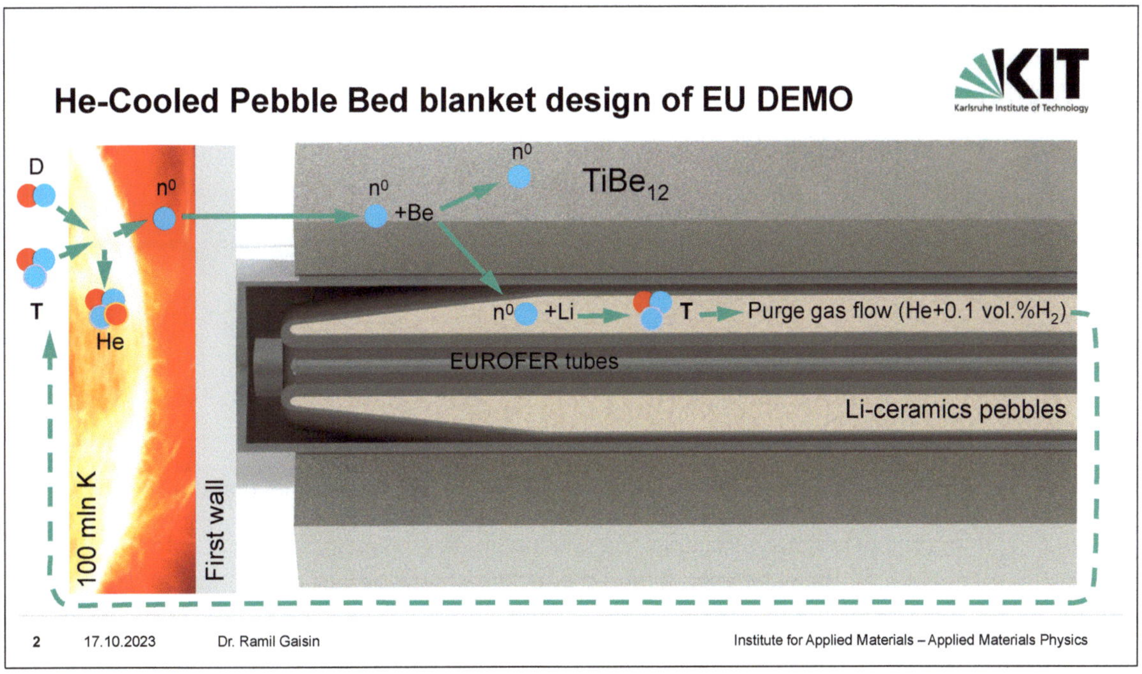
He-Cooled Pebble Bed blanket design of EU DEMO
KIT
Karlsruhe Institute of Technology
D
T
He
n0
100 mln K
First wall
n0
+Be
n0
TiBe12
n0
+Li
T
Purge gas flow (He+0.1 vol.%H2)
EUROFER tubes
Li-ceramics pebbles
2
17.10.2023
Dr. Ramil Gaisin
Institute for Applied Materials – Applied Materials Physics

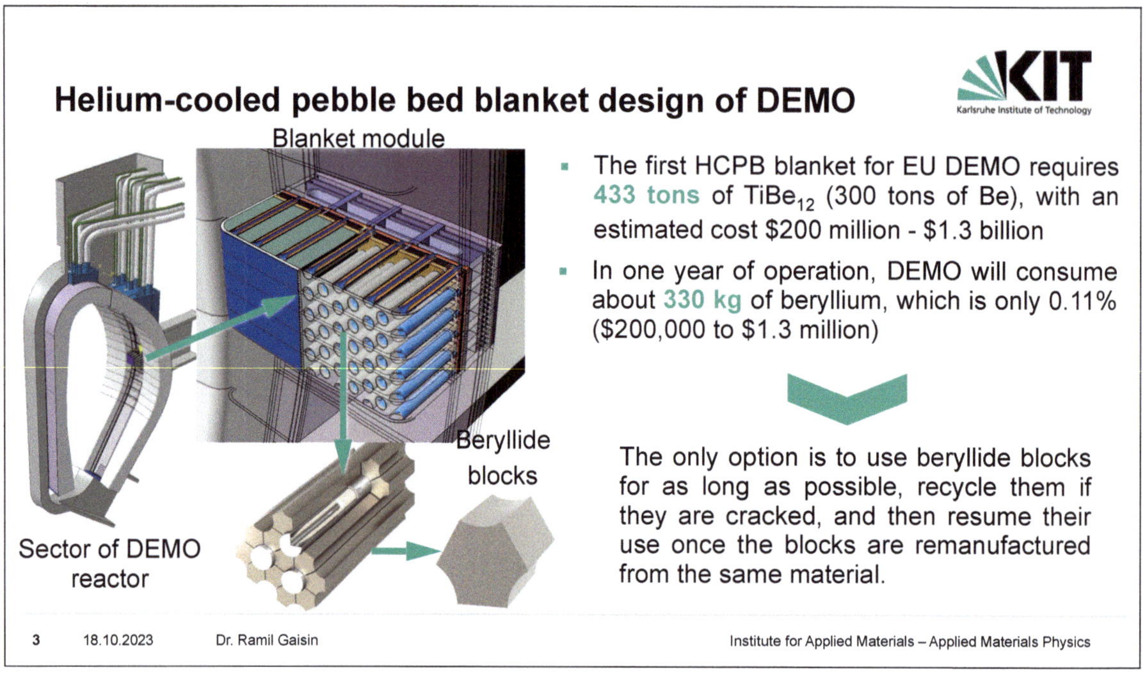
Helium-cooled pebble bed blanket design of DEMO
KIT
Karlsruhe Institute of Technology
Blanket module
Beryllide blocks
Sector of DEMO reactor
The first HCPB blanket for EU DEMO requires 433 tons of TiBe12 (300 tons of Be), with an estimated cost $200 million - $1.3 billion
In one year of operation, DEMO will consume about 330 kg of beryllium, which is only 0.11% ($200,000 to $1.3 million)
The only option is to use beryllide blocks for as long as possible, recycle them if they are cracked, and then resume their use once the blocks are remanufactured from the same material.
3 18.10.2023 Dr. Ramil Gaisin
Institute for Applied Materials – Applied Materials Physics

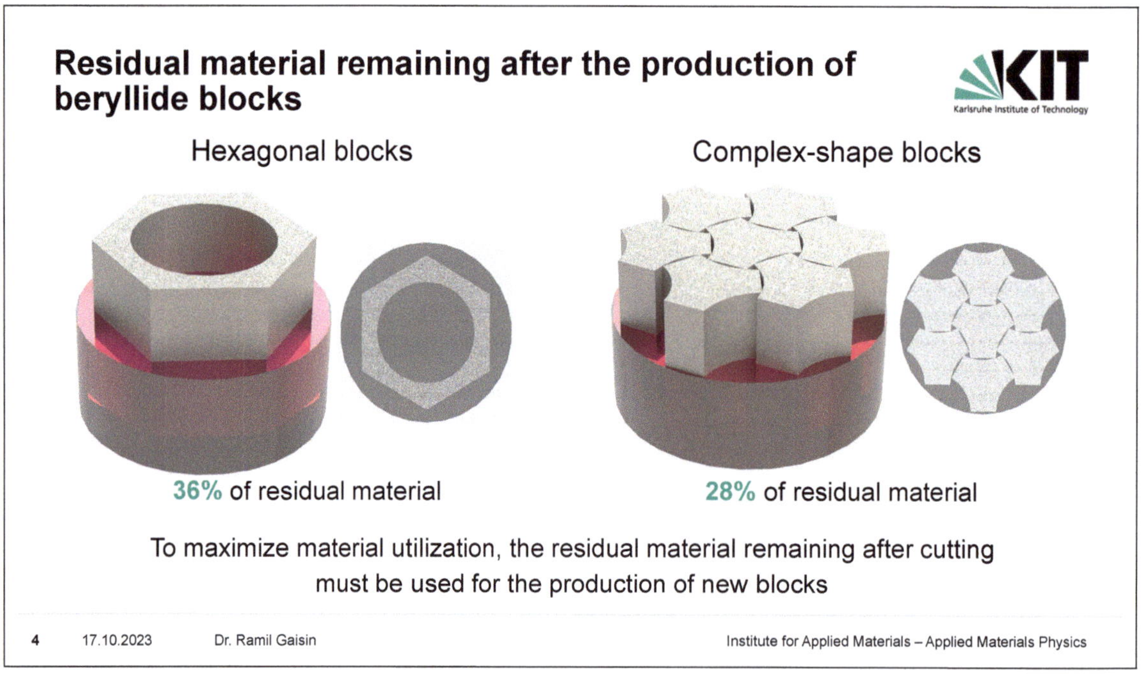
Residual material remaining after the production of beryllide blocks
KIT
Karlsruhe Institute of Technology
Hexagonal blocks
Complex-shape blocks
36% of residual material
28% of residual material
To maximize material utilization, the residual material remaining after cutting must be used for the production of new blocks
4 17.10.2023 Dr. Ramil Gaisin
Institute for Applied Materials – Applied Materials Physics

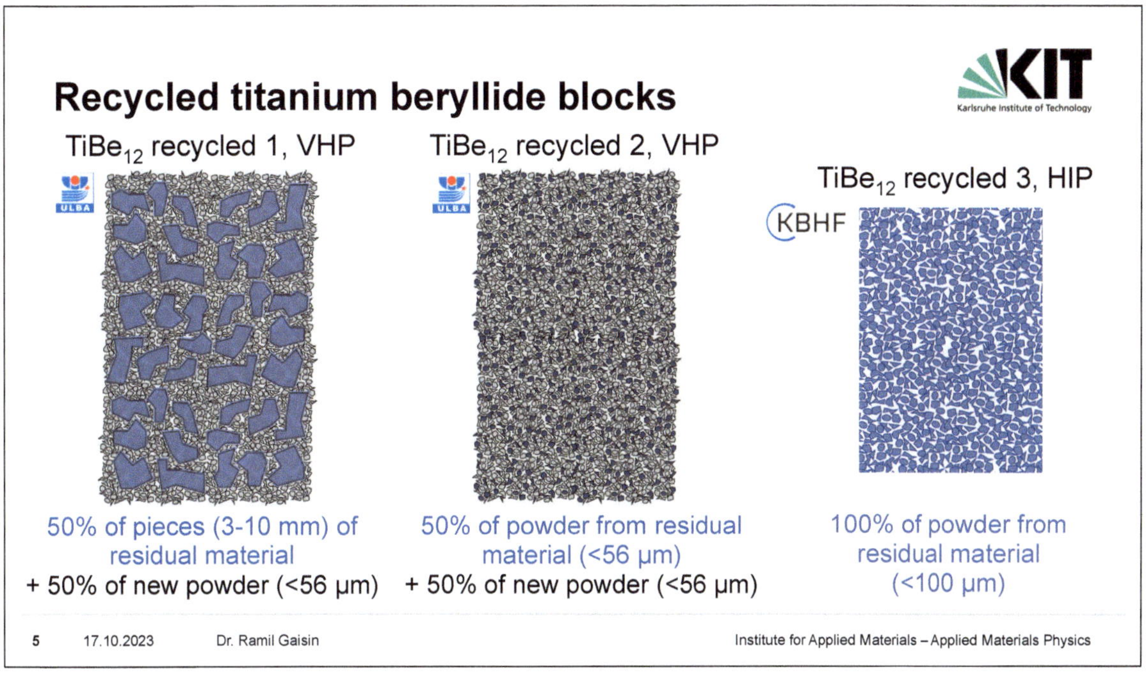
Recycled titanium beryllide blocks
KIT
Karlsruhe Institute of Technology
TiBe12 recycled 1, VHP
TiBe12 recycled 2, VHP
TiBe12 recycled 3, HIP
ULBA
ULBA
KBHF
50% of pieces (3-10 mm) of residual material
+ 50% of new powder (<56 µm)
50% of powder from residual material (<56 µm)
+ 50% of new powder (<56 µm)
100% of powder from residual material (<100 µm)
5
17.10.2023
Dr. Ramil Gaisin
Institute for Applied Materials – Applied Materials Physics

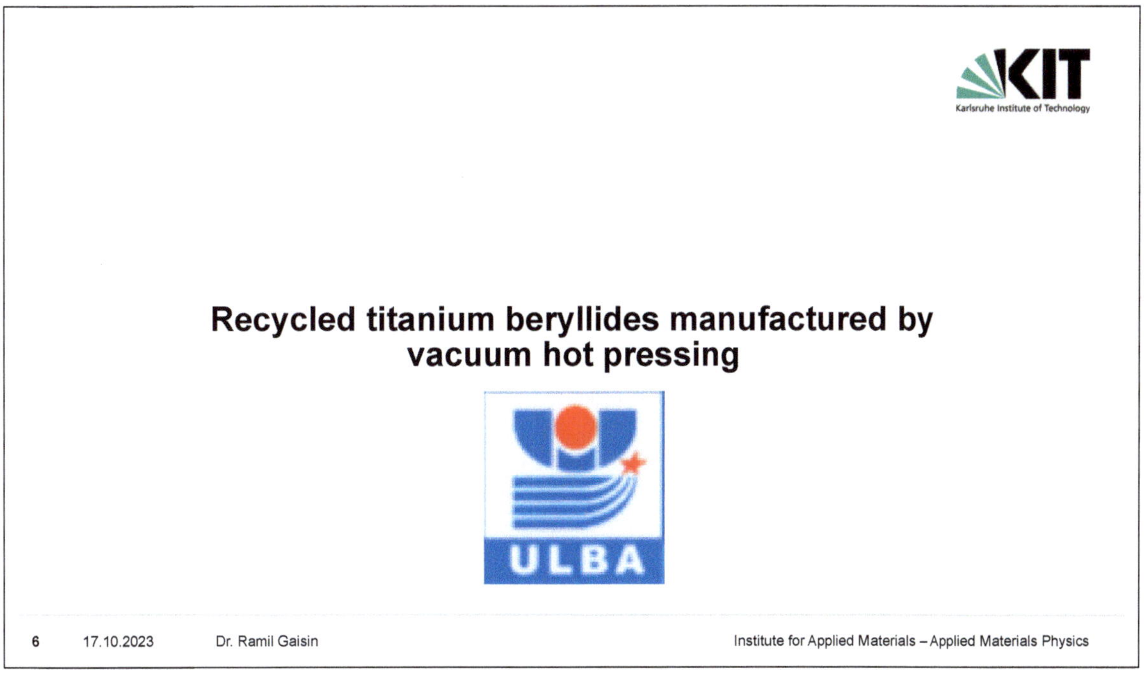
KIT
Karlsruhe Institute of Technology
Recycled titanium beryllides manufactured by vacuum hot pressing
ULBA
6
17.10.2023
Dr. Ramil Gaisin
Institute for Applied Materials – Applied Materials Physics

Recycled blocks manufactured in Ulba

KIT Karlsruhe Institute of Technology

$TiBe_{12}$ recycled 1 | $TiBe_{12}$ recycled 2

Ø51mm×52mm

Density	g/cm³	% of TD
edge	2.254	98.5
center	2.268	99.1

Ø51mm×58mm

Density	g/cm³	% of TD
edge	2.271	99.3
center	2.24	97.9

50% of pieces (3-10 mm) of residual material
+ 50% of new powder (<56 μm)

50% of powder from residual material (<56 μm)
+ 50% of new powder (<56 μm)

7 17.10.2023 Dr. Ramil Gaisin Institute for Applied Materials – Applied Materials Physics

Chemical analysis of the recycled beryllides

KIT Karlsruhe Institute of Technology

Wt. %	$TiBe_{12}$ +7%Be	$TiBe_{12}$	$TiBe_{12}$ recycled 1	$TiBe_{12}$ recycled 2
Be	70.5	69.7	69.7	68.6
C	0.0339	0.038	0.0484	0.0554
N	0.106	0.0801	0.0652	0.0691
O	0.686	0.597	0.74	0.785
Na	< 0.0004	< 0.0001	< 0.0002	< 0.0002
Mg	< 0.00005	< 0.0002	0.00005	0.00013
Al	0.0203	0.0222	0.0206	0.0229
Si	0.0162	0.0213	0.0222	0.0236
Ca	0.015	0.01	0.132	0.137
Ti	27.8	29.6	28.9	28.6
Fe	0.118	0.124	0.11	0.218
U, ppm	0.395	0.51	0.555	0.520

8 17.10.2023 Dr. Ramil Gaisin Institute for Applied Materials – Applied Materials Physics

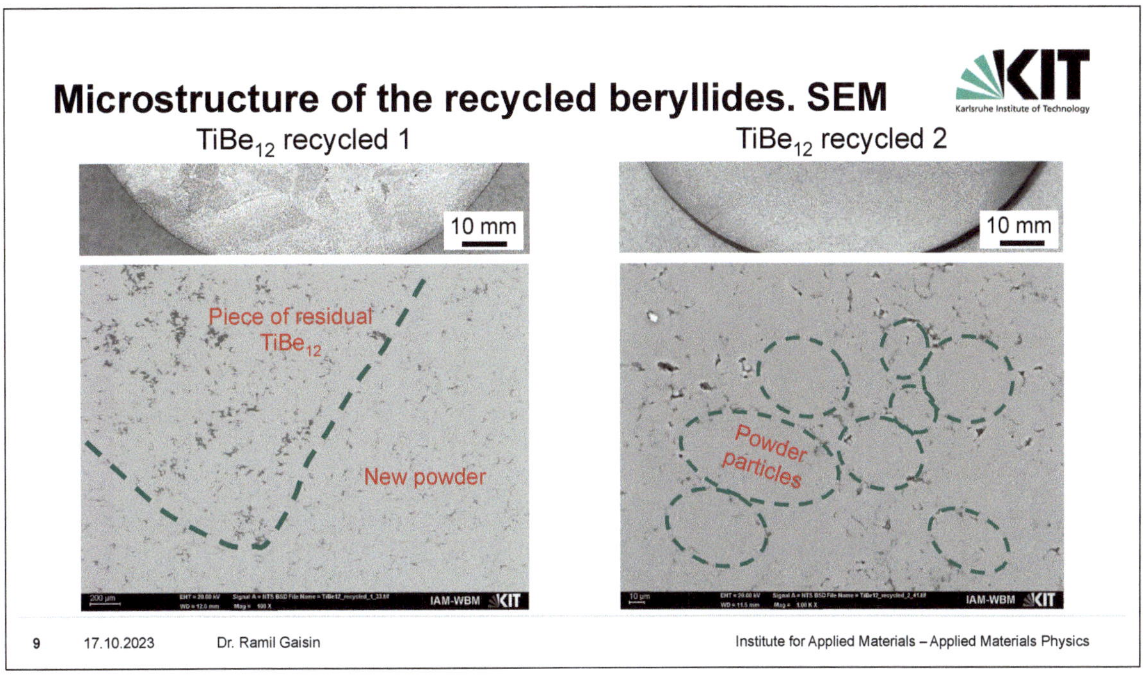
Microstructure of the recycled beryllides. SEM
KIT
Karlsruhe Institute of Technology
TiBe12 recycled 1
TiBe12 recycled 2
10 mm
10 mm
Piece of residual TiBe12
New powder
Powder particles
IAM-WBM
IAM-WBM
9
17.10.2023
Dr. Ramil Gaisin
Institute for Applied Materials – Applied Materials Physics

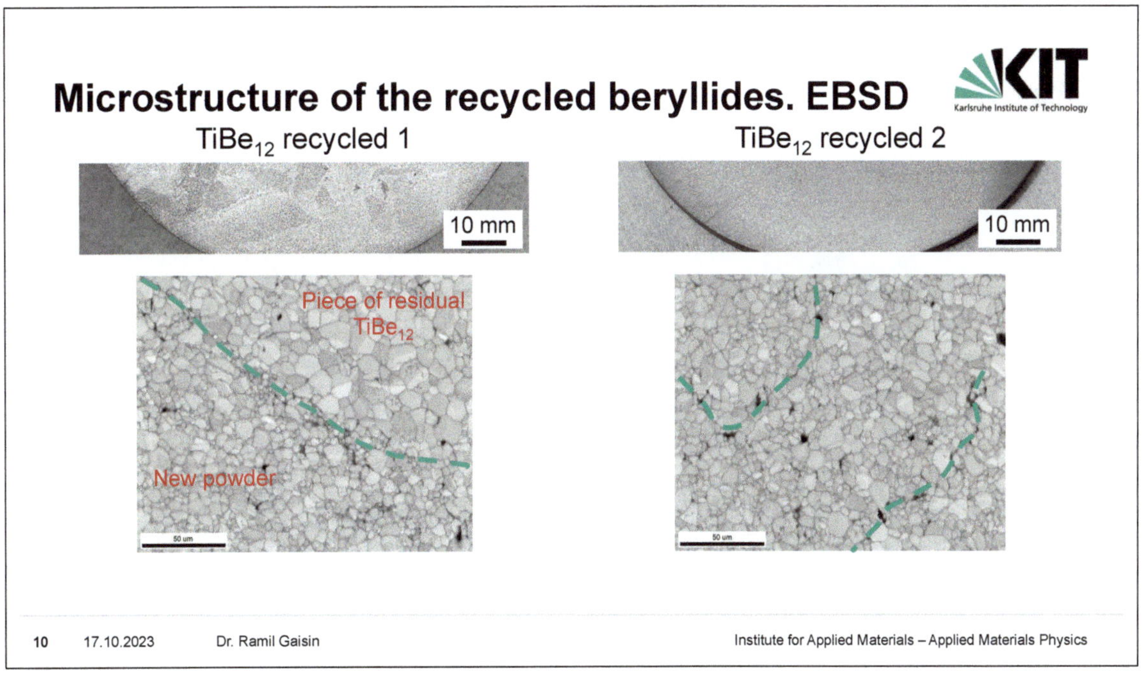
Microstructure of the recycled beryllides. EBSD
KIT
Karlsruhe Institute of Technology
TiBe12 recycled 1
TiBe12 recycled 2
10 mm
10 mm
Piece of residual TiBe12
New powder
50 um
50 um
10
17.10.2023
Dr. Ramil Gaisin
Institute for Applied Materials – Applied Materials Physics

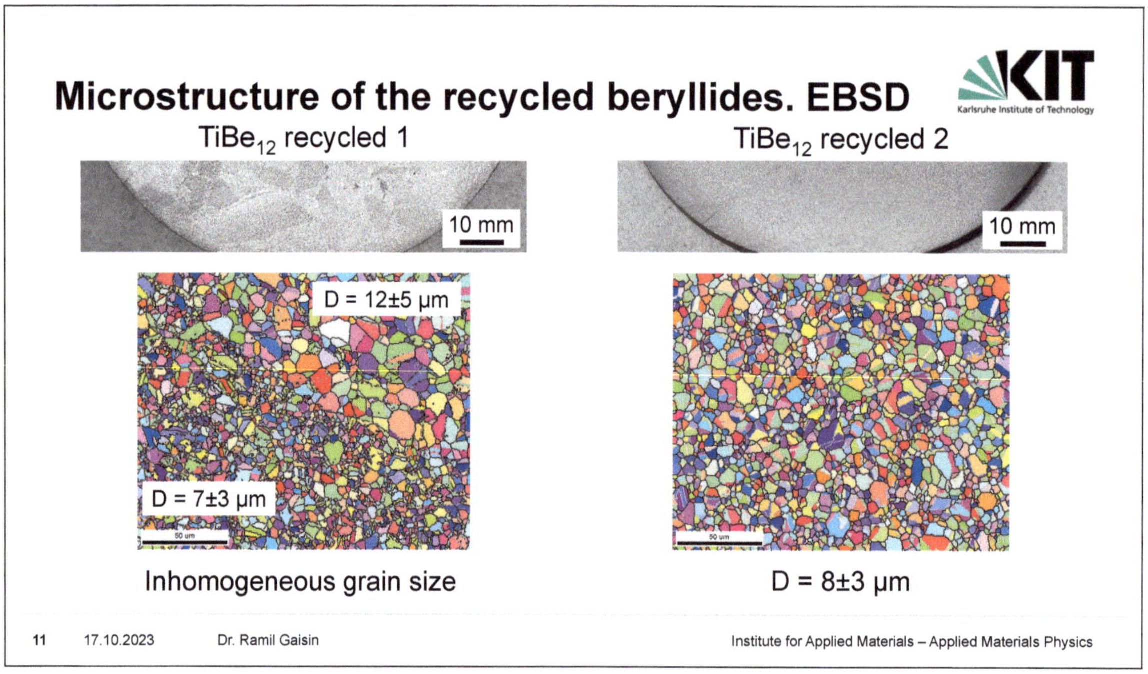
Microstructure of the recycled beryllides. EBSD
KIT
Karlsruhe Institute of Technology
TiBe12 recycled 1
TiBe12 recycled 2
10 mm
10 mm
D = 12±5 µm
D = 7±3 µm
Inhomogeneous grain size
D = 8±3 µm
11
17.10.2023
Dr. Ramil Gaisin
Institute for Applied Materials – Applied Materials Physics

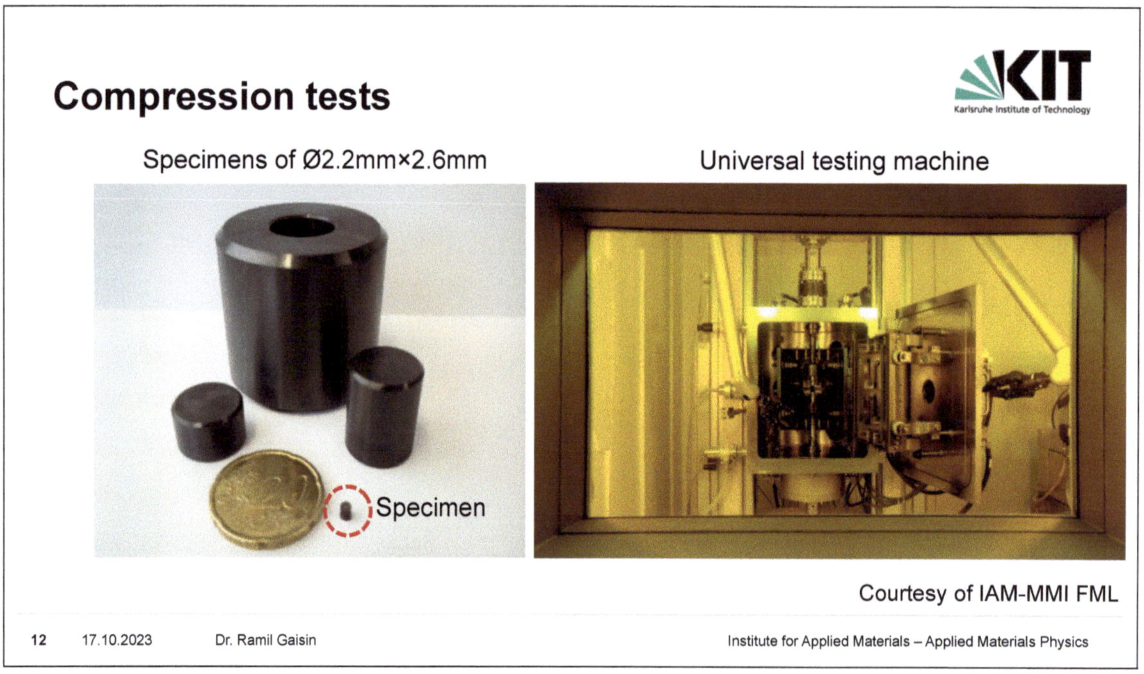
Compression tests
KIT
Karlsruhe Institute of Technology
Specimens of Ø2.2mm×2.6mm
Universal testing machine
Specimen
Courtesy of IAM-MMI FML
12
17.10.2023
Dr. Ramil Gaisin
Institute for Applied Materials – Applied Materials Physics

Mechanical properties of the recycled beryllides
KIT
Karlsruhe Institute of Technology
TiBe12 +7% Be
TiBe12
TiBe12 recycled 1
TiBe12 recycled 2
Compressive strength, MPa
Temperature, °C
0
500
1000
1500
2000
2500
3000
0
200
400
600
800
1000
1200
13
17.10.2023
Dr. Ramil Gaisin
Institute for Applied Materials – Applied Materials Physics

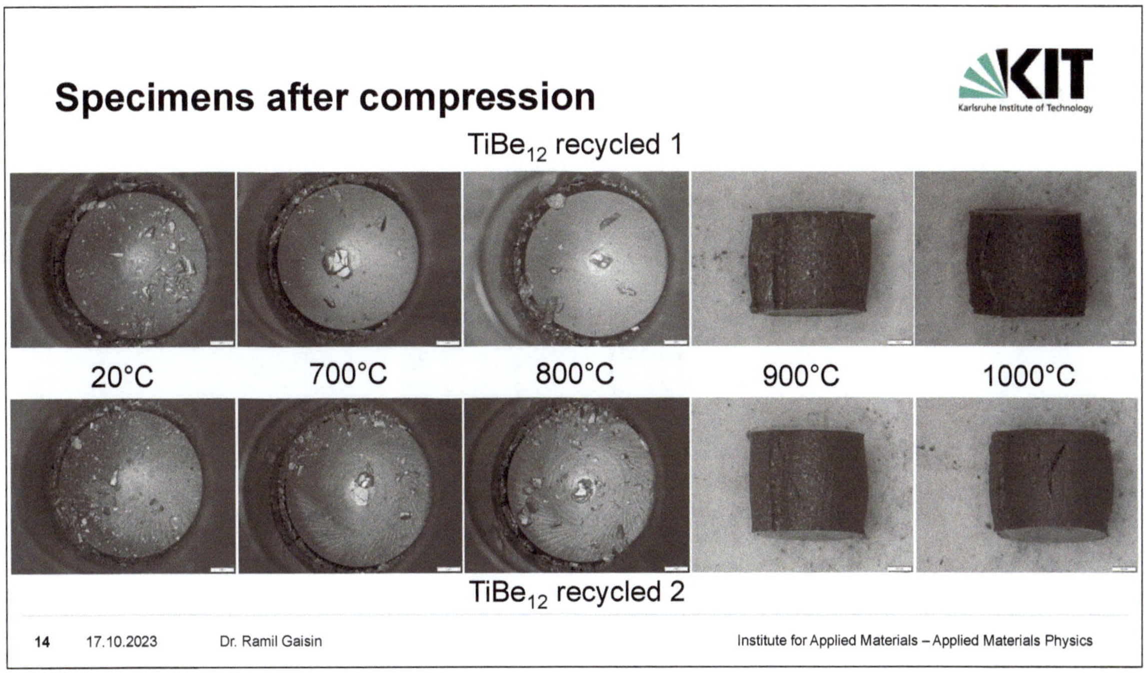
Specimens after compression
KIT
Karlsruhe Institute of Technology
TiBe12 recycled 1
20°C
700°C
800°C
900°C
1000°C
TiBe12 recycled 2
14
17.10.2023
Dr. Ramil Gaisin
Institute for Applied Materials – Applied Materials Physics

Karlsruhe Institute of Technology

Recycled titanium beryllide manufactured by hot isostatic pressing

15 17.10.2023 Dr. Ramil Gaisin Institute for Applied Materials – Applied Materials Physics

Preparation of $TiBe_{12}$ powder from residual materials

Karlsruhe Institute of Technology

Residual $TiBe_{12}$ | Mortar with beryllides | Beryllide powder

16 17.10.2023 Dr. Ramil Gaisin Institute for Applied Materials – Applied Materials Physics

TiBe12 powder from residual materials
Karlsruhe Institute of Technology
Beryllide powder <100 µm
10 µm
EHT = 15.00 kV
I Probe = 6.0 nA
TiBe12_powder_61.tif
Signal A = SE2
Mag = 572 X
WD = 15.1 mm
Width = 200.0 µm
User Name = GAISIN
Date :17 Nov 2022
17 17.10.2023 Dr. Ramil Gaisin
Institute for Applied Materials – Applied Materials Physics

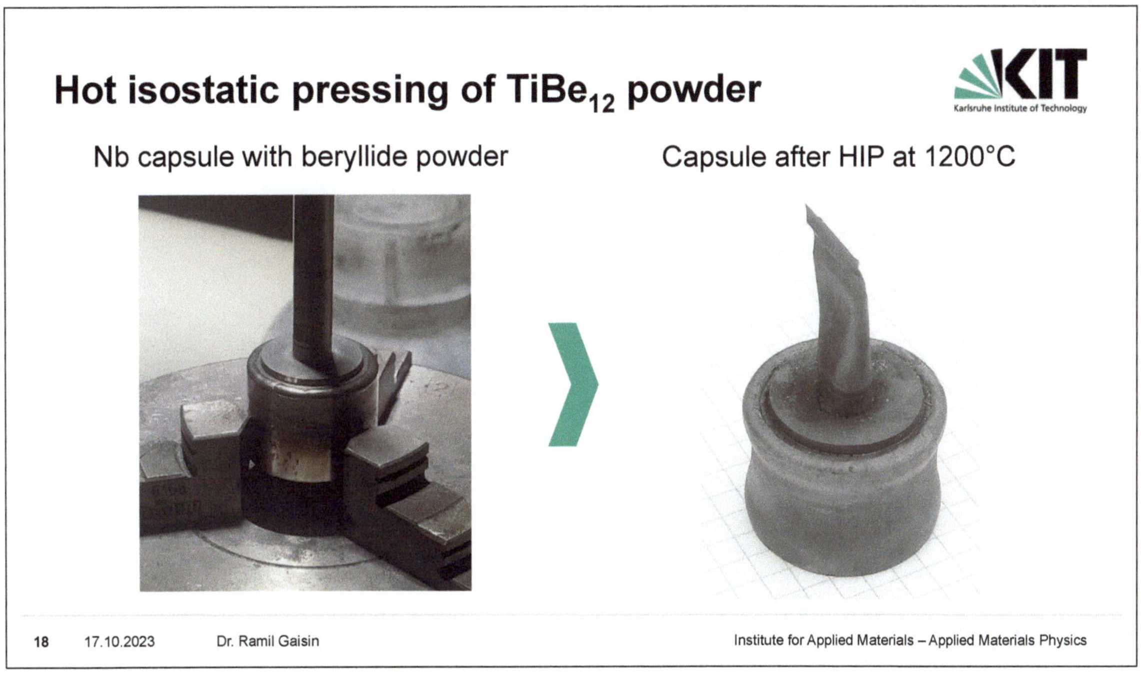
Hot isostatic pressing of TiBe12 powder
Karlsruhe Institute of Technology
Nb capsule with beryllide powder
Capsule after HIP at 1200°C
18 17.10.2023 Dr. Ramil Gaisin
Institute for Applied Materials – Applied Materials Physics

SEM after hot isostatic pressing

Karlsruhe Institute of Technology

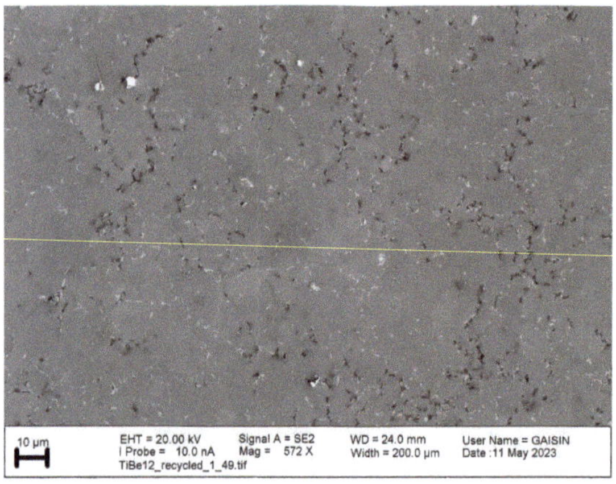

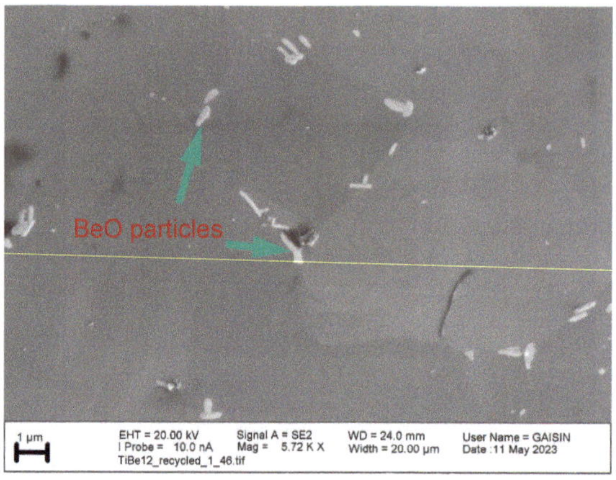

19 17.10.2023 Dr. Ramil Gaisin Institute for Applied Materials – Applied Materials Physics

EBSD after hot isostatic pressing

Karlsruhe Institute of Technology

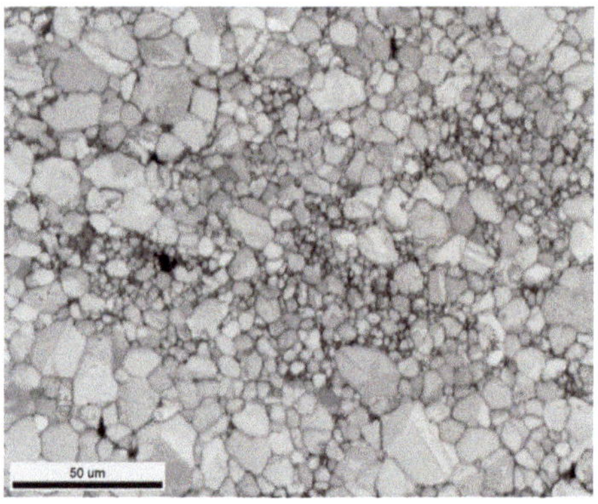

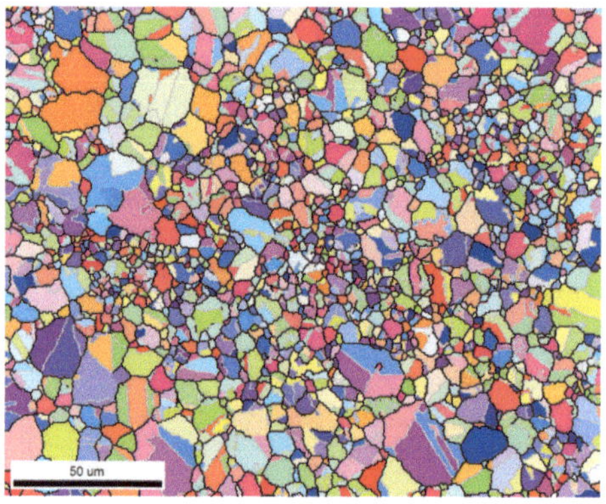

D = 10±6 µm, >25 µm – 5%

20 17.10.2023 Dr. Ramil Gaisin Institute for Applied Materials – Applied Materials Physics

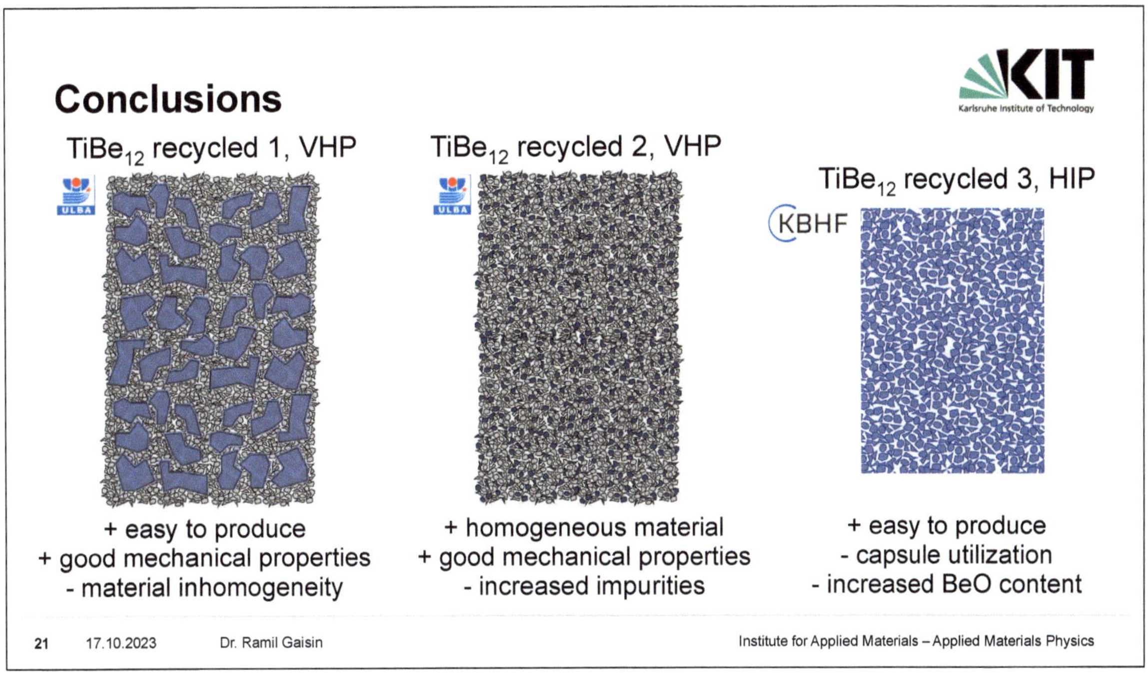
Conclusions
KIT
Karlsruhe Institute of Technology
TiBe12 recycled 1, VHP
ULBA
TiBe12 recycled 2, VHP
ULBA
TiBe12 recycled 3, HIP
KBHF
+ easy to produce
+ good mechanical properties
- material inhomogeneity
+ homogeneous material
+ good mechanical properties
- increased impurities
+ easy to produce
- capsule utilization
- increased BeO content
21 17.10.2023 Dr. Ramil Gaisin
Institute for Applied Materials – Applied Materials Physics

KIT
Karlsruhe Institute of Technology
Thank you for your attention!
22 17.10.2023 Dr. Ramil Gaisin
Institute for Applied Materials – Applied Materials Physics

3.4 Attempts for uranium removing in beryllium using arc plasma

Norihiro Ikemoto, Yuki Minagawa, Kazuo Yonehara

More than 90% of beryllium production is accounted for a company. Because beryllium mines are nearby uranium mine, beryllium is included uranium as an impurity. In fact, the spec sheet of the supplier is mentioned maximum 150 wppm of uranium is included in beryllium. Much amount of beryllium is planning to use in blanket on fusion reactor. Used beryllium is also planning to dispose of. If the uranium concentration in beryllium is high, it is not shallow land burial. To dispose used beryllium shallow land burial, uranium is removed down to 0.85 ppm. In this study, experiments for uranium removing using the difference of the vapor pressure of beryllium and uranium were attempted.

Platinum was used as a simulant material for uranium because the vapor pressure of platinum and uranium is similar. Beryllium-platinum rod was produced by plasma sintering after mixing 30 g of beryllium powder and 5 mg of platinum powder. The rod was inserted in an experimental apparatus. The tungsten cerium electrode of TIG melting was also inserted in it. These were front position. The arc current was set 100 A. The arc plasma was generated in few decade seconds while pressure was kept at 80 kPa. Beryllium fume was produced during arc melting and it was recovered. The beryllium, platinum, and uranium concentration were measured by a ICP for recovered fume, beryllium-platinum rod. Same experiments are carried out for beryllium rod. The platinum concentration after experiment was lower than before experiment. Possibility uranium removing using arc plasma was suggested.

Presenting Author: Norihiro Ikemoto
Presenting Author's Affiliation: KAKEN CO., LTD., Japan
Email: n.ikemoto@kakenlabo.co.jp

KAKEN

Attempts for uranium removing in beryllium using arc plasma

Norihiro Ikemoto, Yuki Minagawa, and Kazuo Yonehara
KAKEN Inc.
e-mail address *n.ikemoto@kakenlabo.co.jp*

1/18

16th International Workshop on Beryllium technology (BeWS-16), 19-20 October 2023, Santander, Spain

Introduction of KAKEN

KAKEN
独創と協創
Since 1978

Headquarter & Chemical Analysis Dept.: Mito Institute
[Environment, Energy, Nuclear, etc.]

- Headquarters office
- Chemical analysis department
  - Radioactivity measurement
  - Environmental pollutant analysis

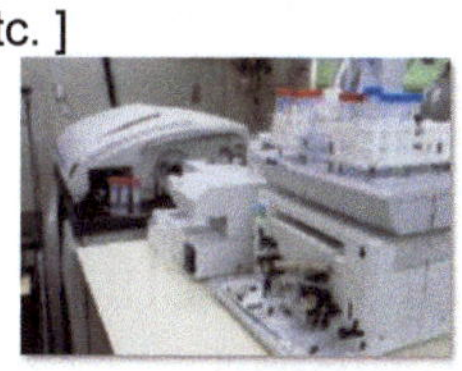

R&D: Hokota Functional-Materials Developmental Center
[Fusion, System & equipment design, etc.]

- Design office
- Laboratory equipment R&D
- R&D of Functional New Materials
- **Fusion R&D**
  - Hydrogen Isotopes **(HDT)** handling and separation
  - Advanced Tritium Breeder **(Li)**
  - **Advanced Neutron Multiplier (Be)**

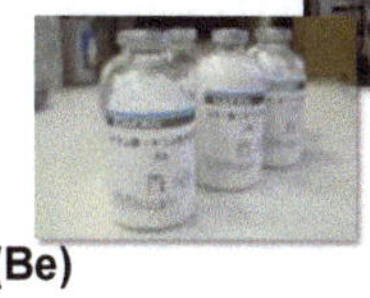

2/18

16th International Workshop on Beryllium technology (BeWS-16), 19-20 October 2023, Santander, Spain

KAKEN fusion activity

KAKEN

Tritium removing system for GB

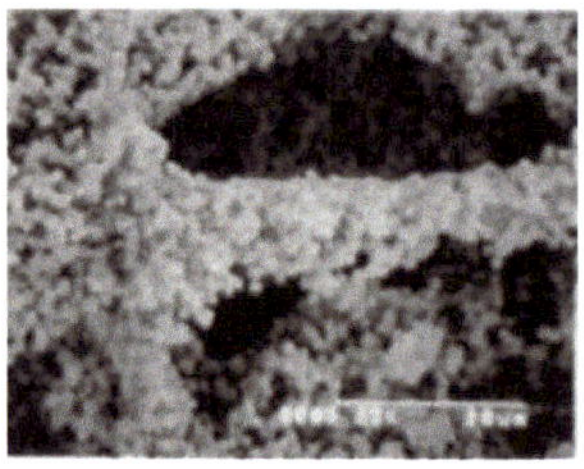

Catalyst for HDT separation

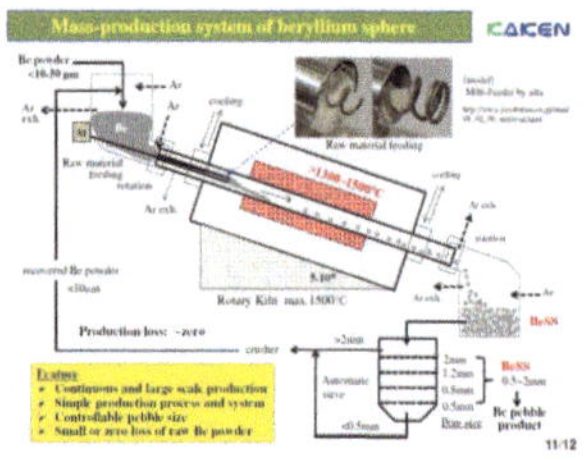

Beryllium pebble production system

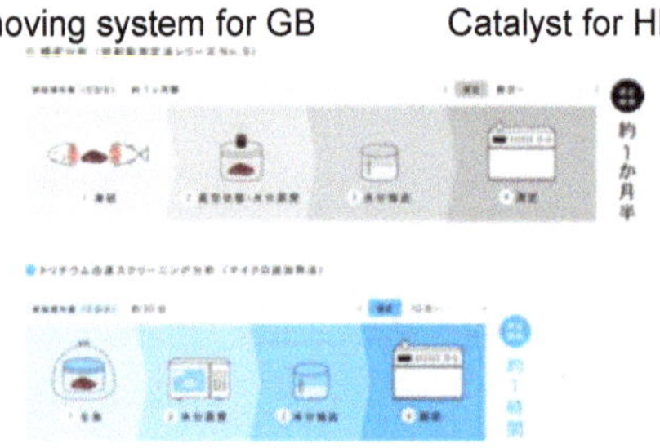

Environmental tritium analysis

Others…
- Tritium removing system for workspace
- Tritium breeder (in developing)
- Tritium sampling device

YouTube link =>

3/18

16th International Workshop on Beryllium technology (BeWS-16), 19-20 October 2023, Santander, Spain

Beryllium facility in KAKEN

KAKEN

KAKEN has beryllium facility (7600x5000x4000 mm)
- ➢ Enclose type hood
- ➢ REP
- ➢ Press machine
- ➢ Polishing machine
- ➢ Sintering device
- ➢ Glove box
- ➢ Electric furnace

Beryllium lab. building

Beryllium handling enclose hood

Beryllium capturing filter

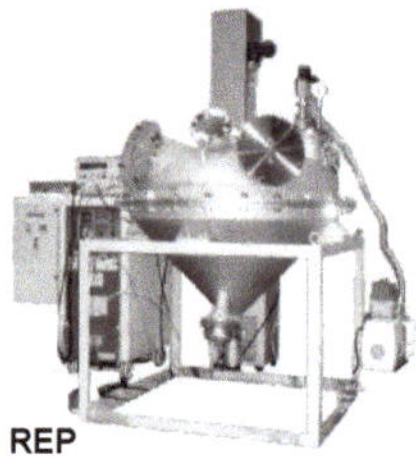

REP

Electric furnace

4/18

16th International Workshop on Beryllium technology (BeWS-16), 19-20 October 2023, Santander, Spain

Beryllium products by KAKEN

KAKEN

Rotating electrode process

Plasma sintering process

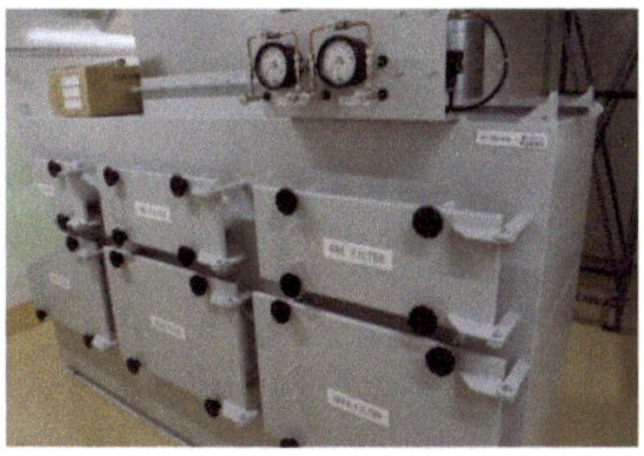

Beryllium dust capturing system

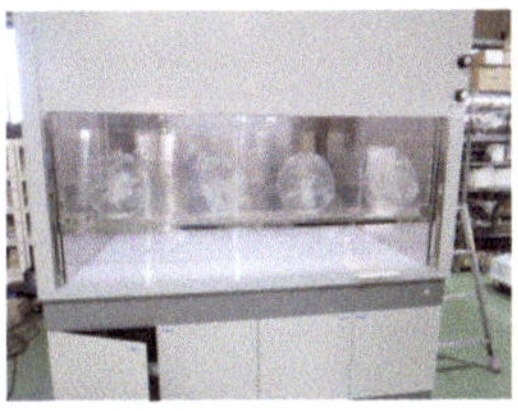

Enclose type hood

Design beryllium laboratory

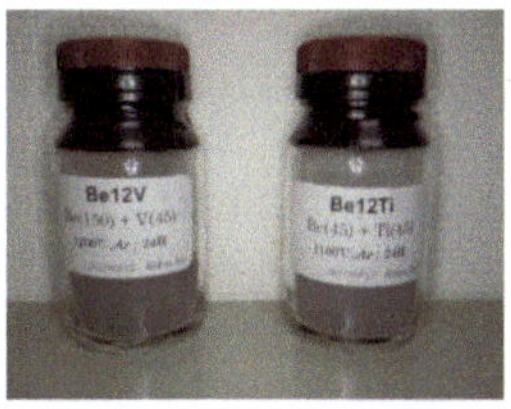

Beryllium alloy powder

5/18

16th International Workshop on Beryllium technology (BeWS-16), 19-20 October 2023, Santander, Spain

KAKEN

Experiments

6/18

Background

- About 90% of beryllium is producing by one company in the world.
- Uranium is containing as impurities in beryllium which maximum concentration is 150ppm by supplier's spec sheet.
- Present examination for fusion reactor, beryllium amount using on a fusion reactor is estimated about 500 tons.
- For shallow disposal of beryllium used in fusion reactor, uranium concentration in beryllium should be decreased(*).
- Uranium can be removed by Ion exchange resin or solvent extraction.
- Wet process requires a reduction process of beryllium ion.
- In this study, we are focus on the difference of the vapor pressure between beryllium and uranium.
- Removing uranium in beryllium was attempted by irradiating arc plasma to beryllium sample.

CHEMISTRY	
Beryllium Assay, % minimum	99.20
Beryllium Oxide, % maximum	0.900
Al, % maximum	0.050
C, % maximum	0.090
Fe, % maximum	0.080
Mg, Cr, each % maximum	0.010
Ni, Cu, Ti, Zr, each, % maximum	0.025
Zn, Mn, Ag, Co, Pb, Ca, Mo, each, % maximum	0.005
Silicon, % maximum	0.045
U, % maximum	0.015
Other Metallic Impurities, each, % maximum	0.040

Supplier's spec sheet

* Someya et al., Fusion Eng. Des., 136B (2018) 1306-1312

7/18

16th International Workshop on Beryllium technology (BeWS-16), 19-20 October 2023, Santander, Spain

Vapor pressure

- Beryllium has much high vapor pressure than uranium.
- Possibility to separate beryllium and uranium using the difference of these vapor pressure.
- Several research exists to remove impurities in metals by arc plasma.
- We attempted to remove uranium in beryllium by irradiating arc plasma.
- In this study, we used platinum instead of uranium because vapor pressure of platinum is similar tendency that of uranium.
- We used a commercial TIG welding machine as arc plasma generator.

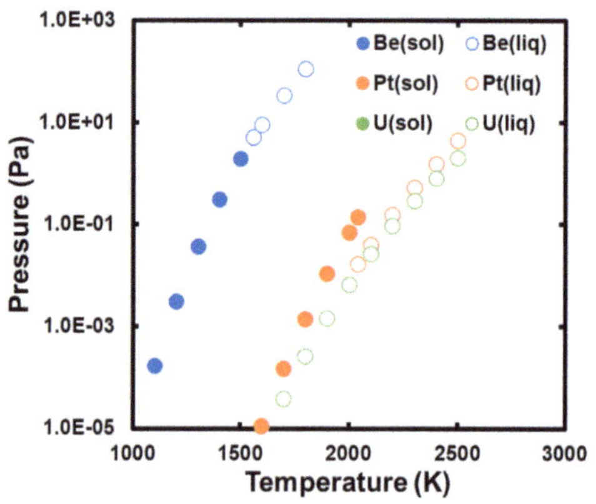

8/18

16th International Workshop on Beryllium technology (BeWS-16), 19-20 October 2023, Santander, Spain

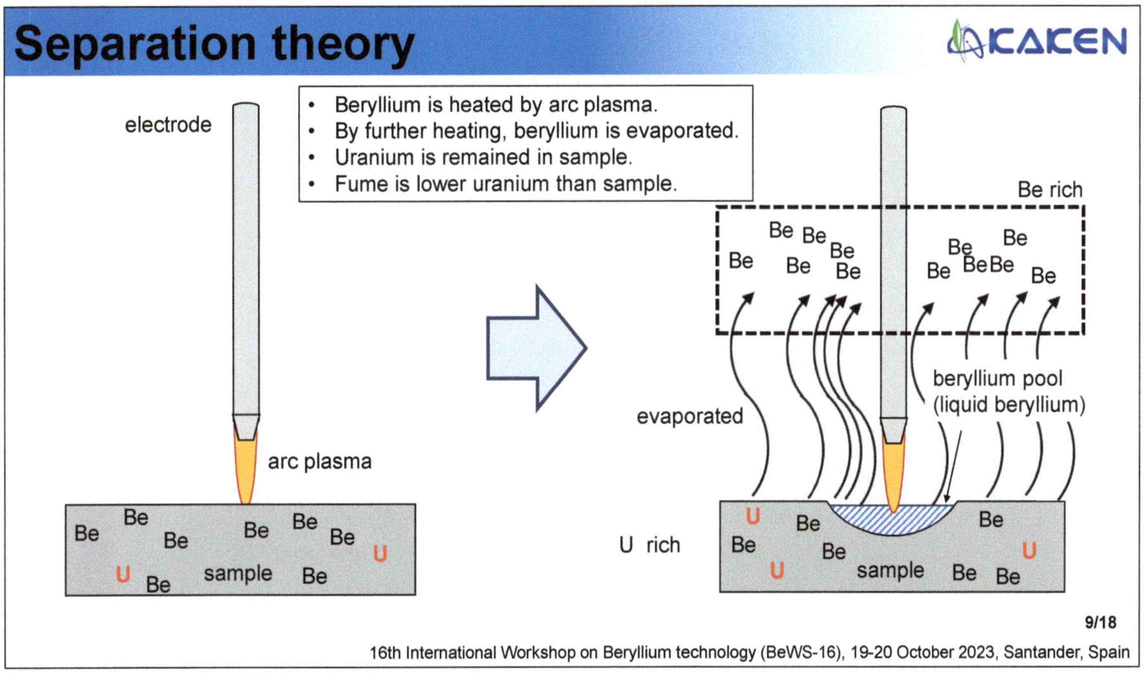
Separation theory
ICAKEN
electrode
Beryllium is heated by arc plasma.
By further heating, beryllium is evaporated.
Uranium is remained in sample.
Fume is lower uranium than sample.
arc plasma
Be Be Be Be Be Be U U Be sample Be
Be rich
evaporated
beryllium pool
(liquid beryllium)
U rich
sample
9/18
16th International Workshop on Beryllium technology (BeWS-16), 19-20 October 2023, Santander, Spain

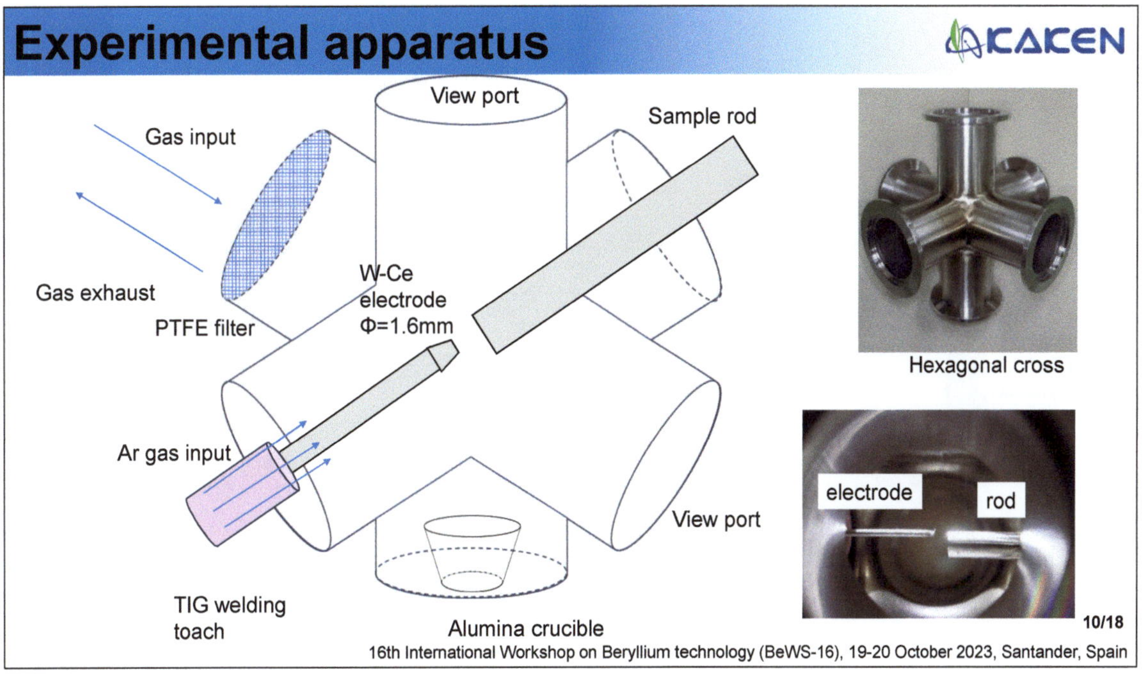
Experimental apparatus
ICAKEN
Gas input
Gas exhaust
PTFE filter
View port
Sample rod
W-Ce
electrode
Φ=1.6mm
Ar gas input
TIG welding
toach
Alumina crucible
View port
Hexagonal cross
electrode
rod
10/18
16th International Workshop on Beryllium technology (BeWS-16), 19-20 October 2023, Santander, Spain

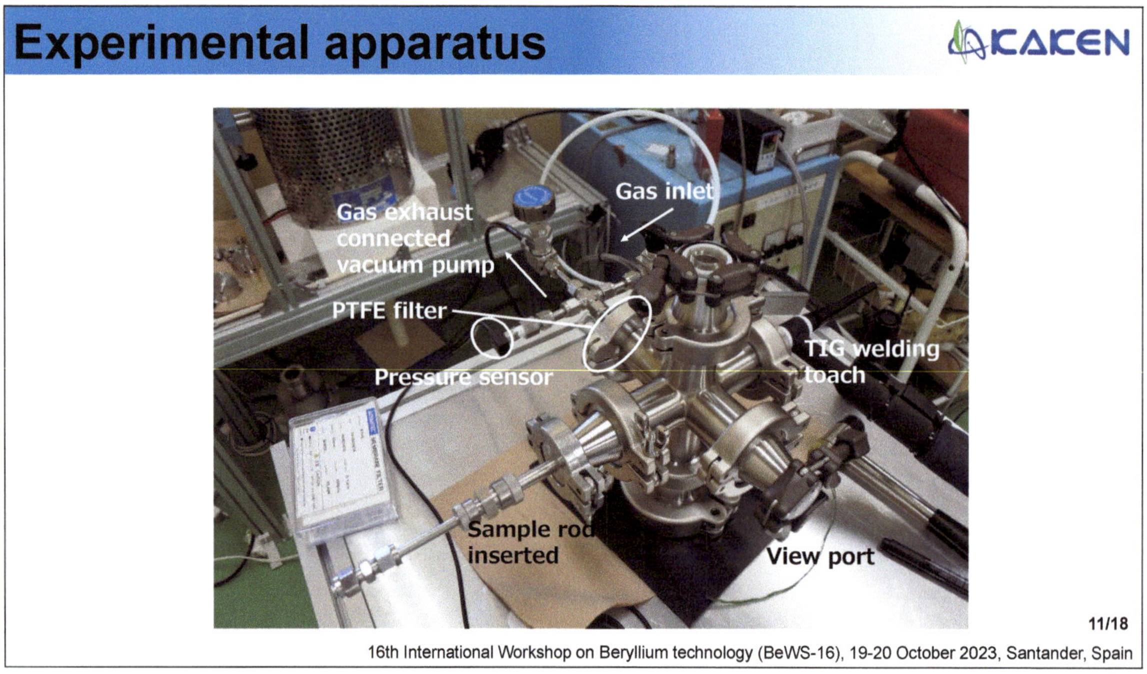
Experimental apparatus
KAKEN
Gas inlet
Gas exhaust
connected
vacuum pump
PTFE filter
Pressure sensor
TIG welding
toach
Sample rod
inserted
View port
11/18
16th International Workshop on Beryllium technology (BeWS-16), 19-20 October 2023, Santander, Spain

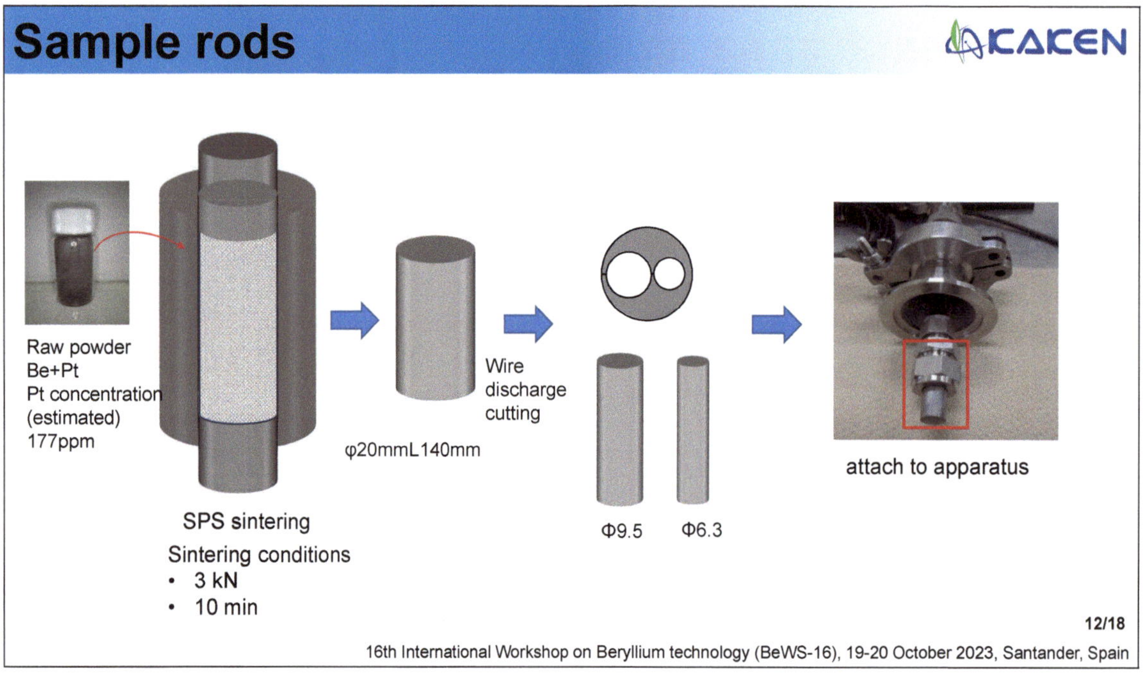
Sample rods
KAKEN
Raw powder
Be+Pt
Pt concentration
(estimated)
177ppm
SPS sintering
Sintering conditions
• 3 kN
• 10 min
φ20mmL140mm
Wire
discharge
cutting
Φ9.5
Φ6.3
attach to apparatus
12/18
16th International Workshop on Beryllium technology (BeWS-16), 19-20 October 2023, Santander, Spain

Experimental procedure

KAKEN

Step 1 gas replacement
Evacuation and argon input were repeated twice.

Step 2 arc discharging
Arc temperature was set 100 A, discharged few decade seconds.

Step 3 analysis
Generated fume and remained rod are analyzed by EDX, and ICP.

13/18

16th International Workshop on Beryllium technology (BeWS-16), 19-20 October 2023, Santander, Spain

Video

KAKEN

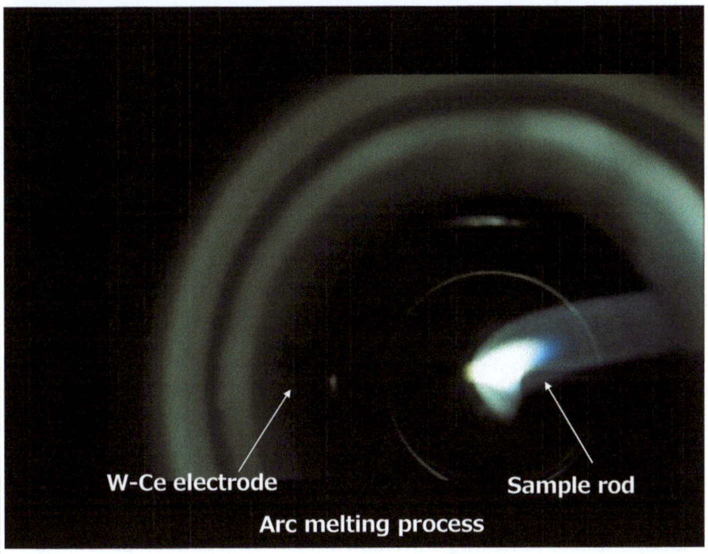

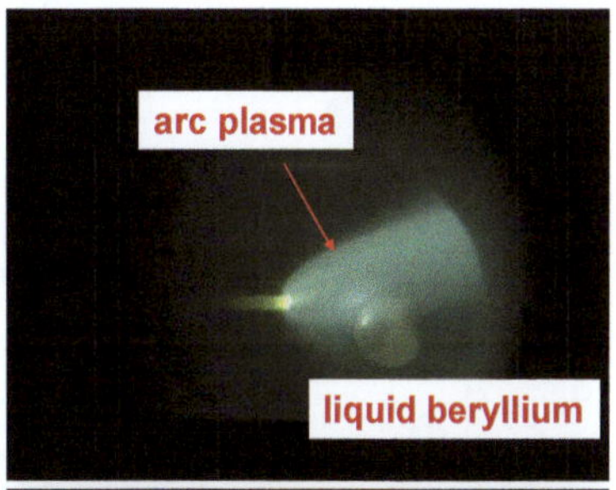

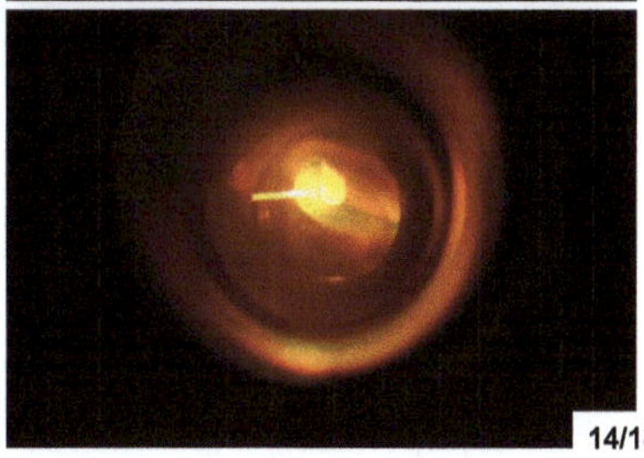

14/18

16th International Workshop on Beryllium technology (BeWS-16), 19-20 October 2023, Santander, Spain

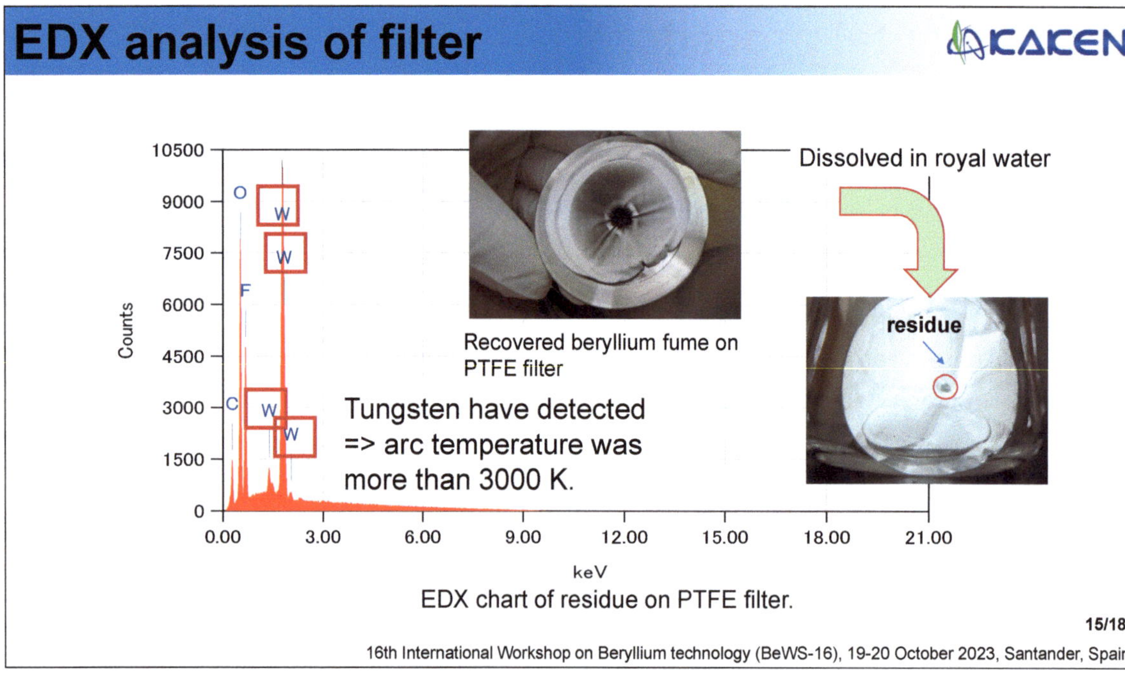
EDX analysis of filter
KAKEN
10500
9000
7500
6000
4500
3000
1500
0
Counts
O
W
W
F
C
W
W
0.00
3.00
6.00
9.00
12.00
15.00
18.00
21.00
keV
Recovered beryllium fume on PTFE filter
Dissolved in royal water
residue
Tungsten have detected
=> arc temperature was
more than 3000 K.
EDX chart of residue on PTFE filter.
15/18
16th International Workshop on Beryllium technology (BeWS-16), 19-20 October 2023, Santander, Spain

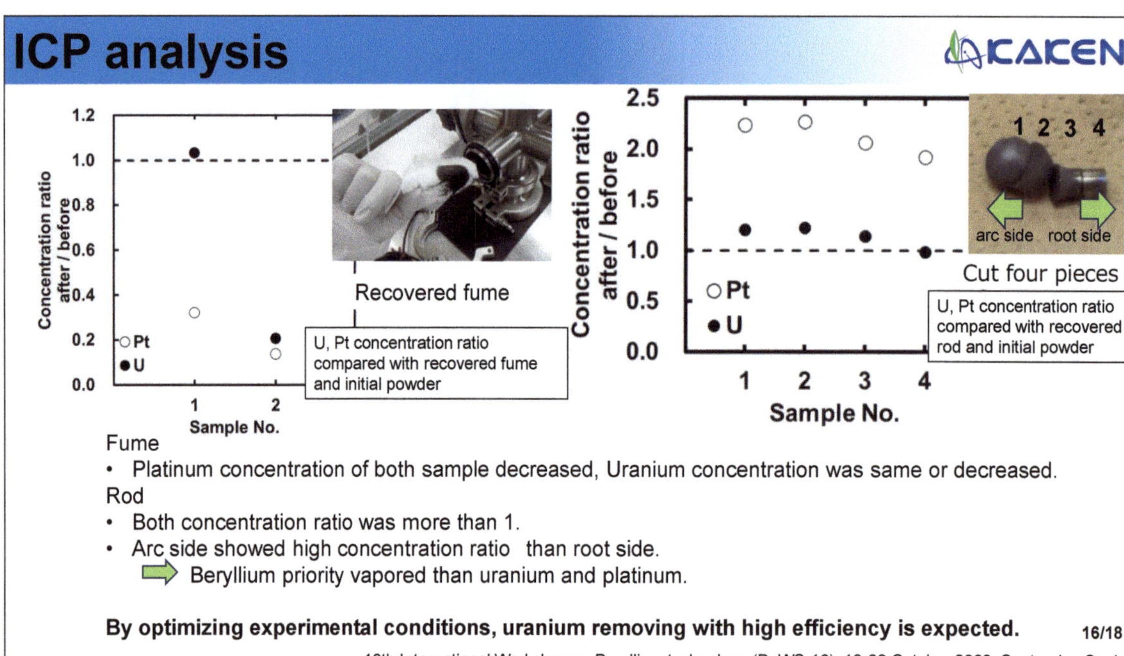
ICP analysis
KAKEN
1.2
1.0
0.8
0.6
0.4
0.2
0.0
Concentration ratio after / before
Pt
U
1
2
Sample No.
Recovered fume
U, Pt concentration ratio compared with recovered fume and initial powder
2.5
2.0
1.5
1.0
0.5
0.0
Concentration ratio after / before
Pt
U
1
2
3
4
Sample No.
1 2 3 4
arc side
root side
Cut four pieces
U, Pt concentration ratio compared with recovered rod and initial powder
Fume
• Platinum concentration of both sample decreased, Uranium concentration was same or decreased.
Rod
• Both concentration ratio was more than 1.
• Arc side showed high concentration ratio than root side.
Beryllium priority vapored than uranium and platinum.
By optimizing experimental conditions, uranium removing with high efficiency is expected.
16/18
16th International Workshop on Beryllium technology (BeWS-16), 19-20 October 2023, Santander, Spain

Apply for other elements removing

ICAKEN

CHEMISTRY	
Beryllium Assay, % minimum	99.20
Beryllium Oxide, % maximum	0.900
Al, % maximum	0.050
C, % maximum	0.090
Fe, % maximum	0.080
Mg, Cr, each % maximum	0.010
Ni, Cu, Ti, Zr, each, % maximum	0.025
Zn, Mn, Ag, Co, Pb, Ca, Mo, each, % maximum	0.005
Silicon, % maximum	0.045
U, % maximum	0.015
Other Metallic Impurities, each, % maximum	0.040

Supplier's spec sheet

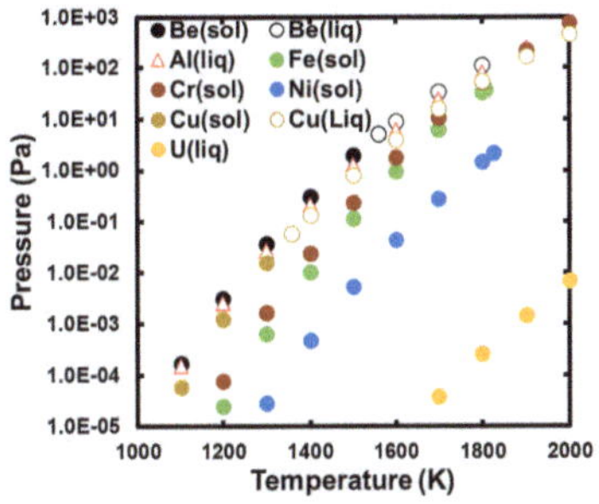

Vapor pressure of beryllium and impurities

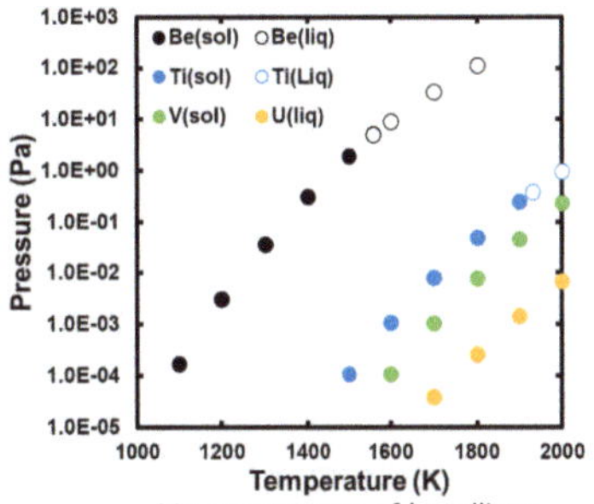

Vapor pressure of beryllium, titanium, vanadium, and uranium

- Considering removing impurity elements in beryllium by this study's method are available or not.
- The vapor pressures of the impurities are not as far apart as beryllium and uranium.
  => difficult removing by our method?
- We think that titanium or vanadium removing is suitable because beryllium is used as not simple but alloy on fusion reactor.
  Vapor pressure of titanium and vanadium are larger difference than impurities.
  =>It has the potential to isolate beryllium from beryllium alloys.

17/18

16th International Workshop on Beryllium technology (BeWS-16), 19-20 October 2023, Santander, Spain

Conclusion

ICAKEN

We focused on uranium containing beryllium, attempted removing by arc plasma.

- platinum concentration in recovered fume was lower than initial powder.
- platinum concentration in the rod after experiment was higher than initial raw powder.

- Similar results are expected to be obtained with uranium
- Considered molten platinum in sample rod diffused to root side.

Future works

1. By optimizing experimental conditions, more effective separation will achieve
   - Current
   - Pressure
   - Electrode-rod distance
2. Do experiments using beryllium-titanium or beryllium-vanadium rod
3. Mass removing system assemble

18/18

16th International Workshop on Beryllium technology (BeWS-16), 19-20 October 2023, Santander, Spain

3.5 Method for processing spent products from titanium beryllide $TiBe_{12}$ after completion of their operation in the DEMO reactor

Ye. Frants, M. Kolmakov, B. Zorin, M. Kylyshkanov, M Podoinikov, S. Udartsev

In the concept of the helium-cooled blanket of the DEMO reactor, it is planned to use blocks made of the $TiBe_{12}$ intermetallic compound. The total weight of products made of titanium beryllide will be at least 300 tons in beryllium equivalent. Due to neutron exposure, by the end of their service life, the products will be radioactive due to the accumulation of ^{3}H, ^{10}Be and other radioactive elements. Also, the consequence of operation will be the loss of shape and integrity of individual blocks. This raises the question of handling waste blocks.

One of the areas of use is the preliminary grinding of blocks into pieces up to 10 mm in size, the extraction of tritium during high-temperature annealing. After the tritium is extracted, the pieces are used to re-manufacture blanks.

The use of this technology will eliminate the cost of disposal of waste material, as well as the manufacture of a new set of blocks. Research conducted at UMP JSC, Ust-Kamenogorsk, made it possible to obtain workpieces with a size of $\varnothing 150\times 200$ mm from titanium beryllide $TiBe_{12}$ by recycling waste generated during manufacturing.

Presenting Author: Sergey Udartsev
Presenting Author's Affiliation: "Ulba Metallurgical Plant" JSC, Ust-Kamenogorsk, Republic of Kazakhstan
Email: UdartsevSV@ulba.kz

Method for recycling spent products from titanium beryllide $TiBe_{12}$ after completion of their operation in DEMO reactor

Ye. Frants, M. Kolmakov,
M. Kylyshkanov, M. Podoinikov, S. Udartsev
Ulba Metallurgical Plant JSC, Ust-Kamenogorsk, Kazakhstan

P. Vladimirov, V. Chakin,
R. Gaisin
Karlsruhe Institute of Technology, Germany

$TiBe_{12}$ blocks during and after operation in DEMO

$TiBe_{12}$ blocks will be subjected to the harsh conditions of rapid temperature change and neutron irradiation

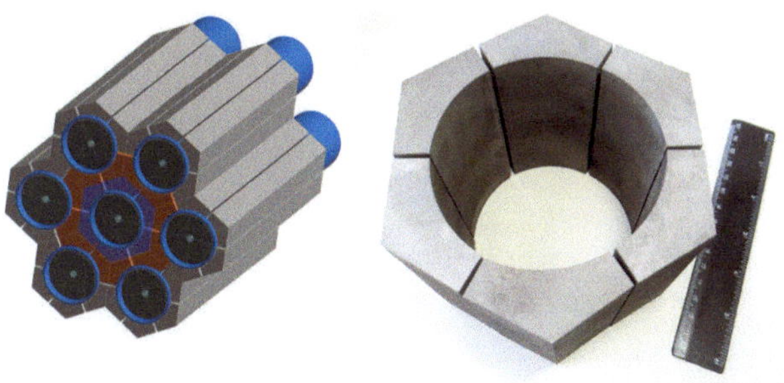

Possible issues

1. Swelling
2. Cracking
3. Fracture
4. Be-burnout
5. ^{3}H, He, ^{10}Be, ^{60}Co, long-lived isotops formation

2 Sergey Udartsev Ulba Metallurgical Plant

Possible ways of $TiBe_{12}$ blocks management after their operation in DEMO

1. Degassing blocks and returning them to reactor
2. Burial
3. Separation Be and Ti and Be ingots or FLiBe manufacturing
4. Repressing $TiBe_{12}$ blocks and returning to reactor

3 Sergey Udartsev Ulba Metallurgical Plant

Possible ways of $TiBe_{12}$ blocks management after their operation in DEMO

1. Degassing blocks and returning them to reactor

4 Sergey Udartsev Ulba Metallurgical Plant

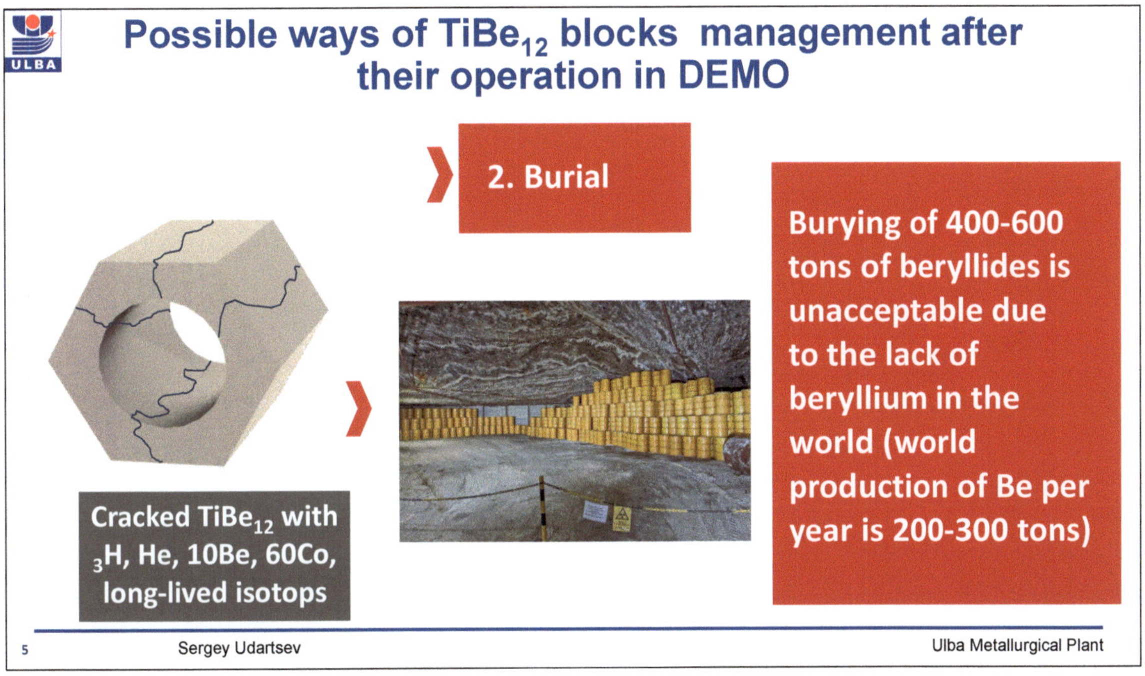
ULBA
Possible ways of TiBe12 blocks management after their operation in DEMO
2. Burial
Burying of 400-600 tons of beryllides is unacceptable due to the lack of beryllium in the world (world production of Be per year is 200-300 tons)
Cracked TiBe12 with 3H, He, 10Be, 60Co, long-lived isotops
5
Sergey Udartsev
Ulba Metallurgical Plant

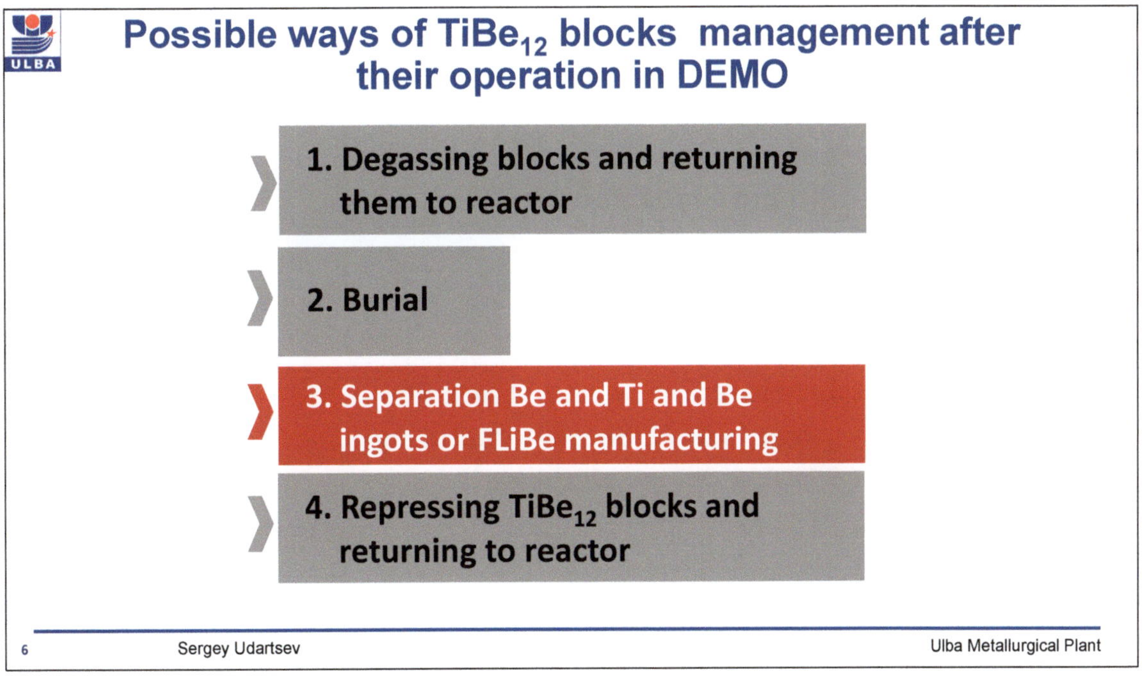
ULBA
Possible ways of TiBe12 blocks management after their operation in DEMO
1. Degassing blocks and returning them to reactor
2. Burial
3. Separation Be and Ti and Be ingots or FLiBe manufacturing
4. Repressing TiBe12 blocks and returning to reactor
6
Sergey Udartsev
Ulba Metallurgical Plant

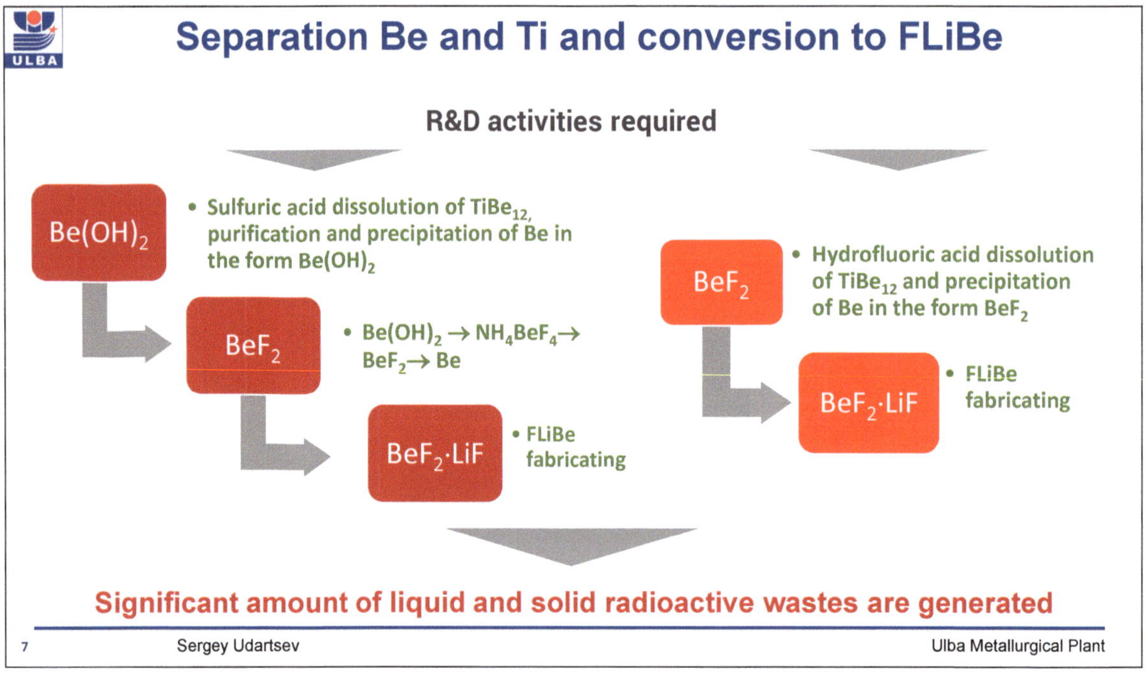
Separation Be and Ti and conversion to FLiBe
R&D activities required
Be(OH)2
• Sulfuric acid dissolution of TiBe12, purification and precipitation of Be in the form Be(OH)2
BeF2
• Be(OH)2 → NH4BeF4→ BeF2→ Be
BeF2·LiF
• FLiBe fabricating
BeF2
• Hydrofluoric acid dissolution of TiBe12 and precipitation of Be in the form BeF2
BeF2·LiF
• FLiBe fabricating
Significant amount of liquid and solid radioactive wastes are generated
7
Sergey Udartsev
Ulba Metallurgical Plant

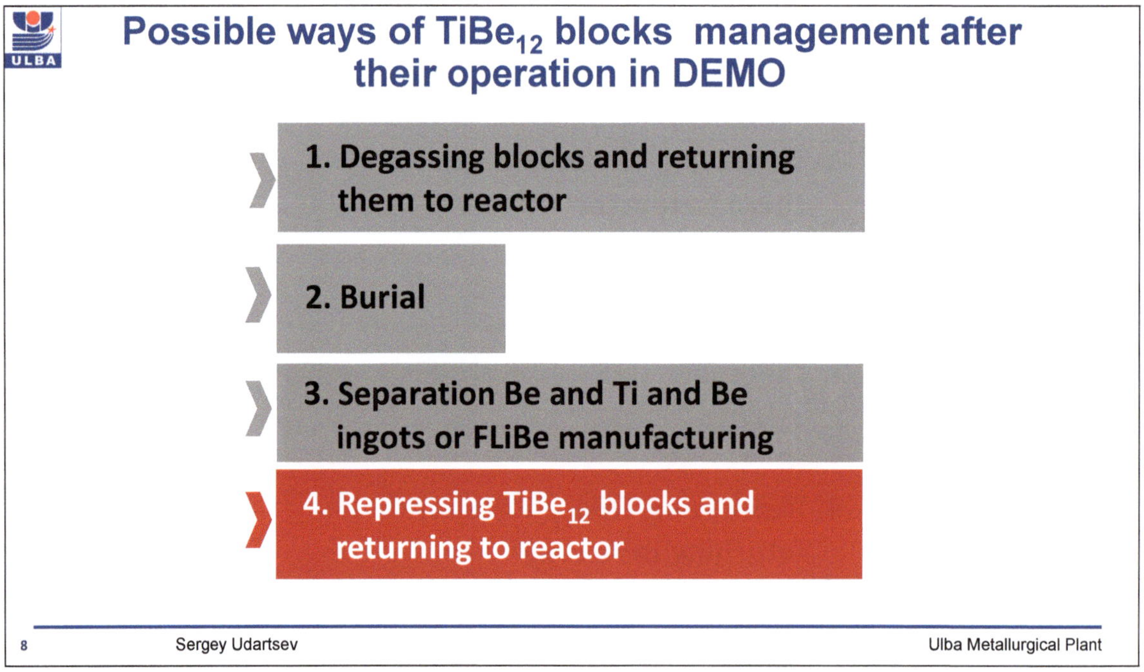
Possible ways of TiBe12 blocks management after their operation in DEMO
1. Degassing blocks and returning them to reactor
2. Burial
3. Separation Be and Ti and Be ingots or FLiBe manufacturing
4. Repressing TiBe12 blocks and returning to reactor
8
Sergey Udartsev
Ulba Metallurgical Plant

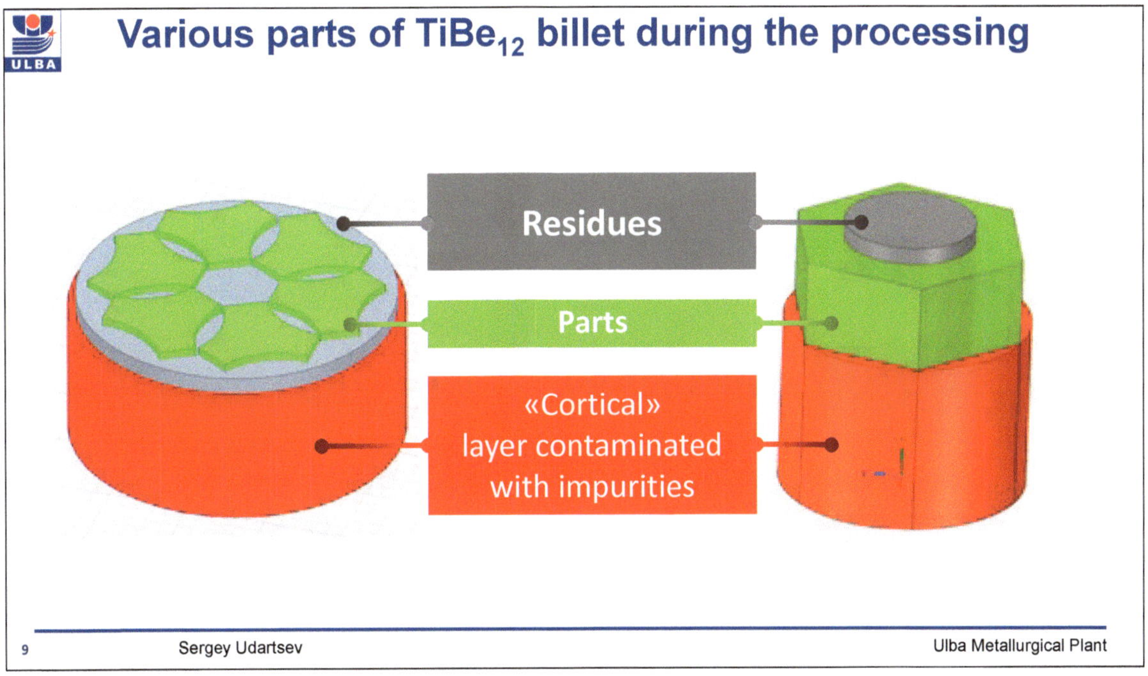
ULBA
Various parts of TiBe12 billet during the processing
Residues
Parts
«Cortical» layer contaminated with impurities
9
Sergey Udartsev
Ulba Metallurgical Plant

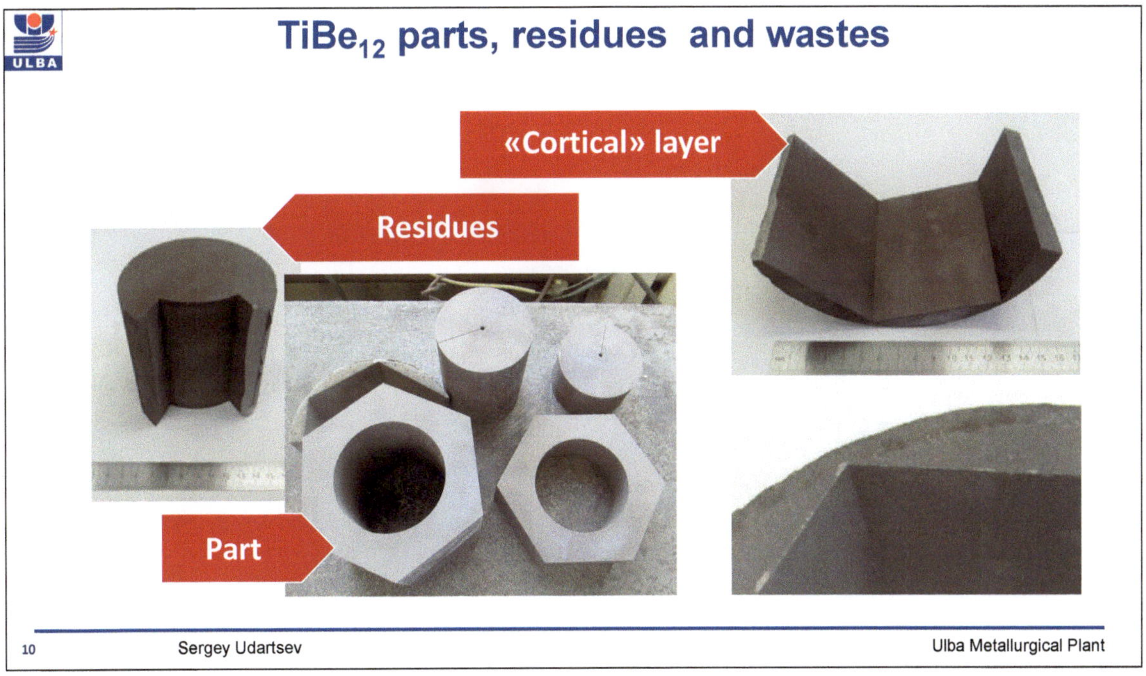
ULBA
TiBe12 parts, residues and wastes
«Cortical» layer
Residues
Part
10
Sergey Udartsev
Ulba Metallurgical Plant

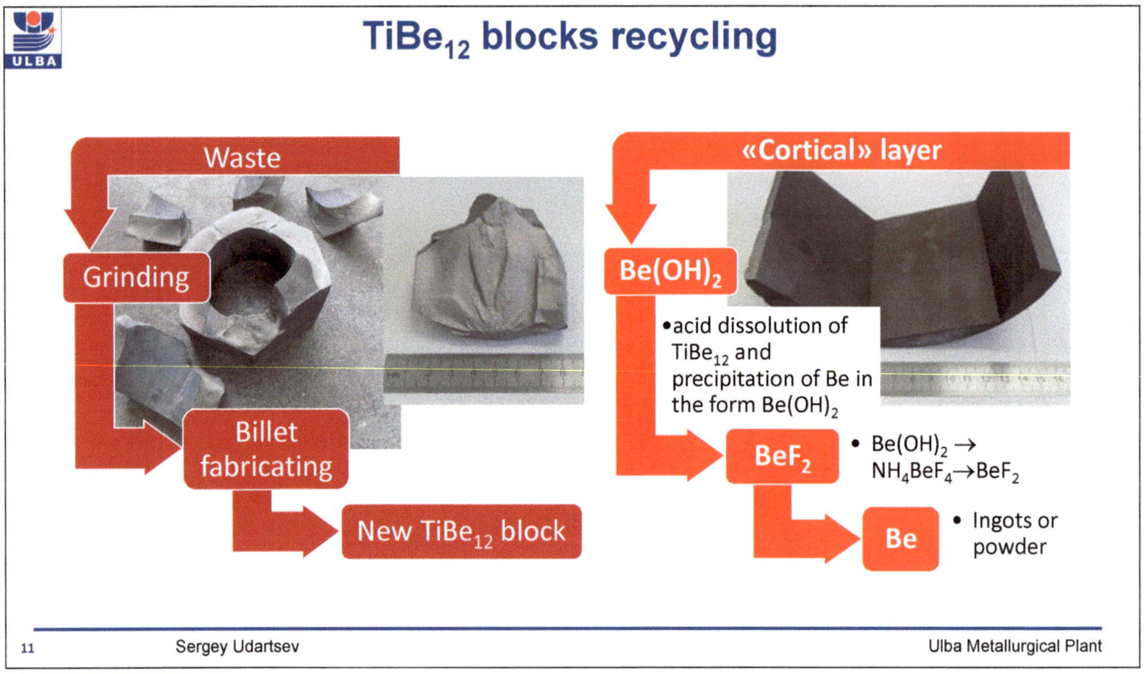
ULBA
TiBe12 blocks recycling
Waste
Grinding
Billet fabricating
New TiBe12 block
«Cortical» layer
Be(OH)2
•acid dissolution of TiBe12 and precipitation of Be in the form Be(OH)2
BeF2
• Be(OH)2 → NH4BeF4→BeF2
Be
• Ingots or powder
11
Sergey Udartsev
Ulba Metallurgical Plant

ULBA
The various fractions of TiBe12 grains after grinding of large pieces
0,315-0,630 mm
0,630-1 mm
1-2 mm
2-3 mm
3-5 mm
12
Sergey Udartsev
Ulba Metallurgical Plant

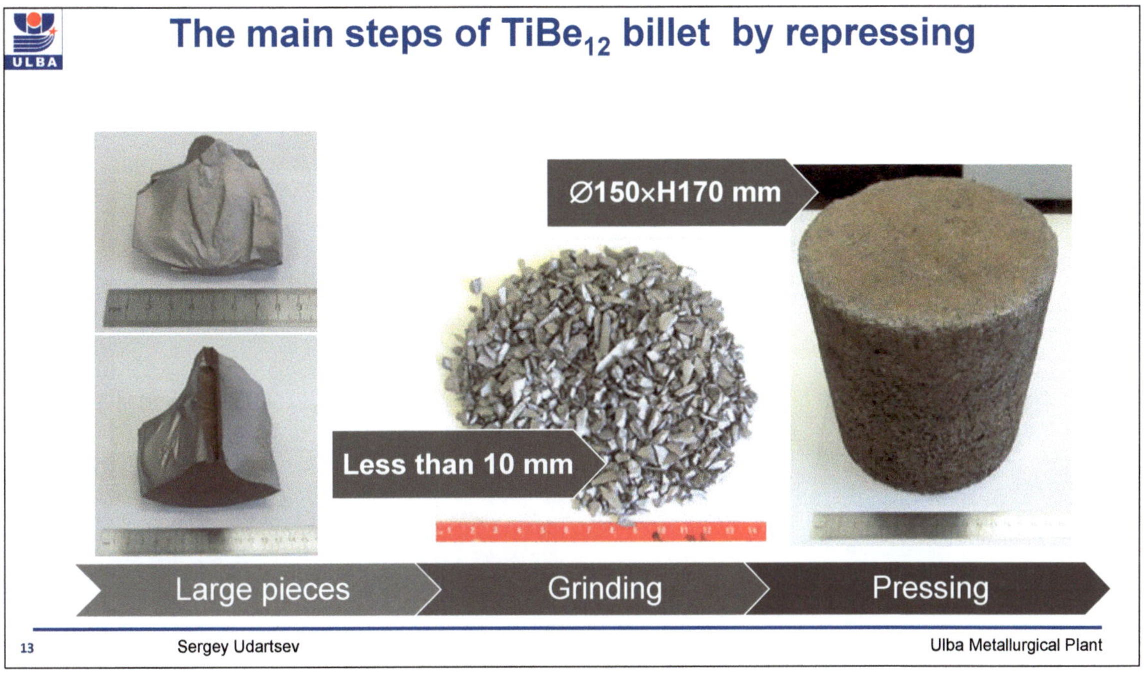
ULBA
The main steps of TiBe12 billet by repressing
Ø150×H170 mm
Less than 10 mm
Large pieces
Grinding
Pressing
13
Sergey Udartsev
Ulba Metallurgical Plant

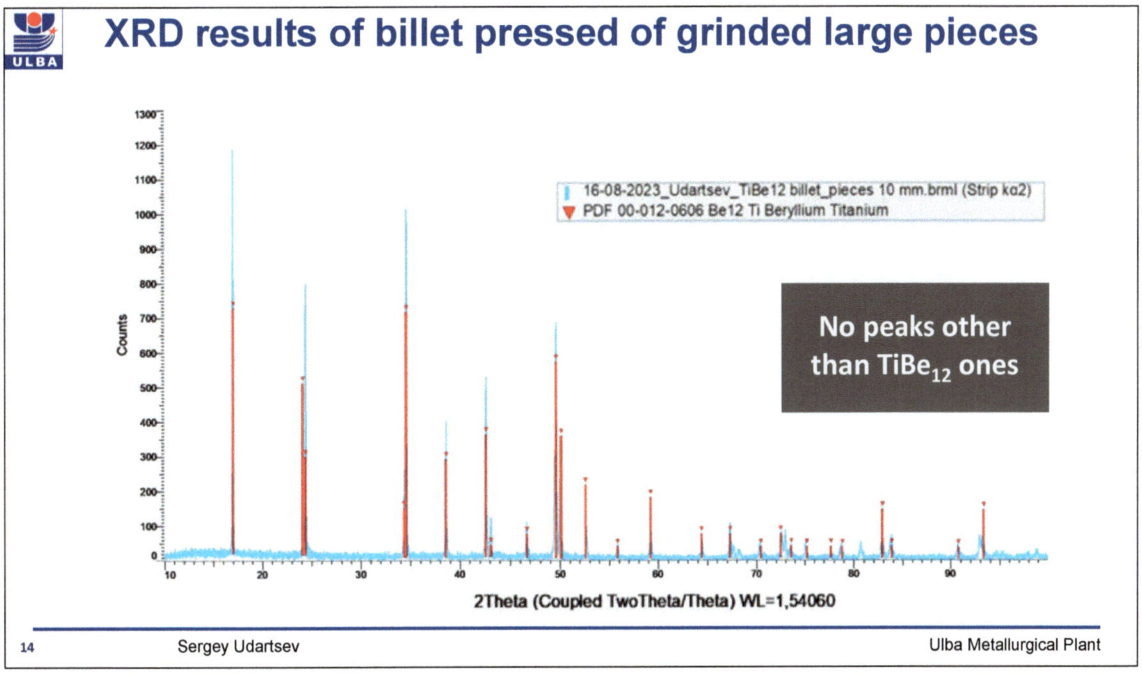
ULBA
XRD results of billet pressed of grinded large pieces
16-08-2023_Udartsev_TiBe12 billet_pieces 10 mm.brml (Strip kα2)
PDF 00-012-0606 Be12 Ti Beryllium Titanium
No peaks other than TiBe12 ones
Counts
2Theta (Coupled TwoTheta/Theta) WL=1,54060
14
Sergey Udartsev
Ulba Metallurgical Plant

Microstructure of the recycled billet

×100

×500

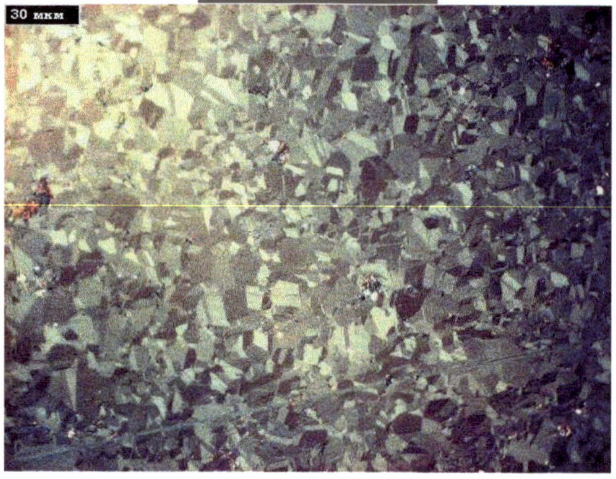

15 Sergey Udartsev Ulba Metallurgical Plant

$TiBe_{12}$ block made from pieces of residual material

∅150×H170 mm
Density >95%
Pressed from
pieces <10 mm

The technology can be adapted for cracked beryllide blocks after their use in DEMO

16 Sergey Udartsev Ulba Metallurgical Plant

ULBA

Conclusions

- $TiBe_{12}$ wastes recycling method is required both for the production of initial blocks and for blocks after completion of operation in DEMO.
  - One of the ways of recycling of blocks after completion of operation in DEMO is FLiBe production. But R&D required.
- Separation of metals by acid dissolution and the subsequent production of FLiBe will lead to the formation of a significant amount of liquid and solid radioactive wastes.
  - UMP researched and successfully obtained billet from $TiBe_{12}$ residues by "dry" recycling. This technology can be used to recycle blocks after completion of their operation in the reactor.

17 Sergey Udartsev Ulba Metallurgical Plant

ULBA

Thank you for your attention!

We invite you to cooperation!

18 Sergey Udartsev Ulba Metallurgical Plant

SESSION 4

Molten salt breeders

4.1 FLiBe Molten Salt as a Tritium Breeder/Coolant for Fusion: Fostering Collaboration for Breakthrough Progress

Richard PEARSON, Paul BARRON, Colin BAUS, Sara FERRY, Ethan PETERSON, Christopher DORN, Bibake UPPAL, Kevin WOLLER

This presentation provides an in-depth analysis of the current Research and Development (R&D) status and the future direction of FLiBe molten salt as an enabling technology for tritium breeding and cooling in fusion reactors. The use of FLiBe presents several advantages for the operation of fusion reactors, including high thermal conductivity, low melting point, and a capacity to generate tritium, which is crucial for self-sustained fusion reactions.

Despite its advantages, FLiBe also poses several challenges that must be overcome, including its corrosive nature, the evolution of corrosive HF gas, the safe handling and storage of tritium, and the need for accurate material property data. This presentation delves into the outcomes of a major community workshop held in North Carolina, U.S., in May 2023, where a diverse group of stakeholders, encompassing laboratories, universities, and industry representatives, convened to evaluate the state of FLiBe technology, as a subset technology option of several breeding blanket types.

The presentation highlights the collaborative efforts of these stakeholders in addressing the challenges by conducting experiments to improve material properties, optimizing reactor designs, developing corrosion-resistant materials, and exploring innovative fabrication methods. It emphasizes the significance of collaboration among stakeholders in tackling these issues and outlines the next steps for advancing FLiBe technology, which include further experimental studies, scaling up reactor designs, and fostering joint research initiatives.

The sustained collaboration and cooperation of stakeholders are paramount for realizing the potential of FLiBe as a key enabling technology in future fusion power plants, ultimately contributing to a clean, abundant, and sustainable energy source for generations to come.

Presenting Author: Richard PEARSON
Presenting Author's Affiliation: Kyoto Fusioneering, Japan
Email: r.pearson@kyotofusioneering.com

KYOTO FUSIONEERING
PSFC Plasma Science and Fusion Center Massachusetts Institute of Technology
FLiBe Molten Salt as a Tritium Breeder/Coolant for Fusion: Fostering Collaboration for Breakthrough Progress
R. PEARSON, S. FERRY, C. DORN, P. BARRON, K. WOLLER, C. BAUS, B. UPPAL & E. PETERSON
16th International Workshop on Beryllium Technology (BeWS-16) Santander, Cantabria, Spain
19th October 2023
BeWS-16
2023
16th International Workshop On Beryllium Technology
Santander, Cantabria, Spain
19-20 October 2023
FUSION for the FUTURE.
©2023 KYOTO FUSIONEERING LTD. ALL RIGHTS RESERVED.

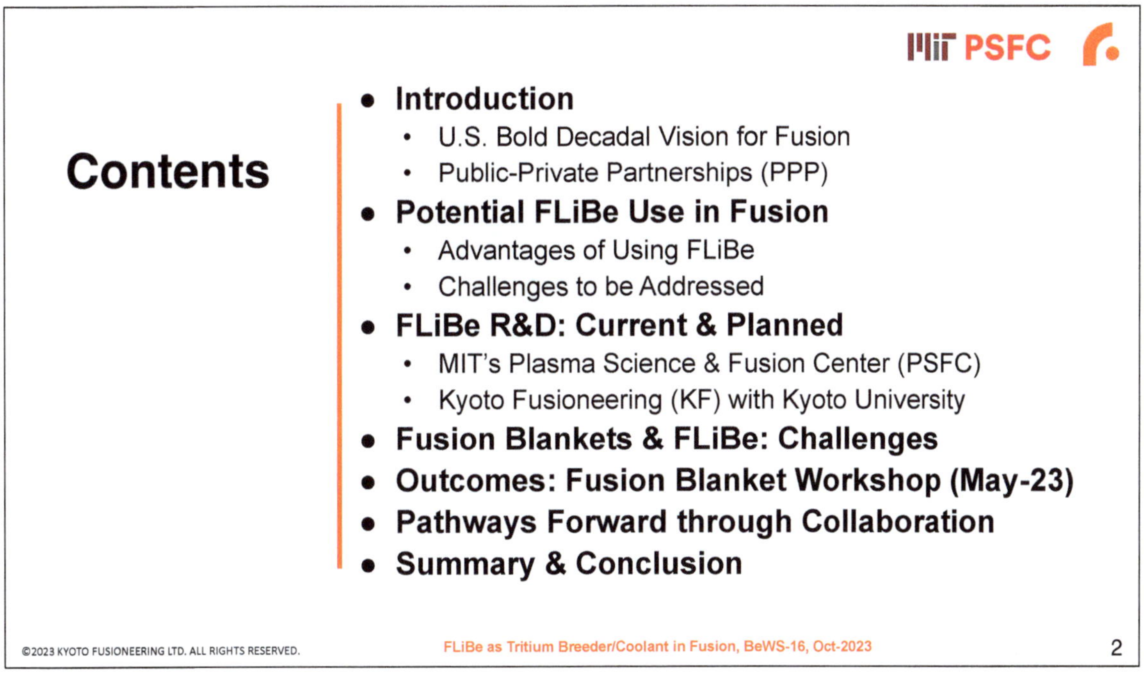
MIT PSFC
Contents
Introduction
U.S. Bold Decadal Vision for Fusion
Public-Private Partnerships (PPP)
Potential FLiBe Use in Fusion
Advantages of Using FLiBe
Challenges to be Addressed
FLiBe R&D: Current & Planned
MIT's Plasma Science & Fusion Center (PSFC)
Kyoto Fusioneering (KF) with Kyoto University
Fusion Blankets & FLiBe: Challenges
Outcomes: Fusion Blanket Workshop (May-23)
Pathways Forward through Collaboration
Summary & Conclusion
©2023 KYOTO FUSIONEERING LTD. ALL RIGHTS RESERVED.
FLiBe as Tritium Breeder/Coolant in Fusion, BeWS-16, Oct-2023
2

Introduction

MIT PSFC

- **U.S. Bold Decadal Vision**
  - Pushed by Climate Change
  - Announced on 15-Mar-2022
  - Purpose: accelerate fusion development
  - Goal: build prototype Fusion Power Plant (FPP) in 10-15 years
- **Public-Private Partnerships**
  - US Department of Energy
    - Milestone-Based Program
    - INFUSE Partnership Grants
  - ARPA-E Workshops on Fusion Fuel Cycle and Blankets (EPRI, May-2023)
  - Symposium on Fusion Energy 2023
    - Satellite Session, Jun-2023, titled "Next Steps to Global Fusion Partnerships"

arpa·e
Advanced Research Projects Agency • ENERGY

INFUSE | Innovation Network for Fusion Energy

SOFE 2023
OXFORD, UK
30TH IEEE SYMPOSIUM ON FUSION ENGINEERING

 FLiBe as Tritium Breeder/Coolant in Fusion, BeWS-16, Oct-2023 3

Potential FLiBe Use in Fusion

MIT PSFC

- **Advantages of using FLiBe**
  - High thermal conductivity
  - Low melting point
  - Tritium generation capability
- **Challenges to be addressed**
  - Corrosion & compatibility
  - Evolution of HF gas
  - Tritium: safe handling & storage
  - Accurate material property data

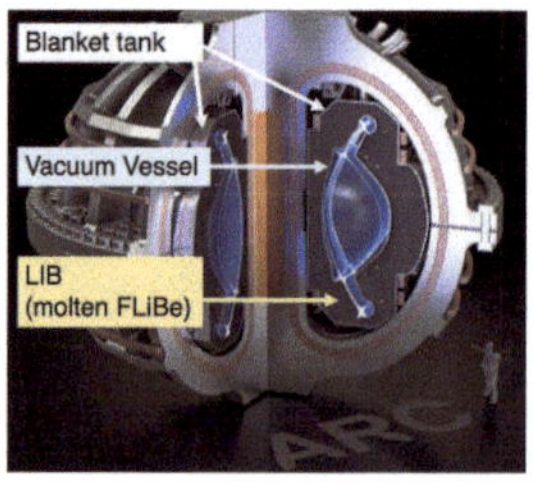

Commonwealth Fusion Systems is pursuing a FLiBe-based liquid immersion blanket for their ARC fusion power plant concept.

Commonwealth Fusion Systems

Rendering of ARC: CFS

XCIMER ENERGY CORPORATION

Xcimer is another US milestone-program company pursuing an IFE approach that uses a FLiBe-waterfall reaction chamber.

Opportunities for fission/ fusion partnership
- Companies like Kairos pursuing FLiBe-based advanced reactor designs
- MSRE operating experience/knowledge

MSR at ORNL 1964-1969

 FLiBe as Tritium Breeder/Coolant in Fusion, BeWS-16, Oct-2023 4

Overview: Current FLiBe R&D at KF & MIT-PSFC

MIT PSFC

FLiBe Principal Research Area	Challenges/ Requirements	Current Activities at KF	Current Activities at MIT - PSFC
Compatibility: Corrosion & Containment	Stability of materials in contact with FLiBe, neutrons, tritium, and heat	Seek out candidate materials based on known properties & other data	FLiBe corrosion R&D: static testing; loop tests at NRL; combined corrosion-irradiation testing
	Longevity and reliability under reactor conditions	Modeling & small-scale testing	Full-scale LIBRA test will allow integrated testing at temperature and under neutron irradiation
	Impact on structural materials and components	Modeling, testing, and design work	Incorporating results from experiments into fuel-cycle systems-level modeling
Tritium Generation & Extraction	Performance related to Tritium Breeding, Release & Extraction	Modeling & small-scale testing	BABY and LIBRA experimental campaigns
	Reactivity of tritium fluoride (TF) and hydrogen fluoride (HF)	Laboratory testing	BABY and LIBRA experimental campaigns; construct LIBRA lab space (Be capable)
Purification Process & Handling	Specifications for and availability of feedstock materials	Supply chain investigation, including make/ buy options	CFS-funded research on salt chemistry
	Impurities in FLiBe increase its reactivity/ corrosivity	Small batch trials conducted	CFS-funded research on salt chemistry; corrosion testing; LIBRA (sectored approach allows chemistry variation)
Commercialization	Beryllium resource supply/demand	Resource & supply chain forecast modelling	N/A
	Non-proliferation and security	Non-proliferation policy development	FLiBe non-proliferation modelling; neutronics and breakout scenarios

 FLiBe as Tritium Breeder/Coolant in Fusion, BeWS-16, Oct-2023 5

FLiBe R&D at MIT-PSFC: LIBRA Prep at Small Scale

MIT PSFC

- LIBRA requires dedicated facility construction (~1 year)
- Want to ramp up experience with FLiBe and tritium handling
- Scale up beryllium safety and respirator use programs
- Practice tritium and neutron detection/accountancy
- Start getting solubility/permeability data
- 100mL test runs completed with surrogate salts; experimental set-up modified

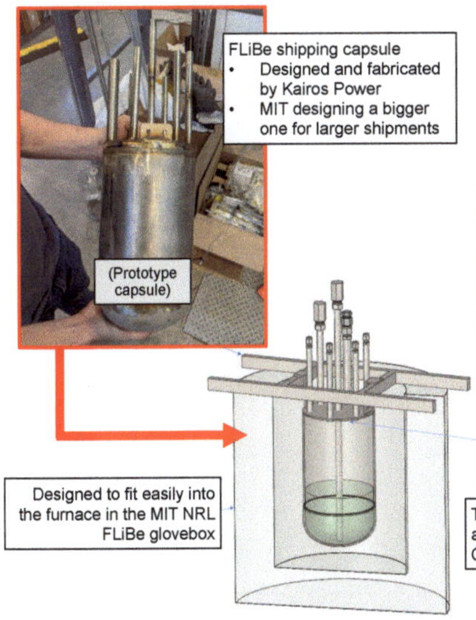

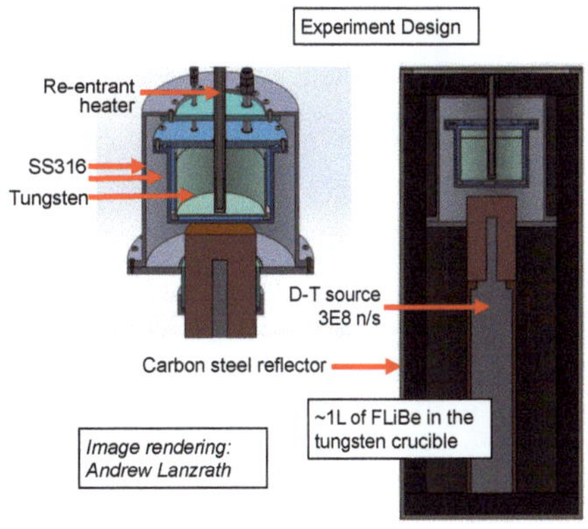

 FLiBe as Tritium Breeder/Coolant in Fusion, BeWS-16, Oct-2023 6

FLiBe R&D at KF: Test Loop for Dynamic Experiments

MIT PSFC

Glovebox set-up at Kyoto University, used for making and purifying laboratory quantities of FLiBe for testing.

Solid beryllium fluoride for making FLiBe in the lab; purification techniques to be developed.

FLiBe Test Loop bench-top experiment at Kyoto University, which will be used to investigate the dynamic performance of the liquid breeder/ coolant.

FLiBe as Tritium Breeder/Coolant in Fusion, BeWS-16, Oct-2023

7

MIT-PSFC: Full-Scale LIBRA Experiment

MIT PSFC

LIBRA = Liquid Immersion Blanket: Robust Accountancy

LIBRA will require a dedicated lab space to accommodate:

- Large volumes of high-temperature salt
- Beryllium and beryllium compounds
- High-energy, high-fluence neutron source
- Significant amounts of tritium being generated

A **new FLiBe lab at MIT PSFC** that will house LIBRA and the MIT NRL molten salt loops (joint fission/fusion venture) is being built.

Key Design Advantage:
LIBRA's four quadrants will function independently, with their own in/out gas-flow systems. That permits data collection on repeatability and also makes it possible to vary parameters between quadrants to see how salt chemistry, etc. affects results.

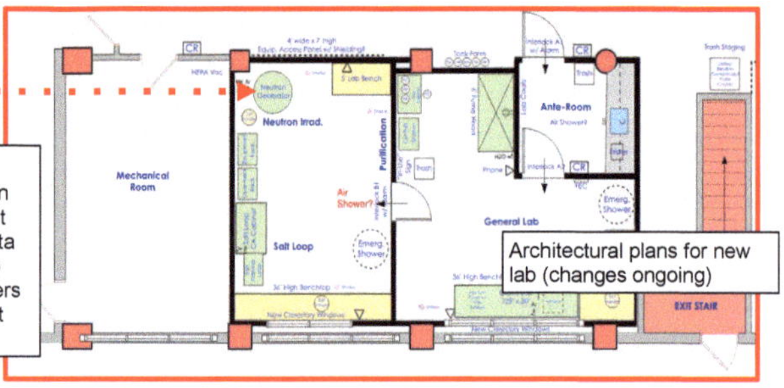

Architectural plans for new lab (changes ongoing)

FLiBe as Tritium Breeder/Coolant in Fusion, BeWS-16, Oct-2023

8

Fusion Blankets & FLiBe: Challenges to Address

MIT PSFC

- **Tritium control**
  - Tritium permeation barriers that are compatible with FLiBe
  - Assessment of how breeder material affects tritium measurement
  - Tritium Extraction Systems (TES)
    - Low Technology Readiness Level (TRL)
    - Down-selection of candidates followed by scale-up is necessary
    - Need to understand impurity effects on TES
  - Understanding of tritium behavior in FLiBe
- **Develop FLiBe blanket designs**
  - Needs facilities that can handle both tritium and beryllium

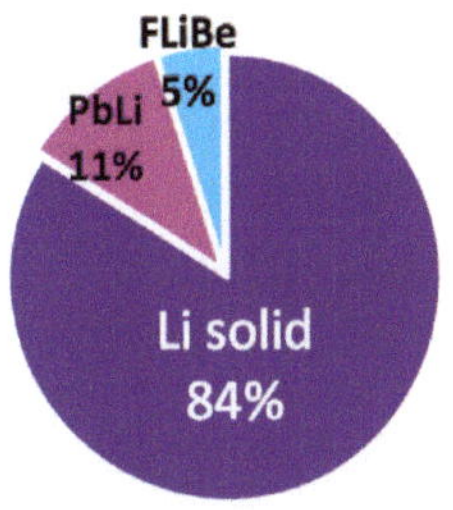

Breakdown of breeder material/neutron irradiation experimental campaigns in the literature by breeder material

Worldwide experience with FLiBe tritium breeding is very limited. This breakdown is based on three (3) papers about FLiBe.

Please Note:
Virtually all of the experimental objectives listed here have associated data and modelling objectives related to acquiring reliable data to feed into well-verified models.

 FLiBe as Tritium Breeder/Coolant in Fusion, BeWS-16, Oct-2023 9

Fusion Blankets & FLiBe: More Challenges

MIT PSFC

- **Materials compatibility with FLiBe**
  - Better corrosion data needed
  - Develop redox and impurity control systems
  - Complicated by evolution of TF during corrosion
  - Good FLiBe reference electrode material required
- **Blanket enabling technologies**
  - Thermal cycle components to be investigated: pumps, primary and secondary coolant loops, heat exchangers, and turbines
  - Lithium enrichment and supply chain must be understood
- **Components need to be robust and maintainable**
- **Efficiency of T-fuel-cycle components**

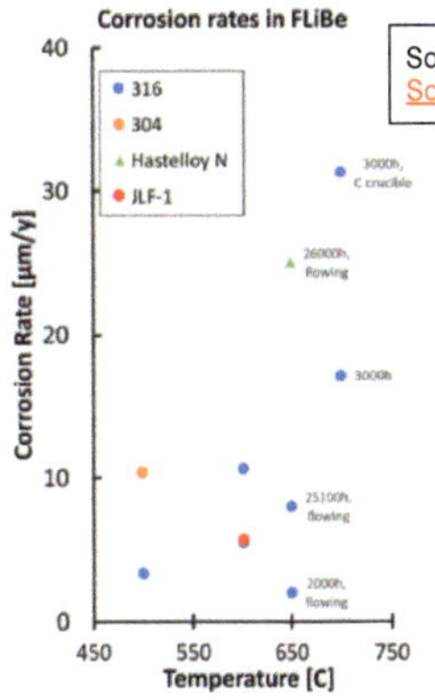

Source: Ferry et al., Fusion Sci & Tech, 2022

Published corrosion data for materials in pure FLiBe are very limited. Exposure time is 1000 h, and tests were static unless otherwise noted. Crucibles were generally made of the same or similar materials to the samples.

Presented at ISFNT-15 (Sep-2023):
Pint et al., *Compatibility assessments for fusion applications: Sn, Li. Pb-Li, and FLiBe*. Four different liquid compatibility tests are currently in progress at Oak Ridge National Lab, including work with SiCf/SiC composites.

 FLiBe as Tritium Breeder/Coolant in Fusion, BeWS-16, Oct-2023 10

Fusion Blankets & FLiBe: Longer-Term Issues

MIT PSFC

- **Activation and waste**
  - Implications of mixed waste disposal (e.g., radioactive materials with Be and T)
  - Effects of activation products on coolant chemistry
- **Maintenance and integration**
  - Ensure fusion companies follow existing Be safety standards
  - Generate new codes, standards, and best practices, specifically for FLiBe and the unique challenges it poses (most previous experience with use of solid beryllium or compounds)
  - Develop monitoring technology (e.g., real-time Be detection)
  - Design blankets for system-wide integration
  - Develop remote-handling strategies for maintenance in toxic (Be) and radioactive environments

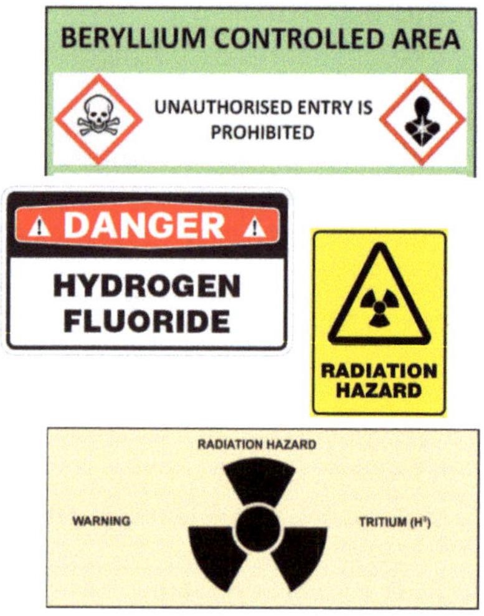

 FLiBe as Tritium Breeder/Coolant in Fusion, BeWS-16, Oct-2023 11

Outcomes: Fusion Blanket Workshop (May-2023)

MIT PSFC

- **2-day workshop on fusion fuel cycle; 2-day workshop on fusion blankets**
- **US blanket program relatively underdeveloped, need to expand experimental capabilities**
  - Tritium capabilities needed for blankets and fuel cycle technology testing
  - Beryllium handling capabilities for FLiBe
- **No plans to down-select a specific breeder at this point: expand R&D on FLiBe, liquid Li (UK partnerships?), Pb-Li, and solid breeders**
  - US will look to support all "milestone winner companies" at this stage (and others, too)
  - Private companies: heavy emphasis on liquid breeders, no solid options being seriously considered
- **Need to quantify and manage waste streams from blankets (underexplored area)**
- **Components need to be robust and maintainable (including qualification)**
- **Efficiency of tritium fuel cycle components that interact with the blanket system (most obviously: tritium extraction system) needs to be improved; need to manage higher throughput than has been experienced with fission or past D-T fusion campaigns**

 FLiBe as Tritium Breeder/Coolant in Fusion, BeWS-16, Oct-2023 12

Pathways Forward through Collaboration

- **Possibilities**
  - Opportunities to collaborate on UNITY?
  - Opportunities to collaborate on scaled-up FLiBe lab at MIT?
  - Corrosion/ materials experiments?

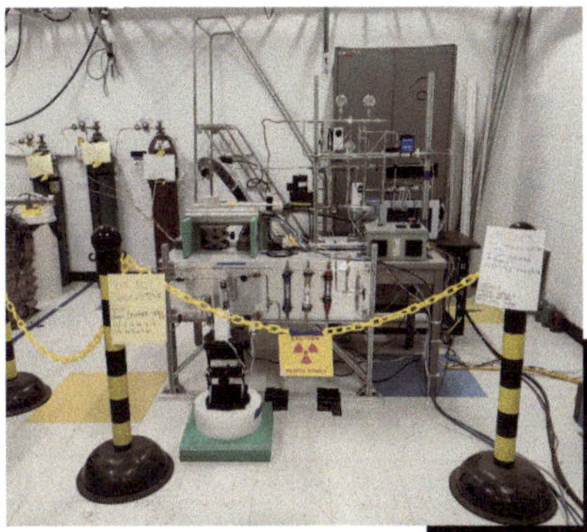

MIT-PSFC set-up for running the small-scale FLiBe tests. The salt crucible is the white cylinder positioned between the two horizontal neutron generators.

Photo credit: Nikola Goleš

PSFC collaborates internally at MIT on FLiBe development with NRL (Nuclear Research Lab), which does fission work.

KF's planned state-of-the-art research test platform, UNITY (UNique Integrated Testing facilitY), which will be the world's first such testing facility for fusion power plant equipment. Electricity generation (proxy heat source; no fusion plasma) with fusion-power plant relevant technologies is planned by end of 2024.

FLiBe as Tritium Breeder/Coolant in Fusion, BeWS-16, Oct-2023

13

Summary & Conclusion

- **Global Fusion Community (Industry) is changing**
  - New US role is underpinned by the Bold Decadal Vision and driven by private investment
  - Private fusion companies are trying to accelerate the timeline
  - Government-funded labs and universities have a critical role to play, including internationally
  - MIT-PSFC and KF are embracing opportunities in this changing industry
- **Collaboration & Partnerships**
  - Public-Private Partnerships (PPPs) are emerging as the dominant model for accelerated fusion development
  - US Government programs targeting specific FPPs to make it easier for national labs, universities, and private companies to work together on common, defined goals
  - Industry-to-industry (incl. fission & fusion synergies?) is an area to be explored further
- **Pathways forward....**
  - Re: beryllium, there is much work for FLiBe to become a workable solution as a fusion breeder, and it may be that some form of solid beryllium or Be compound will still be used for the first FPP
  - KF and MIT-PSFC are proactively seeking collaboration & partnerships, on a global basis

FLiBe as Tritium Breeder/Coolant in Fusion, BeWS-16, Oct-2023

14

PSFC
Plasma Science and Fusion Center
Massachusetts Institute of Technology
KYOTO
FUSIONEERING
BeWS-16
2023
16th International Workshop
On Beryllium Technology
Santander, Cantabria, Spain
19-20 October 2023
Thank You Very Much!
Web: www.psfc.mit.edu
Contact: info@psfc.mit.edu
LinkedIn: linkedin.com/company/plasma-science-and-fusion-center-at-mit/
Web: www.kyotofusioneering.com
Contact: biz@kyotofusioneering.com
Twitter: @kyotofusioneer
LinkedIn: linkedin.com/company/kyoto-fusioneering/

4.2 Prototype verification of tritium breeding with liquid immersion blanket technology using FLIBE

Kevin B. WOLLER, C. CUMMINGS, R. DELAPORTE-MATHURIN, C.S. DUNN, E. EDWARDS, S.E. FERRY, N. GOLEŠ, A.T. LANZRATH, R. LECCACORVI, E.E. PETERSON, S. SEGANTIN, R. VIEIRA, D.G. WHYTE, L. ZHOU, W. ZHOU, M.D. EKLUND, T.F. FUERST, H. GIETL, A.A. RIET, C.N. TAYLOR

Experimental development of liquid FLiBe tritium breeder technology is under way at the Massachusetts Institute of Technology Plasma Science and Fusion Center (PSFC) with modeling and design support from Idaho National Laboratory (INL) Safety and Tritium Applied Research (STAR) Facility. The project goal is to investigate tritium breeding and release from molten FLiBe at up to 700 °C with a source of D-T fusion neutrons. This environment ensures relevant conditions for comparison to Fusion Power Plants (FPP). In order to reach prototypical neutron energy spectrum conditions throughout the breeding blanket, path lengths greater than 30 cm of FLiBe are required. Such a prototypical liquid immersion blanket (LIB) with FLiBe breeder is estimated to use at least 500 L of FLiBe. Neutronics modeling of a prototype blanket of this scale indicates tritium breeding ratios (TBR) approaching unity.

Safe handling and management of FLiBe, taking into account the hazards of corrosive fluoride compounds and beryllium toxicity, are paramount. Efforts to document and share safe practices in addition to scientific results are treated just as importantly. Working directly with high energy D-T fusion neutrons forces experiments to not only account for optimizing tritium breeding, just as would need to be done by FPP developers, but also pushes experimental design to take into account tradeoffs in shielding material versus breeding material.

A high level benefit of LIB technology, especially using FLiBe breeder, is the limited volume of non-breeding, solid materials in the valuable neutron flux zones for tritium breeding blankets, leading to smallest scale breeder blankets [1]. Recent reports [2, 3] highlight the limited supply of tritium for fusion energy development. Smaller scale tritium breeding tests under prototypical conditions are necessary to develop the tritium breeding technology for D-T FPP concepts in parallel to D-T fusion core development. This would help minimize any bottleneck in tritium fuel supply that might inhibit the growth rate of fusion energy adoption.

Facility development, safety and operational review, neutron source characterization, benchtop-scale tritium breeding tests (0.1-1 L), and model comparisons are leading up to and informing full prototype tritium breeding experiment design and planning at the prototypical FPP scale.

ACKNOWLEDGEMENTS

This work is supported by the United States of America Department of Energy (DoE) Advanced Research Projects – Energy (ARPA-E) Agency Award number DE-AR0001542.

REFERENCES

[1] Ferry, S.E., et al. "The LIBRA Experiment: Investigating Robust Tritium Accountancy in Molten FLiBe Exposed to DT Fusion Neutron Spectrum" Fus. Tech. 79.1 (2023): 13-35.
[2] Pearson, R.J., et al. "Tritium supply and use: a key issue for the development of nuclear fusion energy." Fus. Eng. and Design 136 (2018): 1140-1148.
[3] Clery, D., "Out of Gas", Science, Vol 376, Issue 6600.

Presenting Author: Kevin B. Woller
Presenting Author's Affiliation: Plasma Science and Fusion Center, Massachusetts Institute of Technology, USA

Prototype verification of tritium breeding with Liquid Immersion Blanket technology using FLiBe

Kevin B. Woller, kbwoller@mit.edu

19 October 2023

16th International Workshop on Beryllium Technology

This work is supported by
US DOE award AR-0001542

On behalf of the LIBRA team @

MIT PSFC
Plasma Science and Fusion Center

Deuterium and tritium as fusion fuel to heat

Fuse DT in the fusion core:

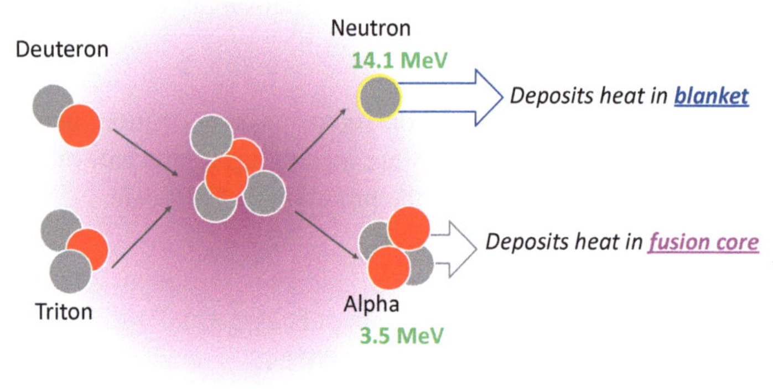

PSFC *Liquid immersion blanket, tritium, and FLiBe* | 19 Oct 2023
Be-WS 16 kbwoller@mit.edu 2

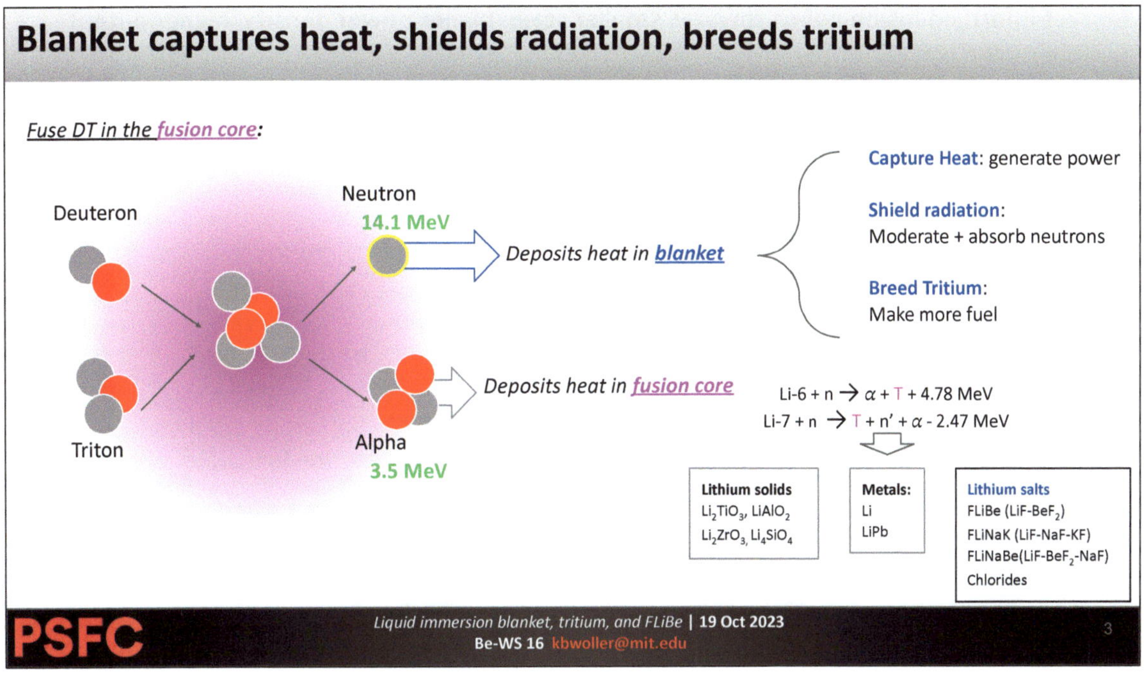
Blanket captures heat, shields radiation, breeds tritium
Fuse DT in the fusion core:
Deuteron
Neutron
14.1 MeV
Deposits heat in blanket
Triton
Alpha
3.5 MeV
Deposits heat in fusion core
Capture Heat: generate power
Shield radiation:
Moderate + absorb neutrons
Breed Tritium:
Make more fuel
Li-6 + n → α + T + 4.78 MeV
Li-7 + n → T + n' + α - 2.47 MeV
Lithium solids
Li2TiO3, LiAlO2
Li2ZrO3, Li4SiO4
Metals:
Li
LiPb
Lithium salts
FLiBe (LiF-BeF2)
FLiNaK (LiF-NaF-KF)
FLiNaBe(LiF-BeF2-NaF)
Chlorides
PSFC
Liquid immersion blanket, tritium, and FLiBe | 19 Oct 2023
Be-WS 16 kbwoller@mit.edu
3

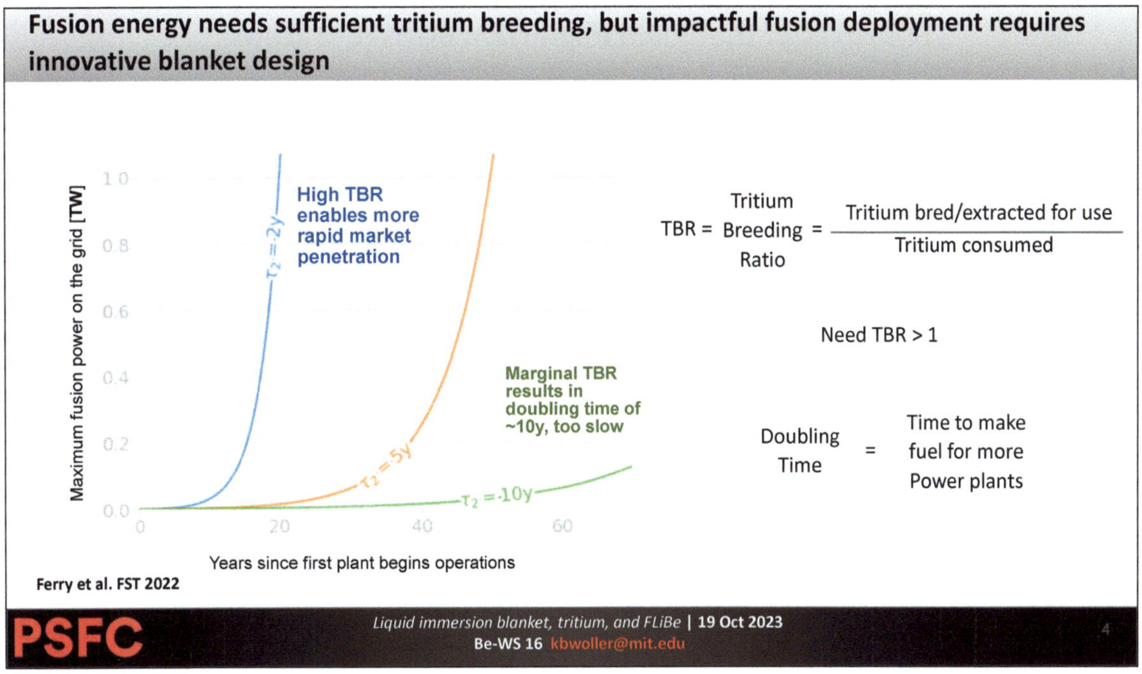
Fusion energy needs sufficient tritium breeding, but impactful fusion deployment requires innovative blanket design
Maximum fusion power on the grid [TW]
1.0
0.8
0.6
0.4
0.2
0.0
0
20
40
60
Years since first plant begins operations
High TBR enables more rapid market penetration
τ2 = 2y
τ2 = 5y
τ2 = 10y
Marginal TBR results in doubling time of ~10y, too slow
Ferry et al. FST 2022
TBR = Tritium Breeding Ratio = Tritium bred/extracted for use / Tritium consumed
Need TBR > 1
Doubling Time = Time to make fuel for more Power plants
PSFC
Liquid immersion blanket, tritium, and FLiBe | 19 Oct 2023
Be-WS 16 kbwoller@mit.edu
4

Market-informed fusion points towards smaller scale, higher power density, higher output temperature

Downside to higher power density is engineering complexity, material degradation

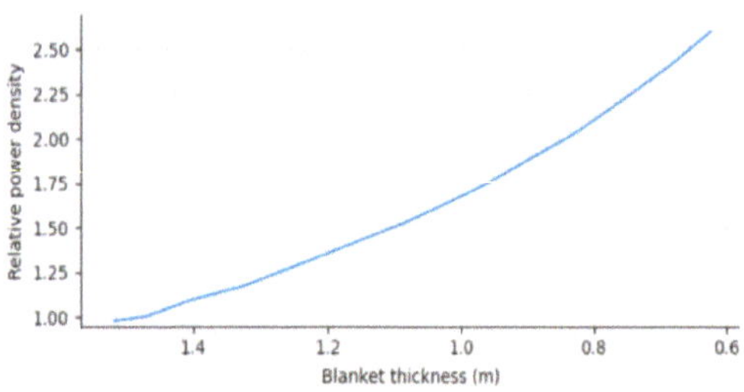

Upside to higher operating temperature is increased efficiency, lower LCOE

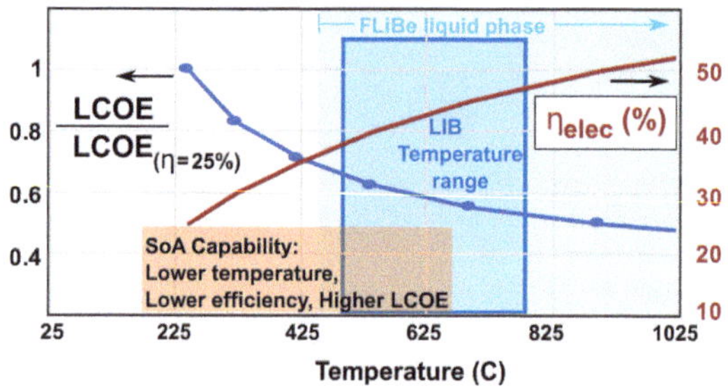

Ferry et al. FST 2022

The Liquid Immersion Blanket (LIB) concept address the market drive for faster, smaller, efficient fusion energy

PSFC *Liquid immersion blanket, tritium, and FLiBe* | **19 Oct 2023**
Be-WS 16 kbwoller@mit.edu 5

Principles of the Liquid Immersion Blanket concept

Magnetic Fusion Energy Power Plant Concept

- **Maximize solid coverage of neutron source to optimize tritium breeding and enhance shielding**
- Molten salt interaction with magnetic fields is **minimal** compared to liquid metal breeders
- Liquid breeder **can be pumped out** for maintenance
- **Online chemistry control** to address corrosion and tritium speciation (T_2, HT over TF)
- Tritium has **low solubility** in molten salts, easing extraction
- Minimize solid structure
- Take advantage of excellent **thermal properties** - good for high power density concepts like ARC

PSFC *Liquid immersion blanket, tritium, and FLiBe* | **19 Oct 2023**
Be-WS 16 kbwoller@mit.edu 6

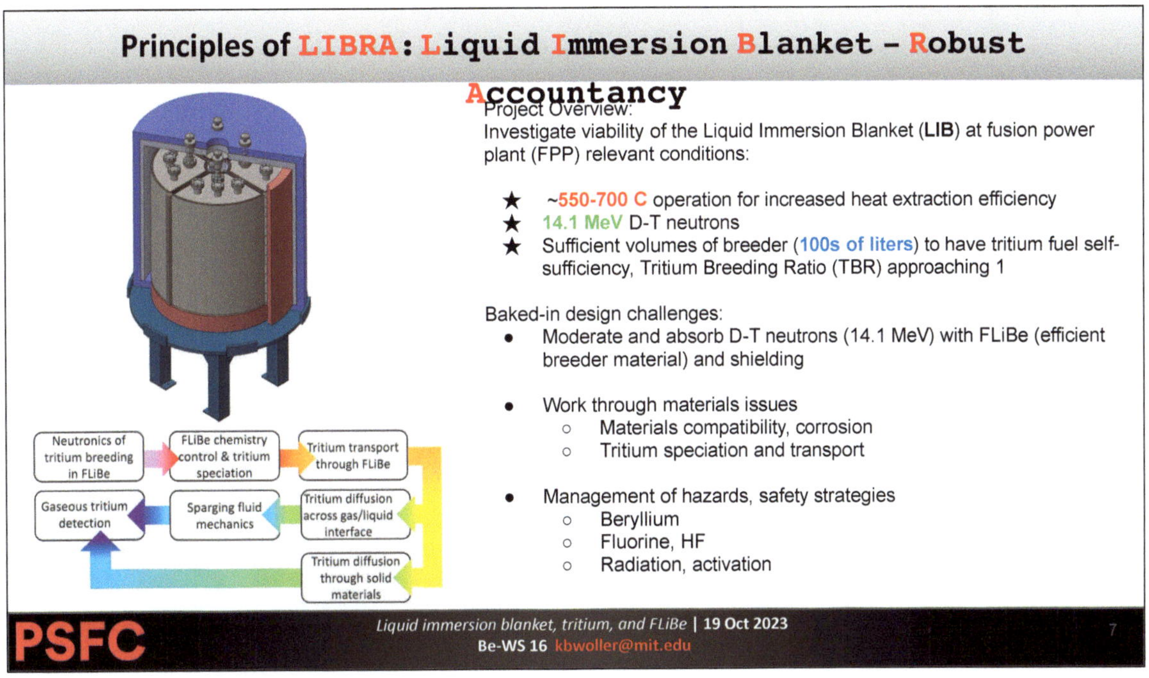
Principles of LIBRA: Liquid Immersion Blanket - Robust Accountancy
Project Overview:
Investigate viability of the Liquid Immersion Blanket (LIB) at fusion power plant (FPP) relevant conditions:
★ ~550-700 C operation for increased heat extraction efficiency
★ 14.1 MeV D-T neutrons
★ Sufficient volumes of breeder (100s of liters) to have tritium fuel self-sufficiency, Tritium Breeding Ratio (TBR) approaching 1
Baked-in design challenges:
• Moderate and absorb D-T neutrons (14.1 MeV) with FLiBe (efficient breeder material) and shielding
• Work through materials issues
○ Materials compatibility, corrosion
○ Tritium speciation and transport
• Management of hazards, safety strategies
○ Beryllium
○ Fluorine, HF
○ Radiation, activation
Neutronics of tritium breeding in FLiBe
FLiBe chemistry control & tritium speciation
Tritium transport through FLiBe
Gaseous tritium detection
Sparging fluid mechanics
Tritium diffusion across gas/liquid interface
Tritium diffusion through solid materials
PSFC
Liquid immersion blanket, tritium, and FLiBe | 19 Oct 2023
Be-WS 16 kbwoller@mit.edu
7

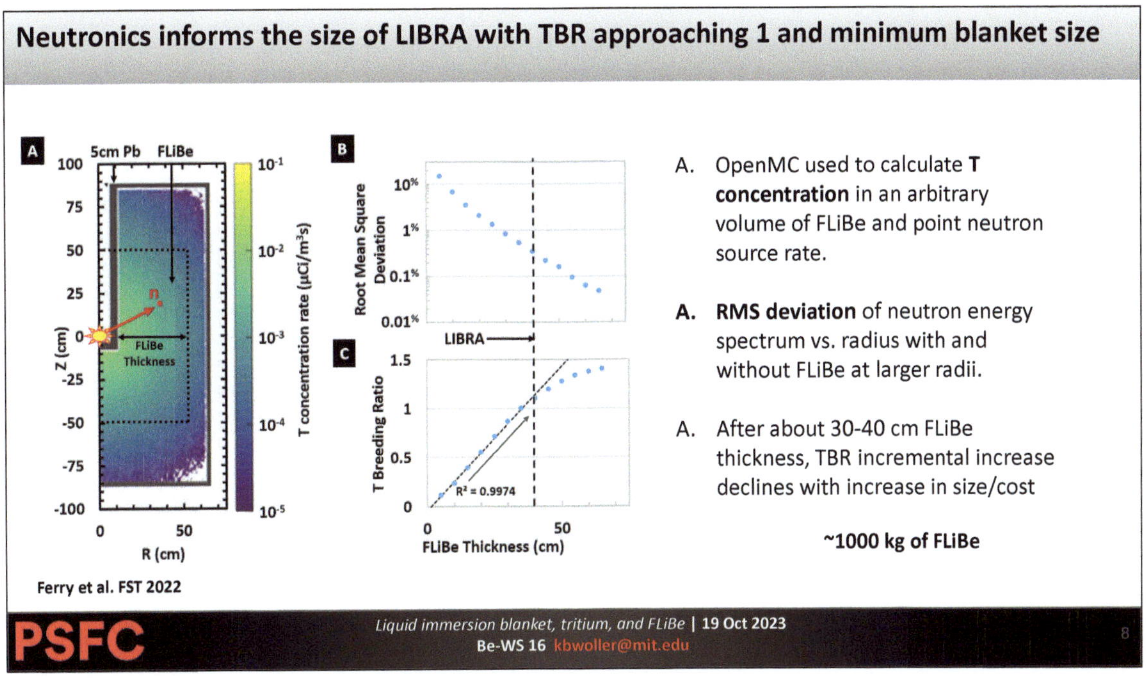
Neutronics informs the size of LIBRA with TBR approaching 1 and minimum blanket size
A
5cm Pb FLiBe
100
75
50
25
0
-25
-50
-75
-100
Z (cm)
n
FLiBe Thickness
0
50
R (cm)
10^{-1}
10^{-2}
10^{-3}
10^{-4}
10^{-5}
T concentration rate (μCi/m³s)
B
Root Mean Square Deviation
10%
1%
0.1%
0.01%
LIBRA
C
T Breeding Ratio
1.5
1
0.5
0
R^2 = 0.9974
0
50
FLiBe Thickness (cm)
Ferry et al. FST 2022
A. OpenMC used to calculate T concentration in an arbitrary volume of FLiBe and point neutron source rate.
A. RMS deviation of neutron energy spectrum vs. radius with and without FLiBe at larger radii.
A. After about 30-40 cm FLiBe thickness, TBR incremental increase declines with increase in size/cost
~1000 kg of FLiBe
PSFC
Liquid immersion blanket, tritium, and FLiBe | 19 Oct 2023
Be-WS 16 kbwoller@mit.edu
8

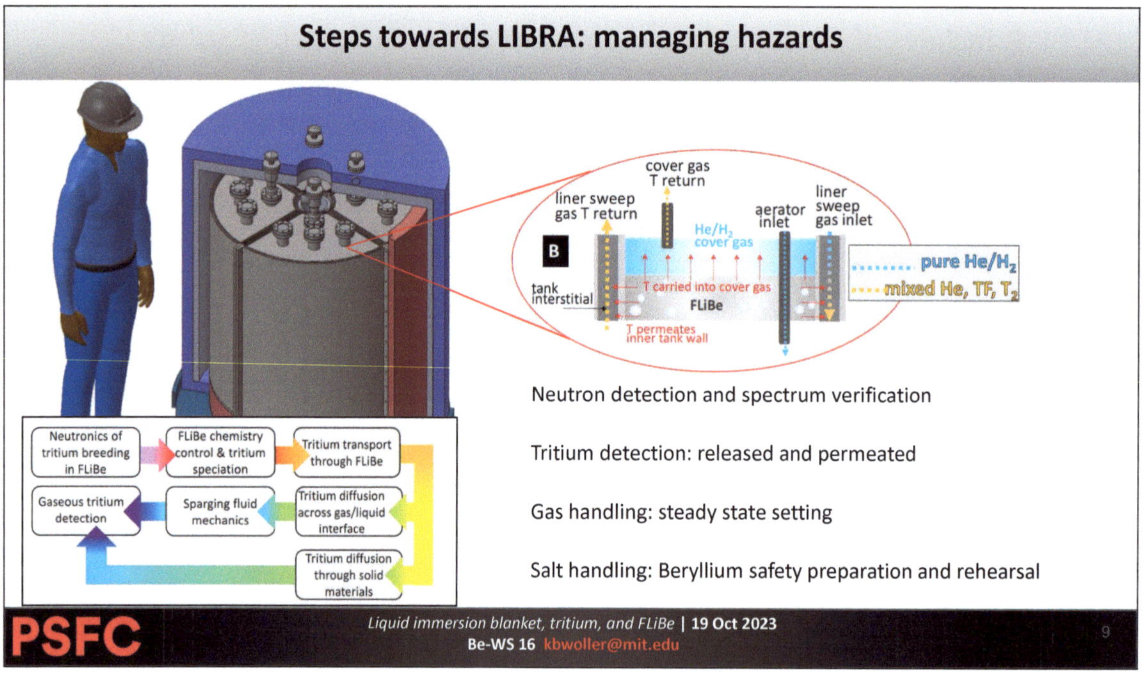
Steps towards LIBRA: managing hazards
cover gas
T return
liner sweep
gas T return
aerator
inlet
liner
sweep
gas inlet
He/H2
cover gas
B
pure He/H2
mixed He, TF, T2
tank
interstitial
T carried into cover gas
FLiBe
T permeates
inner tank wall
Neutron detection and spectrum verification
Tritium detection: released and permeated
Gas handling: steady state setting
Salt handling: Beryllium safety preparation and rehearsal
Neutronics of tritium breeding in FLiBe
FLiBe chemistry control & tritium speciation
Tritium transport through FLiBe
Gaseous tritium detection
Sparging fluid mechanics
Tritium diffusion across gas/liquid interface
Tritium diffusion through solid materials
PSFC
Liquid immersion blanket, tritium, and FLiBe | 19 Oct 2023
Be-WS 16 kbwoller@mit.edu
9

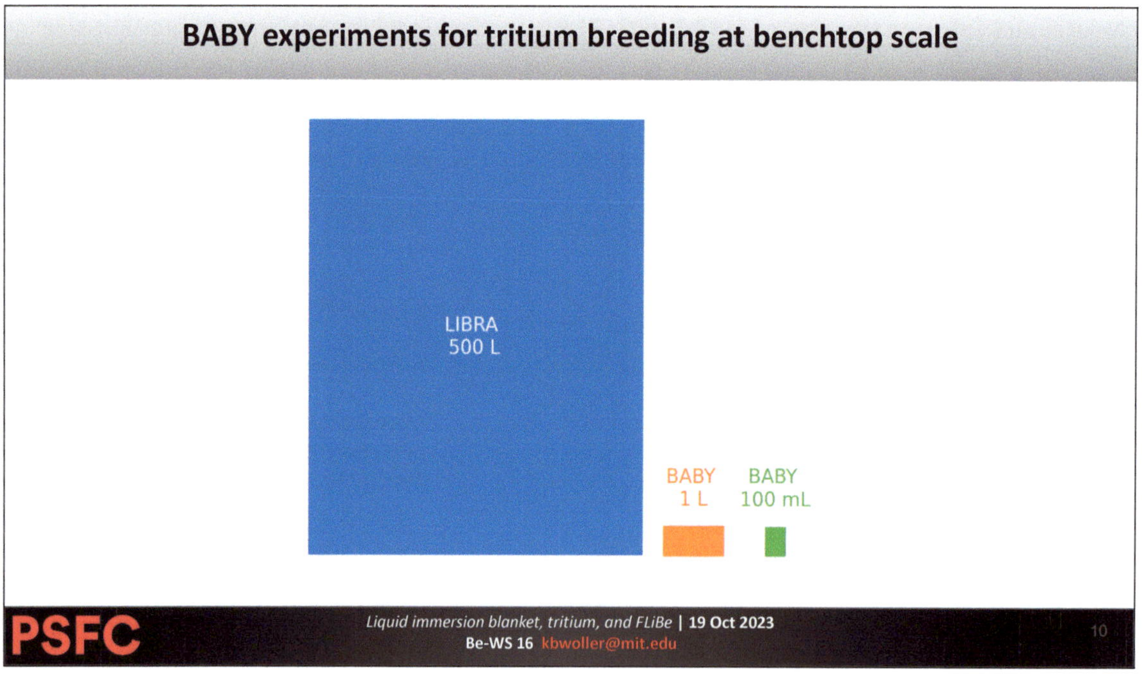
BABY experiments for tritium breeding at benchtop scale
LIBRA
500 L
BABY
1 L
BABY
100 mL
PSFC
Liquid immersion blanket, tritium, and FLiBe | 19 Oct 2023
Be-WS 16 kbwoller@mit.edu
10

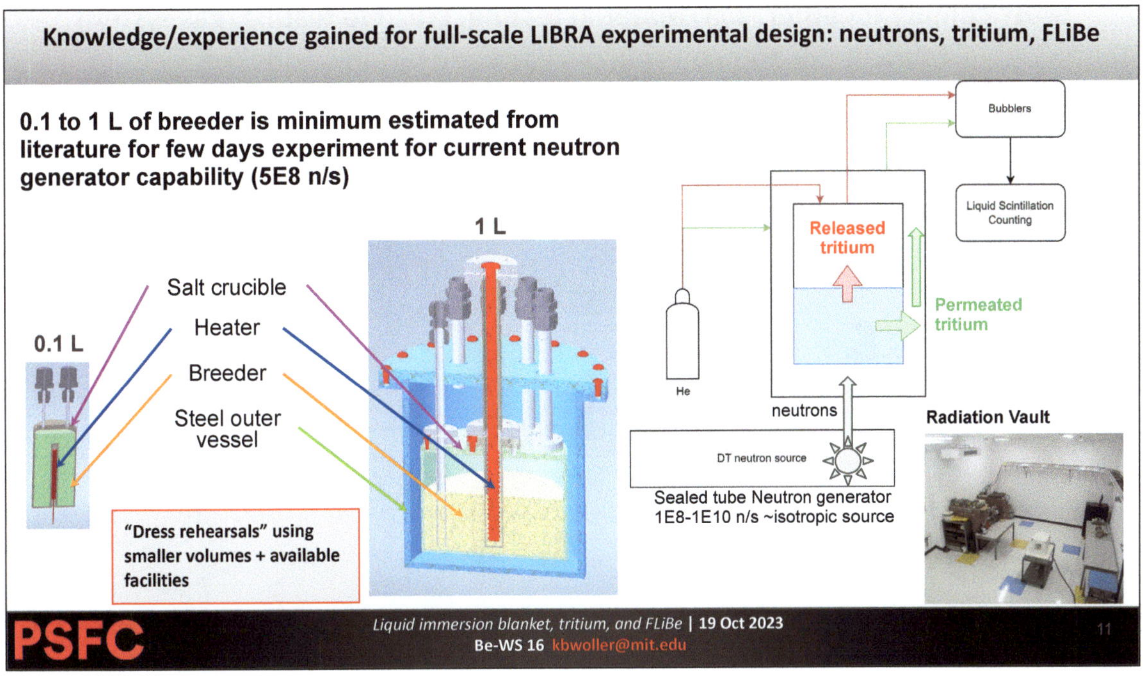
Knowledge/experience gained for full-scale LIBRA experimental design: neutrons, tritium, FLiBe
0.1 to 1 L of breeder is minimum estimated from literature for few days experiment for current neutron generator capability (5E8 n/s)
1 L
0.1 L
Salt crucible
Heater
Breeder
Steel outer vessel
"Dress rehearsals" using smaller volumes + available facilities
Bubblers
Liquid Scintillation Counting
Released tritium
Permeated tritium
He
neutrons
DT neutron source
Sealed tube Neutron generator
1E8-1E10 n/s ~isotropic source
Radiation Vault
PSFC
Liquid immersion blanket, tritium, and FLiBe | 19 Oct 2023
Be-WS 16 kbwoller@mit.edu
11

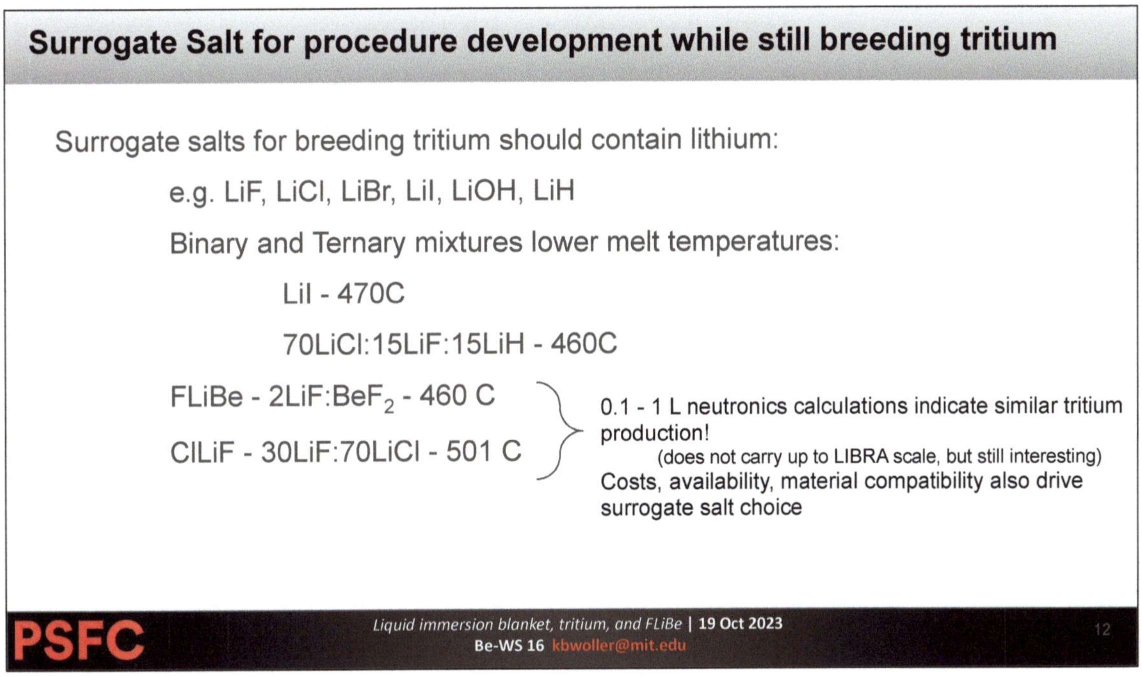
Surrogate Salt for procedure development while still breeding tritium
Surrogate salts for breeding tritium should contain lithium:
e.g. LiF, LiCl, LiBr, LiI, LiOH, LiH
Binary and Ternary mixtures lower melt temperatures:
LiI - 470C
70LiCl:15LiF:15LiH - 460C
FLiBe - 2LiF:BeF2 - 460 C
ClLiF - 30LiF:70LiCl - 501 C
0.1 - 1 L neutronics calculations indicate similar tritium production!
(does not carry up to LIBRA scale, but still interesting)
Costs, availability, material compatibility also drive surrogate salt choice
PSFC
Liquid immersion blanket, tritium, and FLiBe | 19 Oct 2023
Be-WS 16 kbwoller@mit.edu
12

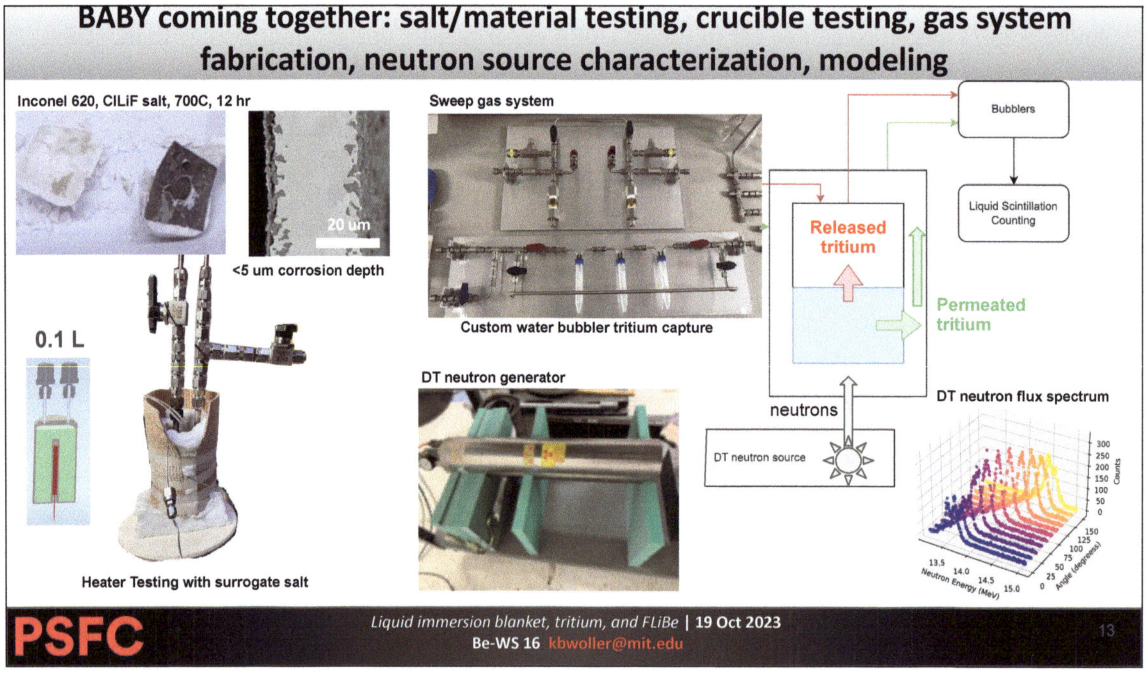
BABY coming together: salt/material testing, crucible testing, gas system fabrication, neutron source characterization, modeling
Inconel 620, ClLiF salt, 700C, 12 hr
20 um
<5 um corrosion depth
0.1 L
Heater Testing with surrogate salt
Sweep gas system
Custom water bubbler tritium capture
DT neutron generator
Bubblers
Liquid Scintillation Counting
Released tritium
Permeated tritium
neutrons
DT neutron source
DT neutron flux spectrum
Counts
Neutron Energy (MeV)
Angle (degrees)
PSFC
Liquid immersion blanket, tritium, and FLiBe | 19 Oct 2023
Be-WS 16 kbwoller@mit.edu
13

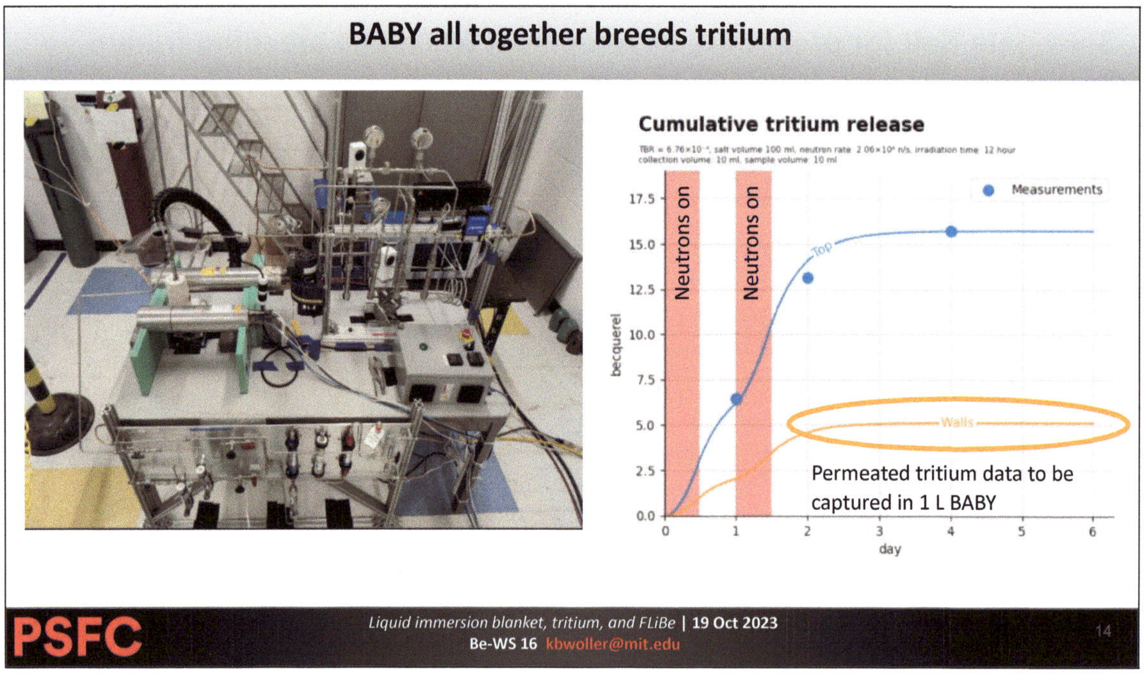
BABY all together breeds tritium
Cumulative tritium release
Measurements
Neutrons on
Neutrons on
Top
Walls
becquerel
day
Permeated tritium data to be captured in 1 L BABY
PSFC
Liquid immersion blanket, tritium, and FLiBe | 19 Oct 2023
Be-WS 16 kbwoller@mit.edu
14

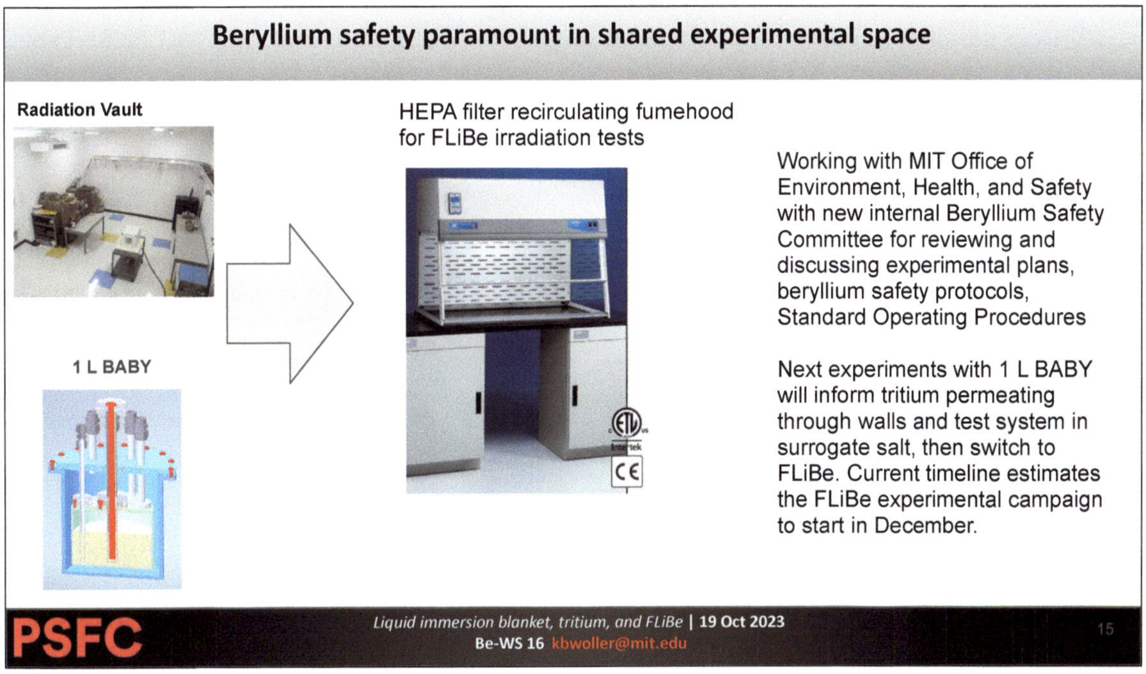
Beryllium safety paramount in shared experimental space
Radiation Vault
HEPA filter recirculating fumehood for FLiBe irradiation tests
1 L BABY
Working with MIT Office of Environment, Health, and Safety with new internal Beryllium Safety Committee for reviewing and discussing experimental plans, beryllium safety protocols, Standard Operating Procedures
Next experiments with 1 L BABY will inform tritium permeating through walls and test system in surrogate salt, then switch to FLiBe. Current timeline estimates the FLiBe experimental campaign to start in December.
PSFC
Liquid immersion blanket, tritium, and FLiBe | 19 Oct 2023
Be-WS 16 kbwoller@mit.edu
15

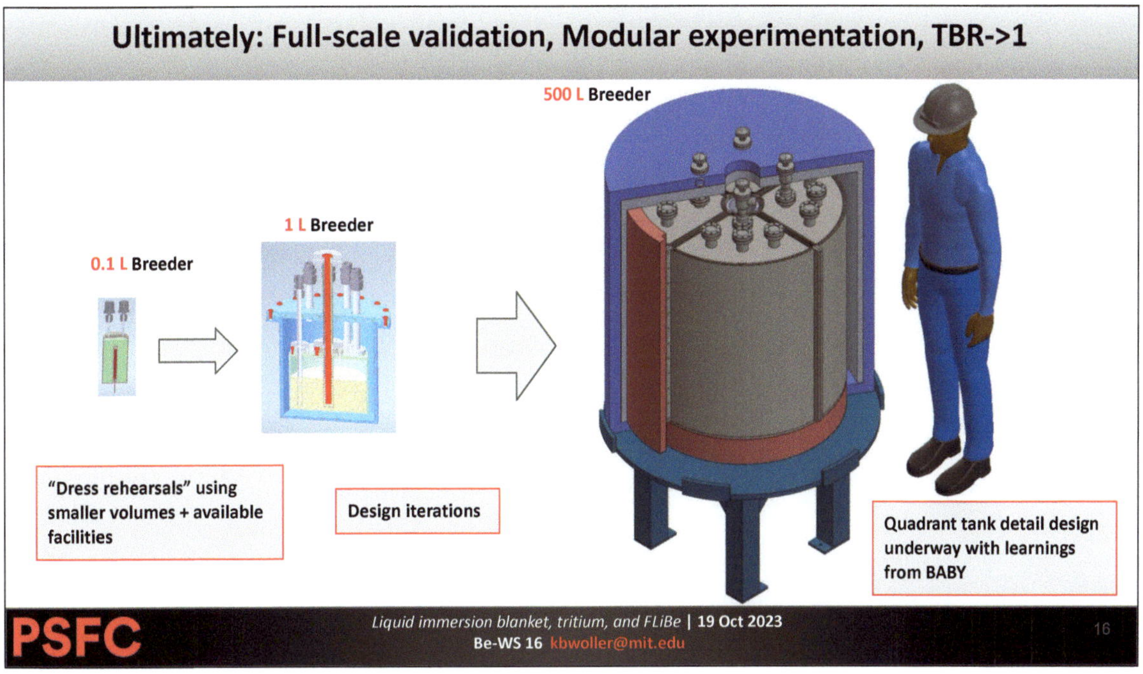
Ultimately: Full-scale validation, Modular experimentation, TBR->1
500 L Breeder
1 L Breeder
0.1 L Breeder
"Dress rehearsals" using smaller volumes + available facilities
Design iterations
Quadrant tank detail design underway with learnings from BABY
PSFC
Liquid immersion blanket, tritium, and FLiBe | 19 Oct 2023
Be-WS 16 kbwoller@mit.edu
16

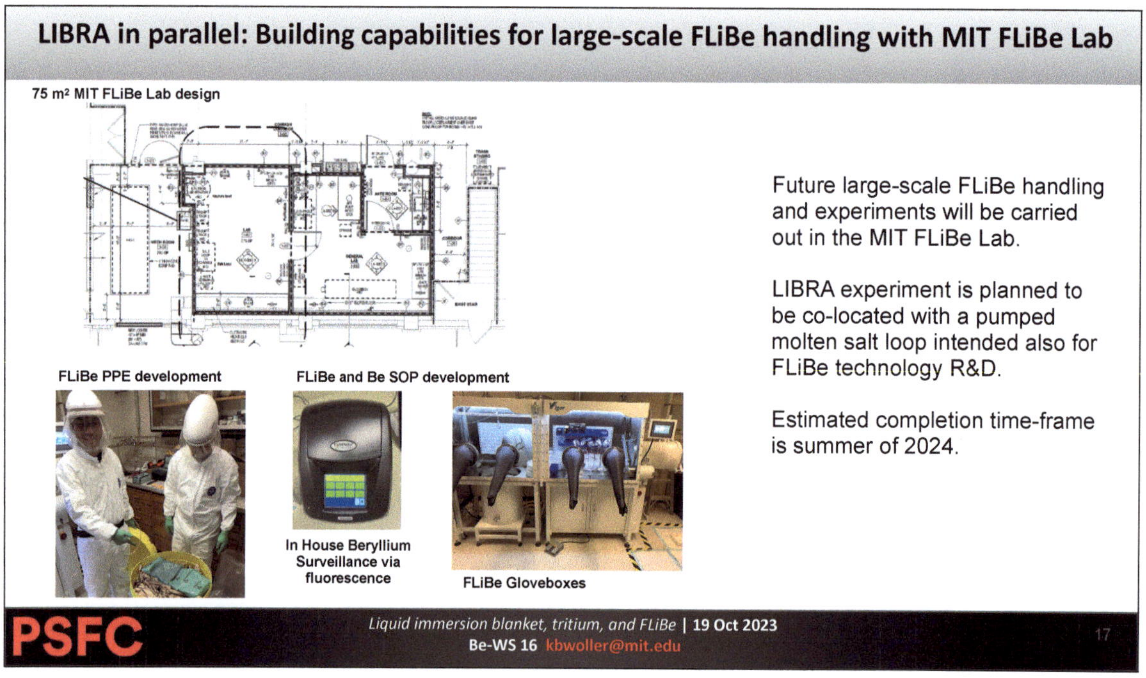
LIBRA in parallel: Building capabilities for large-scale FLiBe handling with MIT FLiBe Lab
75 m² MIT FLiBe Lab design
FLiBe PPE development
FLiBe and Be SOP development
In House Beryllium Surveillance via fluorescence
FLiBe Gloveboxes
Future large-scale FLiBe handling and experiments will be carried out in the MIT FLiBe Lab.
LIBRA experiment is planned to be co-located with a pumped molten salt loop intended also for FLiBe technology R&D.
Estimated completion time-frame is summer of 2024.
PSFC
Liquid immersion blanket, tritium, and FLiBe | 19 Oct 2023
Be-WS 16 kbwoller@mit.edu
17

4.3 Beryllium Worker Protection Model for the Production and Handling of FLiBe

Theodore Knudson, MS, CIH

Lithium beryllium fluoride (Li_2BeF_4), or FLiBe, is a molten salt made from a mixture of lithium fluoride (LiF) and beryllium fluoride (BeF_2) with potential applications as a coolant in molten salt-cooled nuclear reactors and as a liquid blanket for ARC fusion reactors. Beryllium, as either an insoluble or soluble compound, can cause beryllium sensitization (BeS) and chronic beryllium disease (CBD) in susceptible individuals.

Beryllium salts, such as BeF_2 and FLiBe, are soluble compounds. However, most beryllium worker protection efforts are concentrated around operations that process commercially available forms of insoluble beryllium such as beryllium metal, beryllium alloys, and beryllia ceramics. Recently renewed interest in the research and production of FLiBe resulted in the construction of a molten salt purification operation at the Materion Brush Inc. facility in Elmore, OH, where BeF_2 is readily available as a feedstock.

Industrial hygiene design considerations for the production and handling of FLiBe were required to address controls for both BeF_2 salt and molten FLiBe. These efforts included the utilization of existing beryllium worker protection controls used in the production of BeF_2 with the addition of an enclosed design, local exhaust ventilation, administrative procedures for process upsets, and personal protective equipment designed for working with soluble compounds. Initial industrial hygiene beryllium exposure assessments are currently underway.

Presenting Author: Theodore Knudson, MS, CIH
Presenting Author's Affiliation: Materion Corporation, USA
Email: theodore.knudson@materion.com

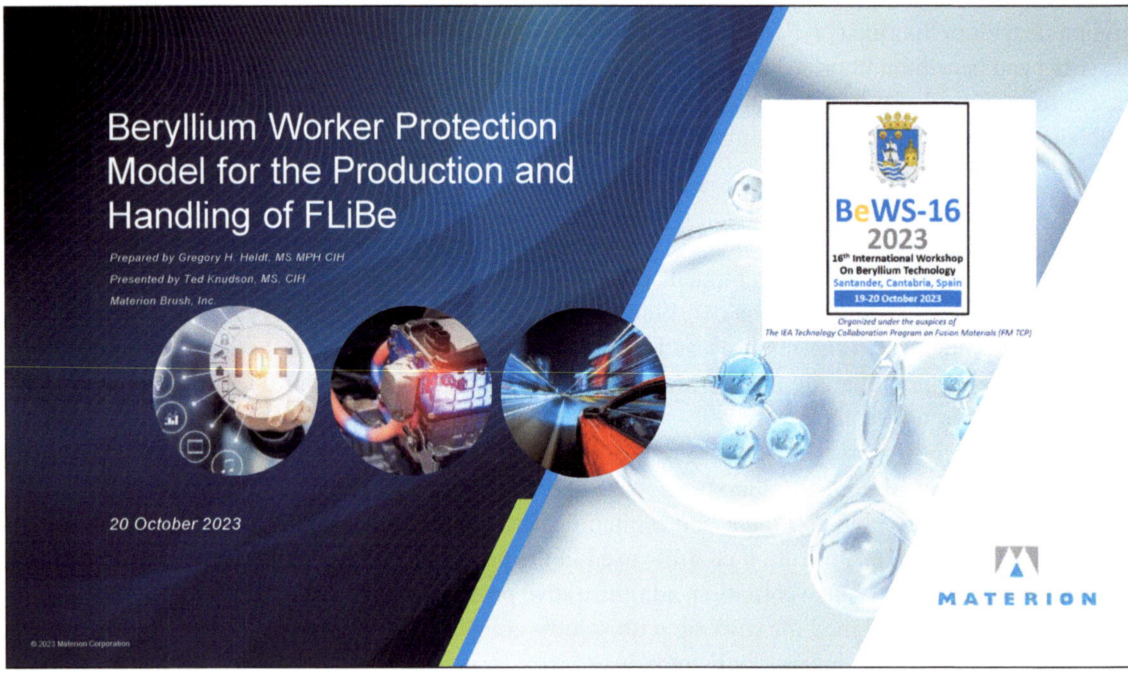

FLiBe

- Molten salt made from a mixture of lithium fluoride and beryllium fluoride
- Applications for both fission and fusion energy technologies

2 © 2023 Materion Corporation

MATERION

Kairos Power and FLiBe

- The Kairos Power fluoride salt-cooled high-temperature reactor (KP-FHR) is a novel advanced reactor technology that leverages TRISO fuel in pebble form combined with a low-pressure fluoride salt coolant. The technology uses a steam cycle to convert heat from fission into electricity.
- Molten fluoride salts, such as FLiBe, have outstanding capability to transfer heat at high temperature, excellent chemical stability, and the ability to retain radioactive fission products that might be released from fuel.

Kairos Power www.kairospower.com

MATERION

3

FLiBe Molten Salt Purification Pilot (MSPP)

- Materion Brush Inc and Kairos Power partnered to produce FLiBe as a coolant for the Kairos Power Fluoride salt-cooled High temperature Reactor (KP-FHR).
- The Molten Salt Purification Pilot (MSPP) facility was commissioned at Materion's Elmore, OH plant in July 2022.

MATERION

4

FLiBe Molten Salt Purification Pilot (MSPP)

- MSPP is the largest FLiBe production facility ever built and has the capacity to generate commercial quantities of material
- Materion operates the facility and provides beryllium fluoride and lithium fluoride as throughput material.

MATERION

Beryllium Worker Protection Model

- Composed of 8 Elements
- Developed specifically for the control of beryllium exposures at very low concentrations (<1 µg/m3)
- Most industrial hygiene concepts are developed for the control of exposures in relatively high concentrations (~mg/m3) and are less effective for controlling beryllium exposure

MATERION

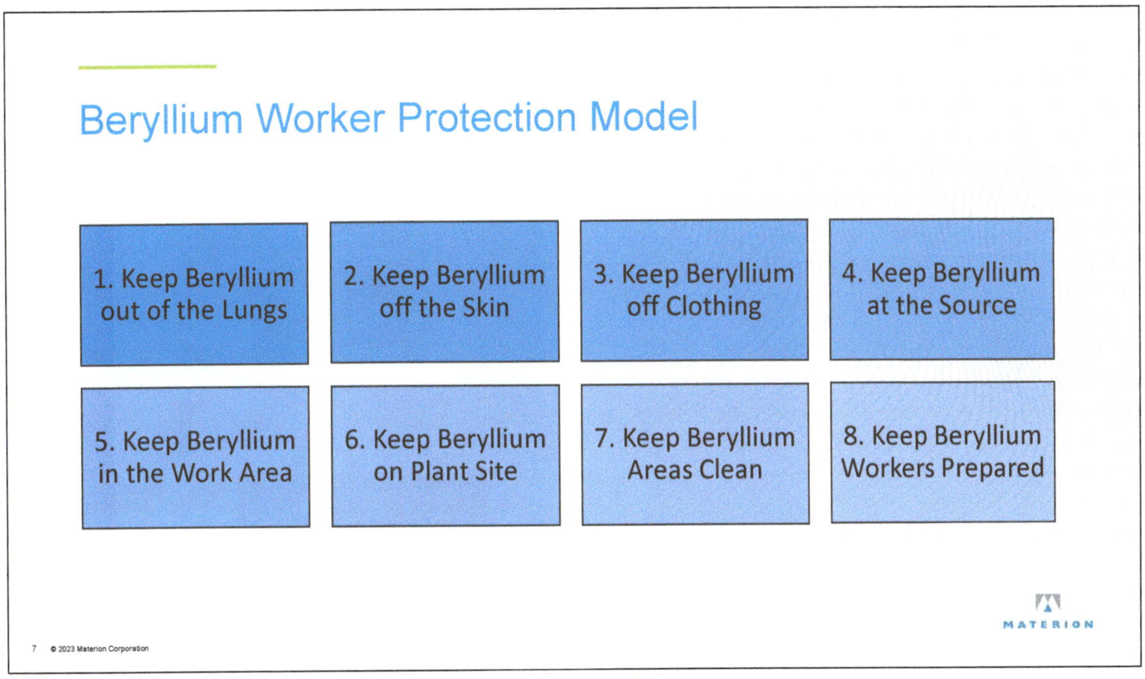
Beryllium Worker Protection Model
1. Keep Beryllium out of the Lungs
2. Keep Beryllium off the Skin
3. Keep Beryllium off Clothing
4. Keep Beryllium at the Source
5. Keep Beryllium in the Work Area
6. Keep Beryllium on Plant Site
7. Keep Beryllium Areas Clean
8. Keep Beryllium Workers Prepared
MATERION
7 © 2023 Materion Corporation

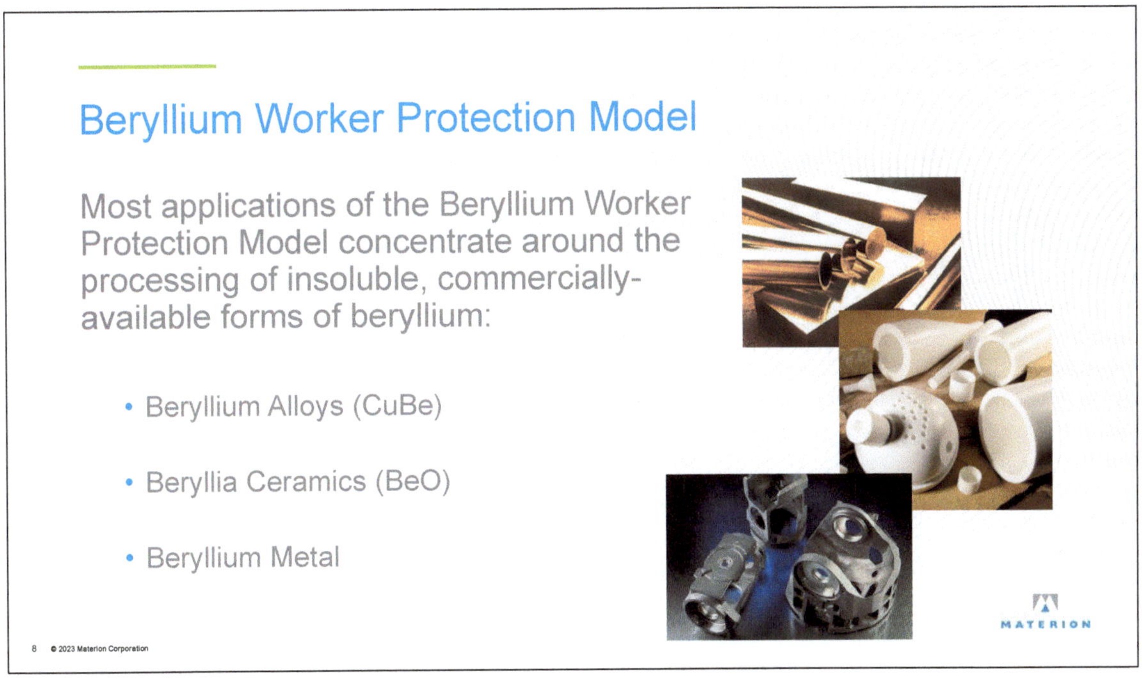
Beryllium Worker Protection Model
Most applications of the Beryllium Worker Protection Model concentrate around the processing of insoluble, commercially-available forms of beryllium:
• Beryllium Alloys (CuBe)
• Beryllia Ceramics (BeO)
• Beryllium Metal
MATERION
8 © 2023 Materion Corporation

Beryllium Worker Protection Model

Insoluble, commercial forms of beryllium are typically processed through casting, rolling, grinding, milling, polishing, and machining.

These activities do not cleanly translate to the processing and handling of FLiBe.

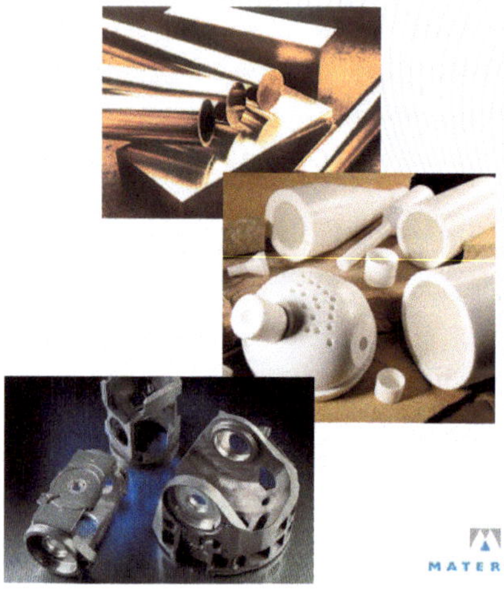

MATERION

9

Beryllium Worker Protection Model

Soluble forms of beryllium, including BeF2 and FLiBe, are generally not commercially available.

Special considerations were required in the design and implementation of the BWPM for the production and handling of soluble molten salt.

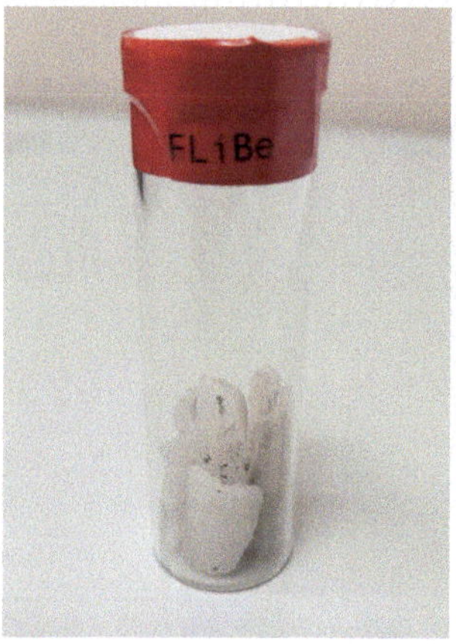

MATERION

10

1. Keep Beryllium out of the Lungs

- Assigned Protection Factor of 1000
- Assumes a worst-case scenario involving the release of BeF_2 or FLiBe
- MSPP is a new process
  - IH beryllium exposure characterization is ongoing

MATERION

2. Keep Beryllium off Skin / 3. Keep Beryllium off Clothing

- All operators entering the MSPP are required to wear:
  - Hooded chemical coveralls
  - Chemical boot covers
  - Two pairs of nitrile gloves
  - Cut-resistant work gloves

MATERION

2. Keep Beryllium off Skin / 3. Keep Beryllium off Clothing

- The selection of chemical overalls is critical to worker protection and should include:
  - High permeation rate (>480 min)
  - Resistance to sweat penetration
  - Welded seams (not stitched)
  - Adhesive closures
- Nitrile gloves are taped at the wrists
- Tight-fitting PAPRs are taped to the hood

MICROCHEM 2000 MICROCHEM 2300+

2. Keep Beryllium off Skin / 3. Keep Beryllium off Clothing

- Operators are provided visual cues designating the color of nitrile glove required to proceed through each part of the plant
- Ensures operators utilize a First On / Last Off glove system to safely doff respiratory protection in transition areas

4. Keep Beryllium at the Source

- The MSPP is designed to isolate operators from directly handling salts
- LiF and BeF2 drums are fitted with flanged cones and inverted for charging
- A HEPA vacuum is used to capture any fugitive material

MATERION

4. Keep Beryllium at the Source

- The MSPP is equipped with general exhaust ventilation on each floor of the skid
- Target is 10-12 air changes per hour (ACH)
- General exhaust ventilation is routed through a pollution control device

MATERION

5. Keep Beryllium in the Work Area

- Access to the MSPP is restricted by using one centralized transition room
- The transition room is HEPA filtered and equipped with downdraft tables for the donning and doffing of respiratory protection

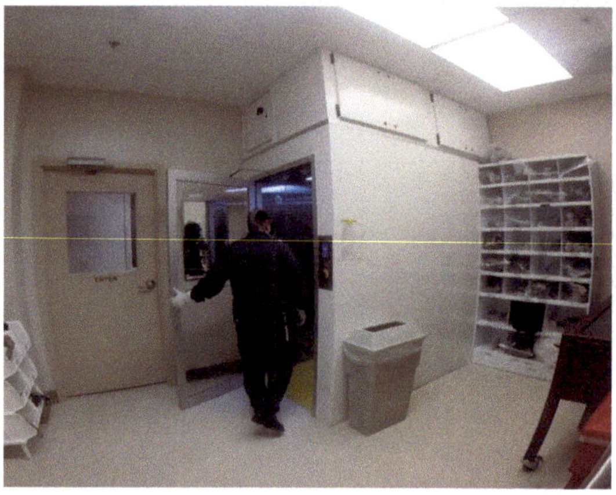

MATERION

5. Keep Beryllium in the Work Area

- To control the migration of beryllium, operators use dedicated footwear when entering production areas

MATERION

5. Keep Beryllium in the Work Area

- To leave production areas, operators must pass through a HEPA-filtered air shower before doffing respiratory protection in the transition room

MATERION

Main Transition Area

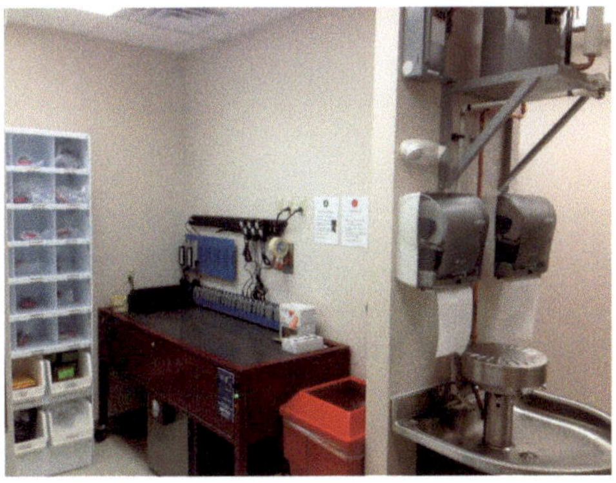

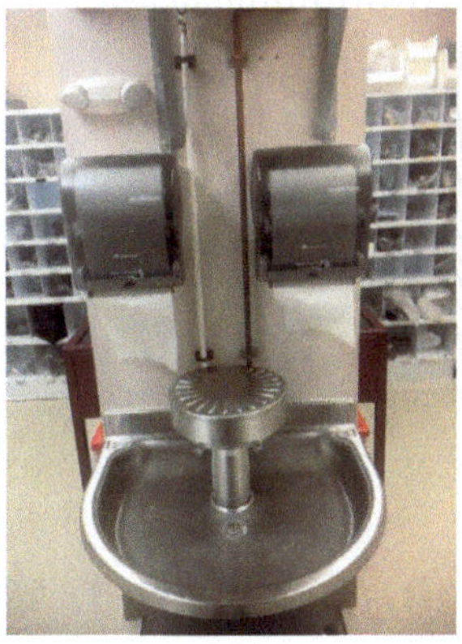

MATERION

6. Keep Beryllium on Plant Site

- Materion utilizes a robust decontamination procedure to ensure transfer vessels are visibly clean and that beryllium remains on-site
- Transfer vessels are wipe sampled after decontamination prior to shipment to assess cleanliness

21 © 2023 Materion Corporation

MATERION

7. Keep Beryllium Areas Clean

- The MSPP is held to the same 5S cleanliness standard as the rest of the manufacturing facility
- Checklists and housekeeping schedules are maintained to help keep the area visibly clean
- Stand-alone HEPA filtration units have been installed increase air changes and help remove fugitive particulate

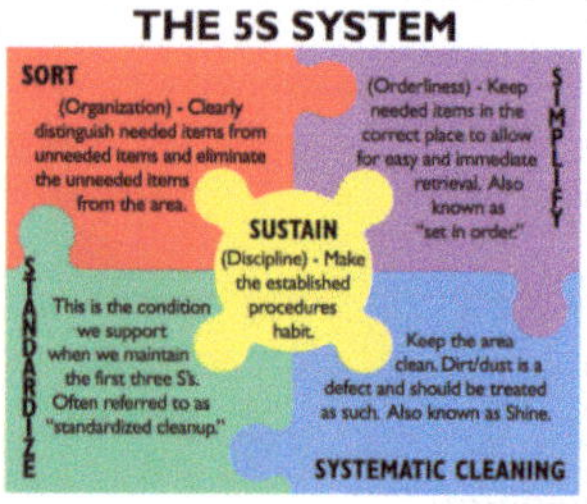

22 © 2023 Materion Corporation

MATERION

8. Keep Beryllium Workers Prepared

- Operators receive annual beryllium hazard awareness training, including the hazards of soluble beryllium
- Each operator must complete a PPE Donning/Doffing Competency Test demonstrating knowledge of PPE usage and removal

Beryllium Work Areas PPE Don/Doff Checklist for Competency			
Employee Name		Clock No.	
Department		Observer Name	
Date		Shift	
Part 1: Donning of PPE for entry into a contaminated work area			
Operator will be wearing prior to commencing this list (at a minimum) blues, a TFPAPR, FOLO gloves, and work boots.			
OBSERVERS MUST READ DONNING SEQUENCE TO THE OPERATOR IN A STEP-BY-STEP FASHION.			
Step 1	Hair length below shoulders must be tied back and secure, remove all jewelry, tools, or any item that may damage the PPE.		
Step 2	Inspect PPE prior to donning.		
Don PPE in the order that follows			
Step 3	Don outer layer of Nitrile gloves		
Step 4	Tyvek hood is donned over head and respirator		
	Step 4a	Hood completely covers ears and neck and is situated on the face piece of the TFPAPR	
	Step 4b	Hood extends past the neck to the shoulders	
Step 5	Don booties		
	Step 5a	Don first layer (grey) booties	
	Step 5b	Don second layer (chemical) booties	
Step 6	Microchem 3000 suit donned		
Step 7	Both suit zippers are zipped completely. Care should be taken to ensure that the Tyvek hood is in place and no skin or hair is visible between the hood and the respirator		
Step 8	The following should be taped: (1) Inner sleeve to outer Nitrile with 50/50 overlap (2) TFPAPR to hood of Microchem 3000 (3) Outer zipper in neck area to prevent sliding of zipper.		
	Step 8a	When conducting wet tasks a complete seal is required around the Microchem hood and TFPAPR, this should be done with yellow tape.	
Step 9	Work or chemical gloves should be underneath the outer sleeve of the Microchem 3000.		
	Step 9a	When conducting wet tasks the outer sleeve for Microchem 3000 should be taped at the wrist to chemical gloves.	
Step 10	Swiffer is prepped for entry into the work area.		
Step 11	Integrity of PPE is verified by the observer and the lack of visible hair or skin is observed as well.		
Step 12	Place Swiffer in Hot Zone prior to beginning work		

MATERION

23

Early Insights – IH Sampling Results

- IH exposure data results are limited due to process stability
  - (e.g., unexpected starts, stops)
- A full beryllium exposure characterization will be completed once the process is in control
- Some opportunities for improvement have been identified

MATERION

24

Early Insights – Local Exhaust Ventilation

- Operators currently use a central vacuum system to provide point ventilation for charging activities and maintenance line breaks
- There are opportunities to reduce worker exposure using permanent local exhaust ventilation collars at points where designated line breaks generate particulate
- Design is already underway

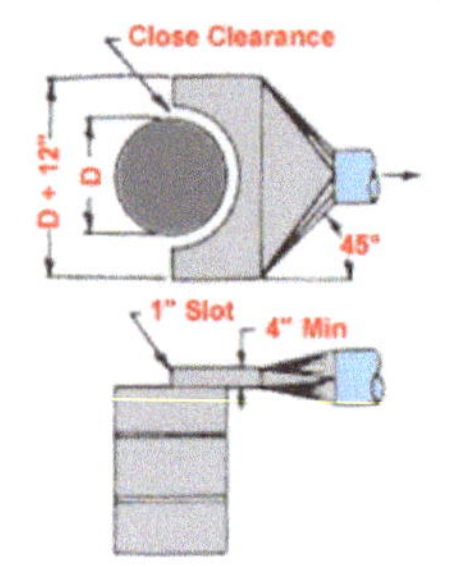

- Q = 100 cfm/sq ft barrel top min
- Duct velocity = 3500 minimum
- Entry loss = 0.25 vp + 1.78 slot vp
- Manual loading

MATERION

25

Early Insight – PPE and Housekeeping

- Efforts to implement a robust 5S cleaning schedule have been limited by short stay times dictated by PPE usage
- Even in ideal conditions stay times are limited to 90 minutes
- Stay times are likely to shorten during upcoming summer months
- There has been some success using standalone air conditioning units to cool the internal temperature of the skid
- Increased local and general exhaust ventilation may improve housekeeping

MATERION

26

Materion enables what's next™
Materion Corporation Product Stewardship
+1 800.862.4118
http://www.materion.com
Connect With Us
MATERION
© 2023 Materion Corporation

SESSION 5

Properties of Be & Beryllides

5.1 Screening tests to qualify thermo-mechanical properties of beryllium pebble beds with non-spherical pebbles

Jörg Reimann, E. Cilingir, A. Goraieb, S. Udartsev, P. Vladimirov

In many ceramic breeder blankets, pebble-shaped beryllium is used as a neutron multiplier. Fairly spherical pebbles are considered as a candidate material, however, non-spherical, NP, particles are of economic interest because production costs are much lower. Yet, thermo-mechanical pebble bed data do merely exist for these beryllium grades, and the blanket relevant potential of these grades cannot be judged.
In the past, screening pebble bed experiments were performed [1] with different grades of NP beryllium pebbles, produced by crushing beryllium blocks and subsequent grinding. Packing experiments, uniaxial compression tests, UCTs, and measurements of the effective thermal pebble bed conductivity *k* were performed.

In the present report, similar investigations are described with NP beryllium pebble beds, produced by Ulba, Kasachstan, where the grinding step was not included during the beryllium particle production. The differences between present and previous results [1] are fundamental:

1. present packing fractions for vibrated pebble beds are about 0.49 compared to 0.62,
2. large compressive strains are obtained; $\approx 6\%$ for a stress of $\approx$ 4MPa compared to 1.7%, and
3. the beryllium-characteristic increase of k with compression shifts to such large strain values which are not expected to occur in blankets.

The important message from the present investigations is that the grinding step is important; further experiments with modified NP beryllium pebble beds should be performed.

REFERENCES

[1] Reimann et al, FED 98-99 (2015) 1851-1854

Presenting Author: Jörg Reimann
Presenting Author's Affiliation: Karlsruhe Beryllium Handling Facility, Germany
Email: reimann-langhans@web.de

Karlsruhe Institute of Technology

Screening tests to qualify thermo-mechanical properties of Be pebble beds with non-spherical particles

J. Reimann[1], E. Cilingir[2], N. Zimber[1], A. Goraieb[2], S. Udartsev[3], P. Vladimirov[1]

[1] Karlsruhe Institute of Technology, Institute for Applied Materials, Hermann-von-Helmholtz-Platz 1, 76344 Eggenstein-Leopoldshafen, Germany

[2] Karlsruhe Beryllium Handling Facility (KBHF GmbH), Hermann-von-Helmholtz-Platz 1, 76344 Eggenstein-Leopoldshafen, Germany

[3] Ulba Metallurgical Plant JSC, Ust-Kamenogorsk, Republic of Kasakhstan

Outline

- Introduction
  - Previous investigations on Be pebble beds, pbs, with non-spherical particles, NPs
  - Modelling parameters of pebble bed thermo-mechanics
  - Experimental set-up
- Results
  - A) Optical comparisons of grinded/non-grinded NPs
  - B) Packing fractions
  - C) Uniaxial stress-strain dependence
  - D) Effective thermal conductivity of pebble beds
  - E) Pebble bed-structure-mechanical interaction (BSMI)
- Discussion/Conclusions

Small database for Be pbs consisting of NPs

- In present ceramic breeder blankets, pebble-shaped beryllium is mostly used as multiplier. As candidate material, fairly spherical pebbles with diameters of d ≈ 1mm are considered and a wide data base of thermo-mechanical properties exists [1], at least for the BOL.
- A Demo reactor requires about 300 tons of Be pbs. The pebble production by the rotating electrode process, REP, is a cost driver. Non-spherical particles, NPs, can be produced considerably cheaper and, therefore, are of significant economic interest.
- First screening results on pbs consisting of NPs were reported in [2]. These NPs were produced by crushing sintered beryllium blocks and a subsequent grinding step.
- The present NPS were produced without this subsequent grinding step.

[1] Reimann et al, FED 81 (2006) 449-454
[2] Reimann et al. FED 98-99 (2015) 1851-1854

3 19.10.2023 Reimann et al., Thermo-mechanics of non-spherical Be pebble beds Department of Metallic Alloys, Institute for Applied Materials - Applied Materials Physics

Main parameters for the modelling of the thermo-mechanical behavior of pbs

BOL:

- packing fractions
- stress-strain relationship $\sigma = f(\varepsilon, T,\ldots)$,
- pebble bed effective thermal conductivity $k = f(\varepsilon, T, \ldots)$,
- thermal creep ε_{cr}.

EOL: Irradiation effects,

- Same parameters as above, no data exist for pbs with NPs.

4 19.10.2023 Reimann et al., Thermo-mechanics of non-spherical Be pebble beds Department of Metallic Alloys, Institute for Applied Materials - Applied Materials Physics

Experimental Setup

KIT Karlsruhe Institute of Technology

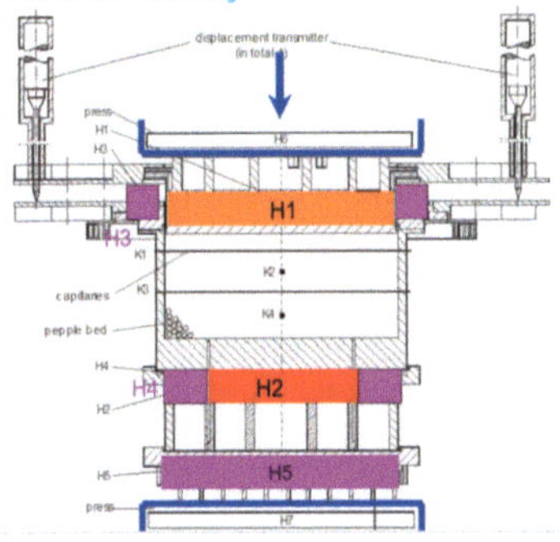

HECOP facility [1]

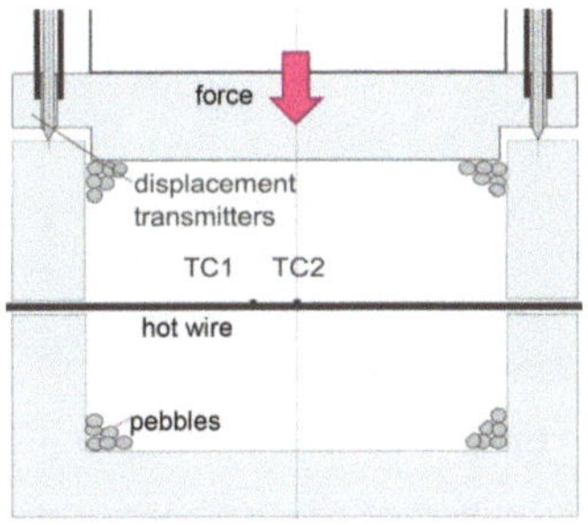

Simplified Setup for T_a [2] : Uniaxial Compression Test facility combined with a Hot Wire, HW

Modification for $T>T_a$ [3]: container with additional heating on cylindrical part and use of the the heating system of HECOP [1]

Parameter range:

- max. pressure (uniaxial stress): 6MPa
- max. T: 500°C

[3] Reimann et al, BeWS15

5 19.10.2023 Reimann et al., Thermo-mechanics of non-spherical Be pebble beds Department of Metallic Alloys, Institute for Applied Materials - Applied Materials Physics IAM

A) Comparison of grinded/non-grinded NPs

KIT Karlsruhe Institute of Technology

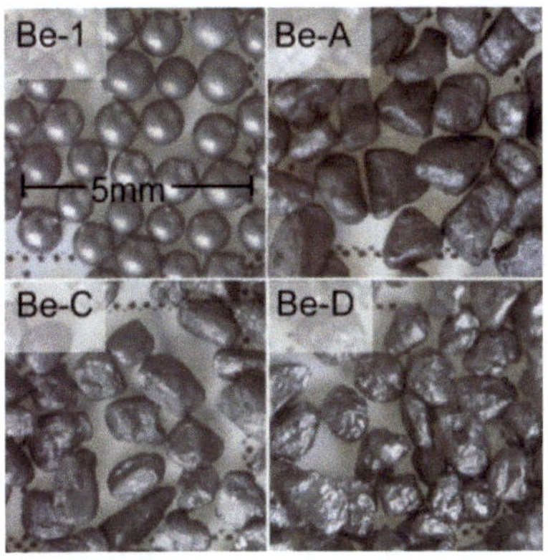

Be grades from [2]:
Be-1: NGK (spherical)
Be-A,C: Bochvar, RR
Be-D: Materion, USA

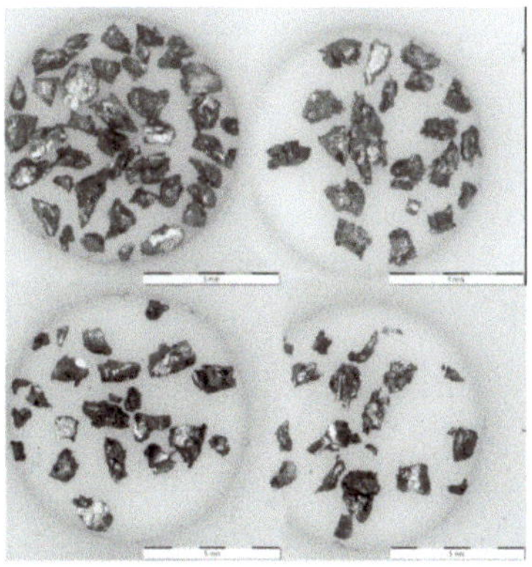

Present NPs from Ulba, " Be-U"

Compared to grinded NPs, non-grinded NPs are much more irregular

6 19.10.2023 Reimann et al., Thermo-mechanics of non-spherical Be pebble beds Department of Metallic Alloys, Institute for Applied Materials - Applied Materials Physics IAM

A) Comparison of grinded/non-grinded NPs

KIT Karlsruhe Institute of Technology

Be-1 Be-D Be-U

Vibrated pebble beds at the bottom of a cylindrical container

- Be-1: pebbles are mostly in contact with the bottom plate
- Be-D: NPs appear to be aligned at the wall with their largest sides (increased packing fractions close to walls?, increased heat transfer coefficient?)
- Be-U: an alignment of the NPs at the bottom wall is hardly observed.

Tomography investigations could determine characteristic void fraction fluctuations because of ordering effects, see e.g. [4].

[4] Reimann et al, IJMR MK111814 (2019)

7 19.10.2023 Reimann et al., Thermo-mechanics of non-spherical Be pebble beds Department of Metallic Alloys, Institute for Applied Materials - Applied Materials Physics IAM

B) Packing fractions

KIT Karlsruhe Institute of Technology

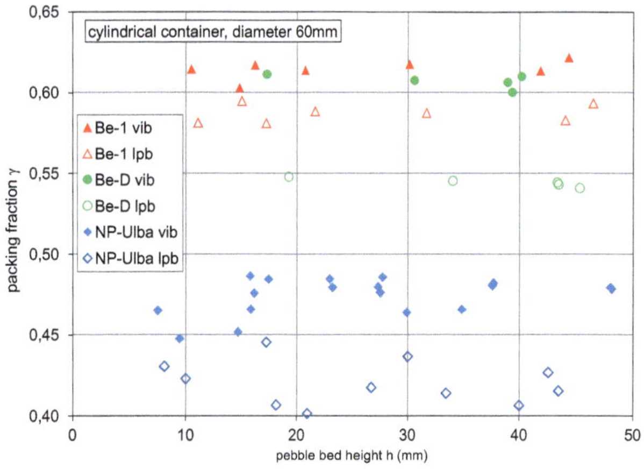

- For the rather spherical Be-1 pebbles γ is close to 0.6 even for loosely filling
- γ for vibrated Be-U pebble beds is still below 0.5. because of the spiky and irregular shape of the NPs.
- The impact of the γ on blanket design will be discussed later

8 19.10.2023 Reimann et al., Thermo-mechanics of non-spherical Be pebble beds Department of Metallic Alloys, Institute for Applied Materials - Applied Materials Physics IAM

C) Uniaxial stress-strain dependence

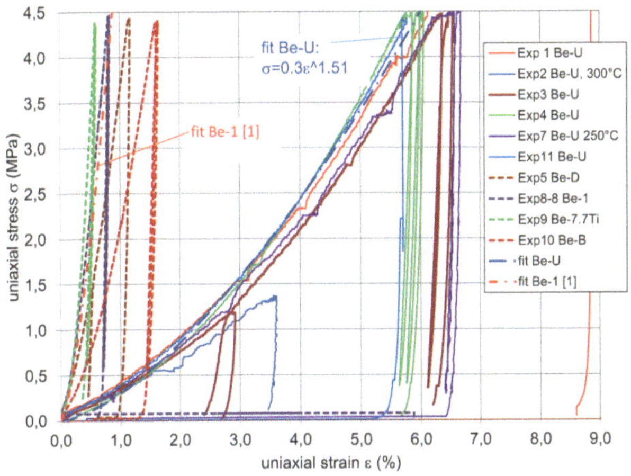

- Be-U pbs are significantly softer than the other Be grades: that is, for a given σ, strains ε are much larger.
- The stiffest pbs (smallest ε for a given σ) are obtained for the rather smooth spherical pebbles from Be-1[1] and Be-7.7Ti [3].
- A distinct temperature effect of the σ–ε curve is not observed, consistent with previous findings [1,3].

D) Effective thermal conductivity k of pbs, k = f(ε)

KIT Karlsruhe Institute of Technology

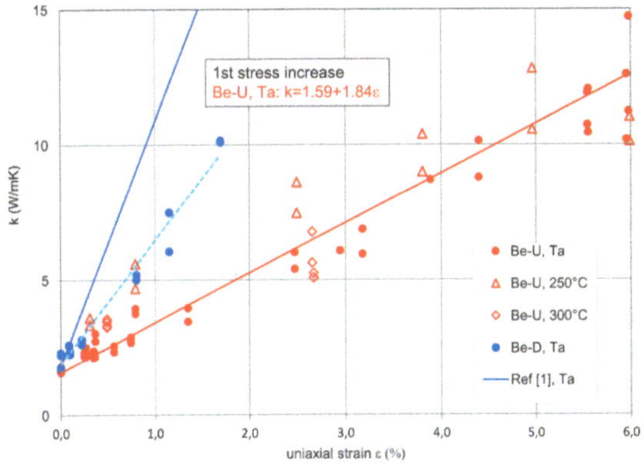

Effective thermal pebble bed conductivity k as a function of strain ε

Similar to previous investigations [1,2,3], a fairly linear relationship is obtained.

Large differences are observed between Be-U pbs and the other Be grade pbs because of differing σ–ε relationships.

D) Effective thermal conductivity k of pbs, k = f(ε)

KIT Karlsruhe Institute of Technology

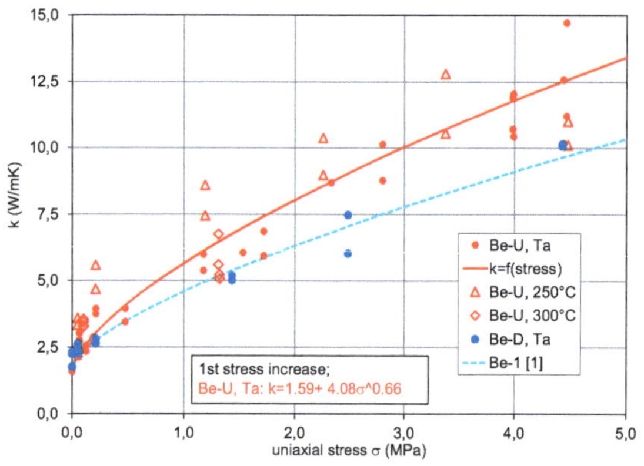

k as a function of stress σ

- For Be-U, k increases even stronger with increasing σ than Be-1.
- But: **what is the relevant parameter for blankets? Stress σ or strain ε ???**

11 19.10.2023 Reimann et al., Thermo-mechanics of non-spherical Be pebble beds Department of Metallic Alloys, Institute for Applied Materials - Applied Materials Physics IAM

D) Bed-structure-mechanical interaction (BSMI)

KIT Karlsruhe Institute of Technology

- During operation of pb blankets, thermal stresses arise because of both different temperature levels and thermal expansion coefficients of the pebble beds and the surrounding cooled structural materials (bed-structure-mechanical interaction (BSMI)).
- The BSMI is predicted by finite element codes with appropriate pebble bed models for BOL (stress-strain, thermal conductivity, thermal creep), and for EOL, with appropriate models considering irradiation effects.
- Although pebble deformations are caused by stresses imposed on the pebbles, the stress is not considered as primary parameter but the refrained strains which result in the stress.
- Therefore, for Be pbs, the softer the pbs, the smaller the build-up of σ but also of increased k.

12 19.10.2023 Reimann et al., Thermo-mechanics of non-spherical Be pebble beds Department of Metallic Alloys, Institute for Applied Materials - Applied Materials Physics IAM

Summary and Conclusions

Karlsruhe Institute of Technology

- For economical reasons, Be pebble beds, pbs, consisting of non-spherical particles, NPs, are of interest.
- In previous exps [2], pbs consisting NPs, were used where after crushing of Be blocks, a grinding step was added.
- Now, NP Be pb exps are reported where no subsequent grinding step was used.
- These NPs have spiky, extremely irregular shapes which results in:

  a) Packing fractions for vibrated pebble beds are in the range of 0.5 compared to ≈ 0.63.

  b) very soft pb behavior (small stress build-up),

  c) Crushing of the irregular pebbles during cycling and because of irradiation effects.

13 19.10.2023 Reimann et al., Thermo-mechanics of non-spherical Be pebble beds Department of Metallic Alloys, Institute for Applied Materials - Applied Materials Physics IAM

Summary and Conclusions

Karlsruhe Institute of Technology

- Consequences for blanket designs:
  - a) Small packing fractions: larger blanket volumes, smaller TBRs expected.
  - b) Soft pbs: no remarkable increase of the pb thermal conductivity.
  - c) Crushing: pb stability endangered.
- In summary: a grinding step (or another process route) is required in order to obtain more regular particles without sharp edges and a less spiky appearance. By optimizing pb densification procedures, NPs similar to those previously investigated [2], might be even superior to spherical pebbles e.g. in respect to packing fraction and heat transfer coefficient.

14 19.10.2023 Reimann et al., Thermo-mechanics of non-spherical Be pebble beds Department of Metallic Alloys, Institute for Applied Materials - Applied Materials Physics IAM

5.2 Study of Mechanical Properties of Beryllium Intermetallic Compounds Fabricated by Plasma Sintering

Taehyun HWANG, Jae-Hwan KIM, Yutaka SUGIMOTO, Yoshiaki AKATSU , Suguru NAKANO, Shota YOKOHAMA, Masaru NAKAMICHI

Advanced neutron multipliers in the Japanese Demonstration (JA DEMO) reactors are expected to be exposed to high temperatures (600–900 °C) and neutron doses (20,000 appm He, 50 dpa) during operation. Therefore, instead of beryllium, Beryllium intermetallic compounds (beryllides) have been considered as a promising candidate material for advanced neutron multipliers in JA DEMO reactors due to their higher stability and lower reactivity with water vapor. Also, block shaped beryllides especially are being considered for use as an advanced neutron multiplier, in place of beryllium/beryllides pebbles, due to their lower swelling properties and the expectation of high Tritium Breeding Ratio (TBR) for the JA DEMO blanket.

It is essential to establish a database of mechanical properties for beryllides with block-shaped for the advanced blanket design. Therefore, the purpose of this study is to evaluate the mechanical properties of beryllides processed by the plasma sintering method. A mixed powder of Be-7.7at.% Ti ($Be_{12}Ti$) was prepared with homogenization treatment at 1200°C for 24 hours in an argon atmosphere. Then, plasma sintering (KE-PAS II, manufactured by Kaken) was performed in the range of 1050°C to 1200°C to evaluate the effect of sintering temperature on strength. Experimental tests for the mechanical strength were conducted at room temperature (RT) and at high temperatures.

A detailed results regarding mechanical properties of beryllides will be reported.

Presenting Author: Taehyun HWANG
Presenting Author's Affiliation: National Institutes for Quantum and Radiological Science and Technology, Japan
Email: hwang.taehyun@qst.go.jp

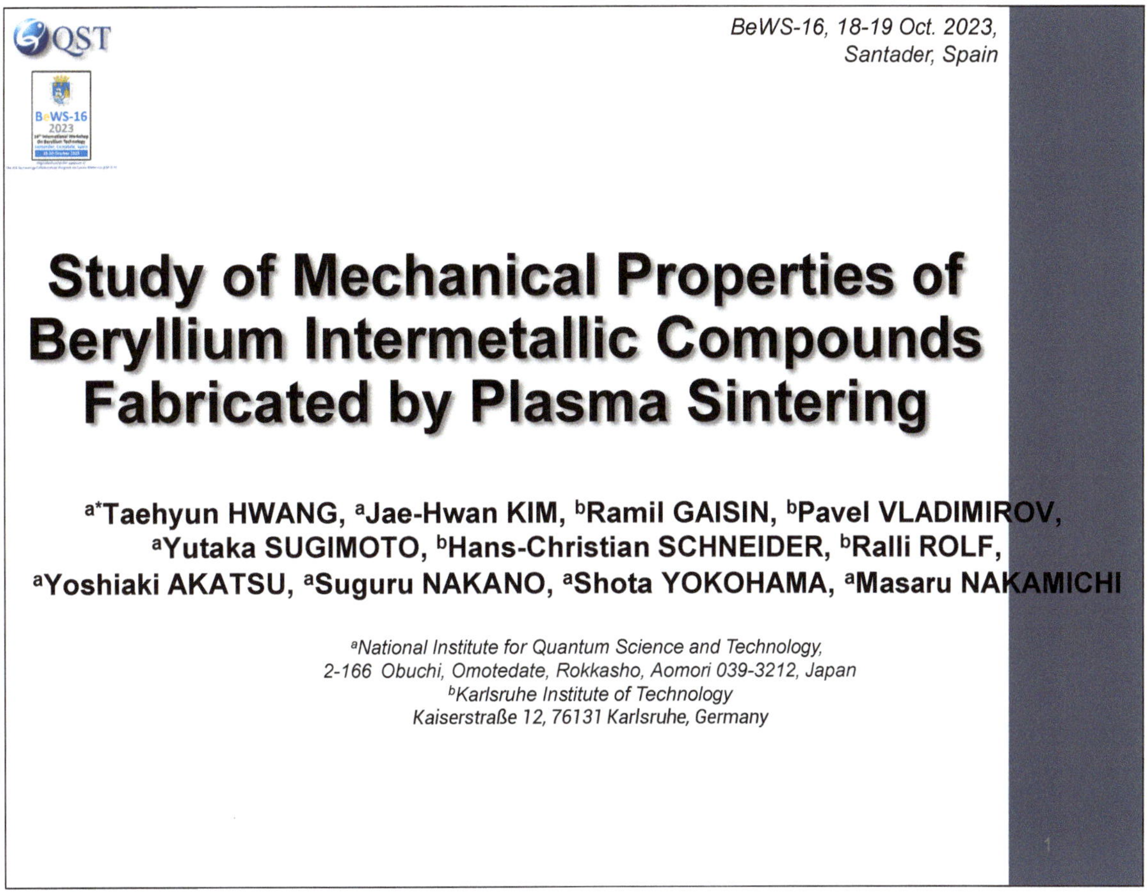
QST
BeWS-16 2023
BeWS-16, 18-19 Oct. 2023, Santader, Spain
Study of Mechanical Properties of Beryllium Intermetallic Compounds Fabricated by Plasma Sintering
a*Taehyun HWANG, aJae-Hwan KIM, bRamil GAISIN, bPavel VLADIMIROV, aYutaka SUGIMOTO, bHans-Christian SCHNEIDER, bRalli ROLF, aYoshiaki AKATSU, aSuguru NAKANO, aShota YOKOHAMA, aMasaru NAKAMICHI
aNational Institute for Quantum Science and Technology, 2-166 Obuchi, Omotedate, Rokkasho, Aomori 039-3212, Japan
bKarlsruhe Institute of Technology
Kaiserstraße 12, 76131 Karlsruhe, Germany
1

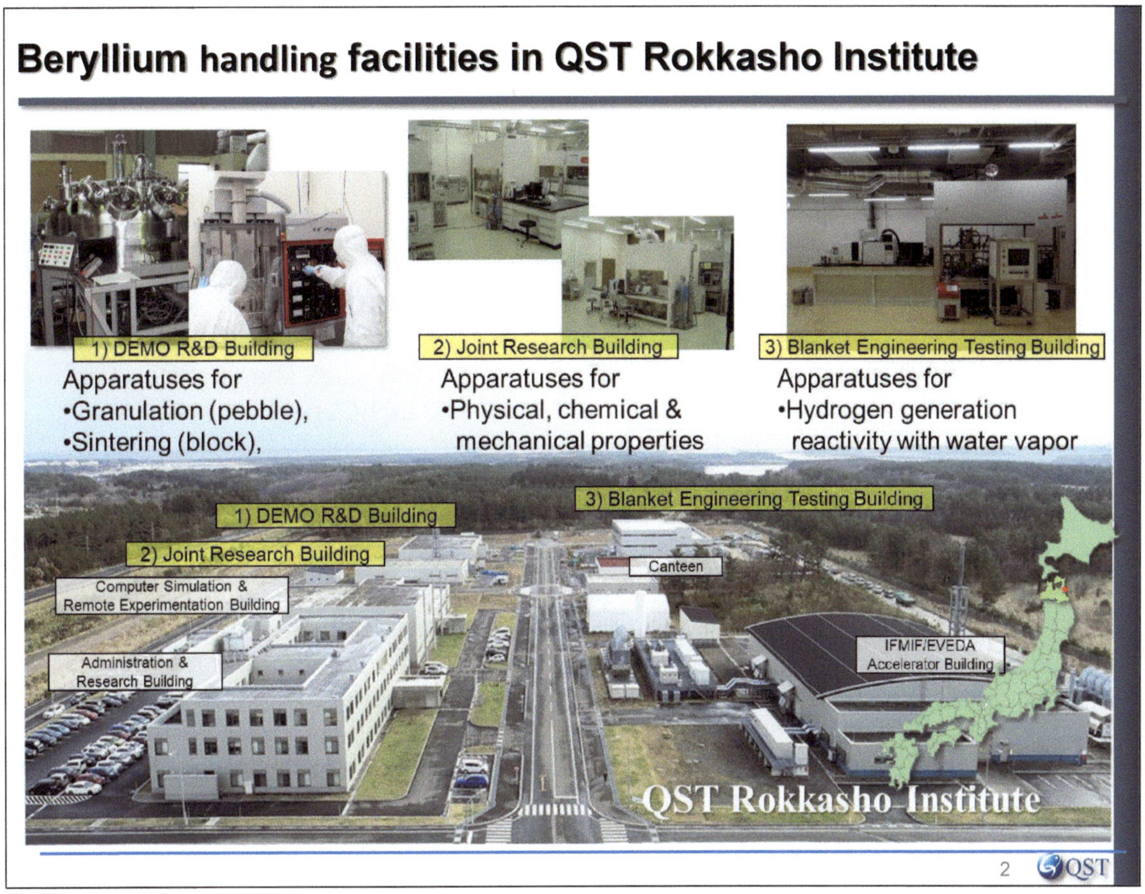
Beryllium handling facilities in QST Rokkasho Institute
1) DEMO R&D Building
Apparatuses for
•Granulation (pebble),
•Sintering (block),
2) Joint Research Building
Apparatuses for
•Physical, chemical & mechanical properties
3) Blanket Engineering Testing Building
Apparatuses for
•Hydrogen generation reactivity with water vapor
1) DEMO R&D Building
2) Joint Research Building
3) Blanket Engineering Testing Building
Canteen
Computer Simulation & Remote Experimentation Building
Administration & Research Building
IFMIF/EVEDA Accelerator Building
QST Rokkasho Institute
2
QST

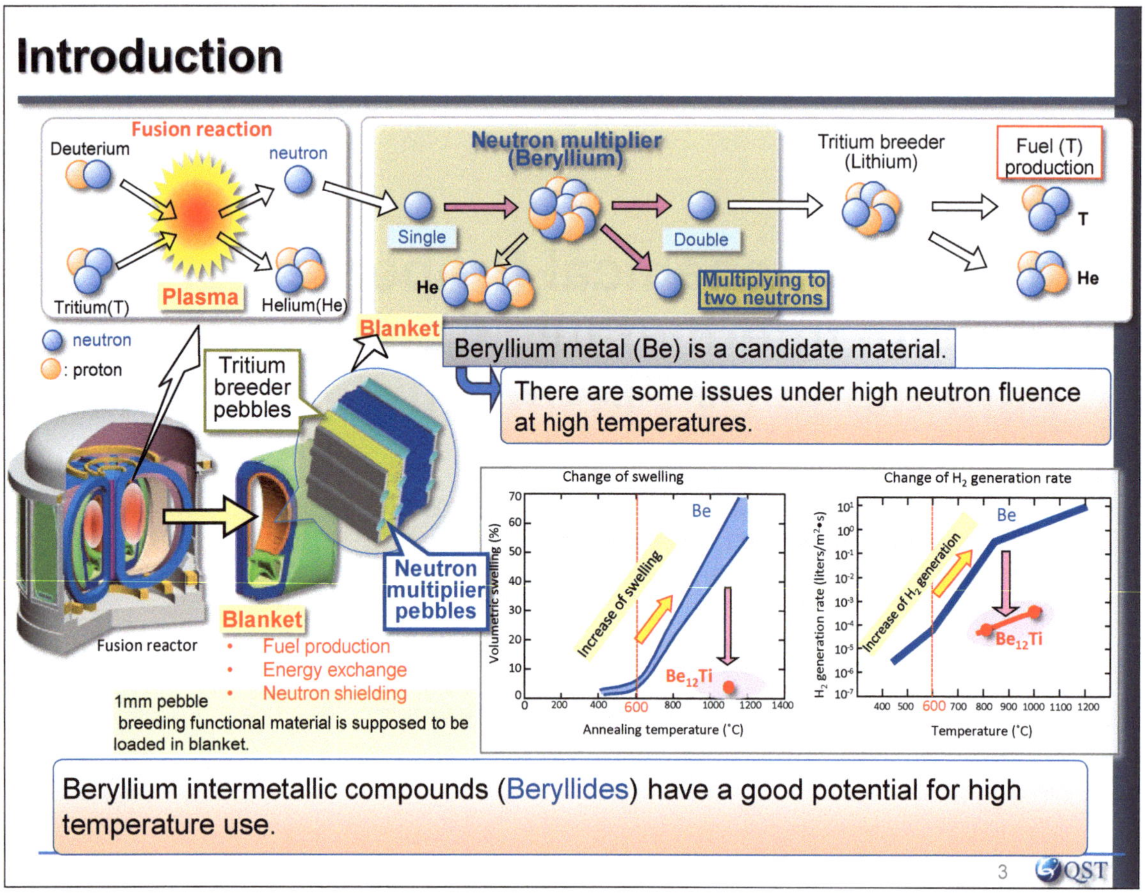
Introduction
Fusion reaction
Deuterium
neutron
Tritium(T)
Plasma
Helium(He)
neutron
: proton
Neutron multiplier
(Beryllium)
Single
Double
He
Multiplying to two neutrons
Tritium breeder
(Lithium)
Fuel (T) production
T
He
Blanket
Beryllium metal (Be) is a candidate material.
There are some issues under high neutron fluence at high temperatures.
Tritium breeder pebbles
Neutron multiplier pebbles
Fusion reactor
Blanket
• Fuel production
• Energy exchange
• Neutron shielding
1mm pebble
breeding functional material is supposed to be loaded in blanket.
Change of swelling
Volumetric swelling (%)
Increase of swelling
Be
Be12Ti
Annealing temperature (˚C)
Change of H2 generation rate
H2 generation rate (liters/m2•s)
Increase of H2 generation
Be
Be12Ti
Temperature (˚C)
Beryllium intermetallic compounds (Beryllides) have a good potential for high temperature use.
3
QST

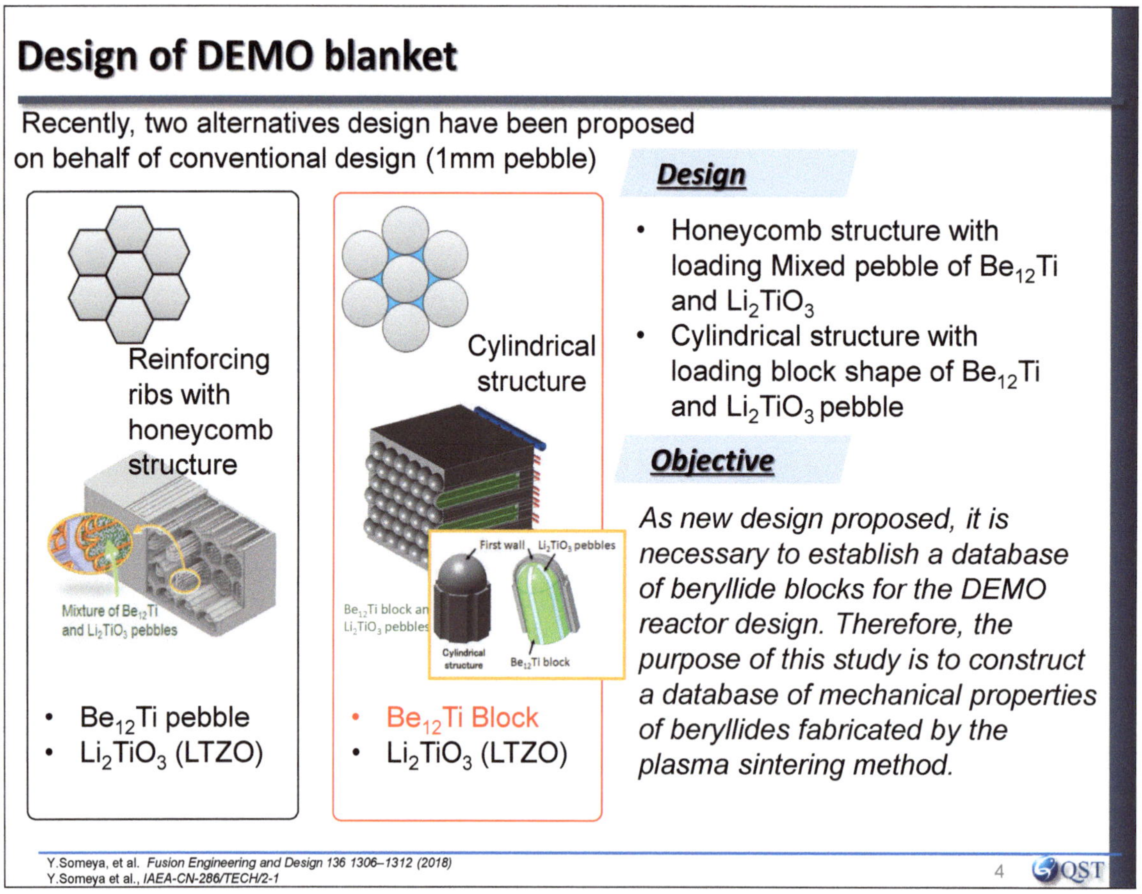
Design of DEMO blanket
Recently, two alternatives design have been proposed
on behalf of conventional design (1mm pebble)
Reinforcing ribs with honeycomb structure
Mixture of Be12Ti and Li2TiO3 pebbles
• Be12Ti pebble
• Li2TiO3 (LTZO)
Cylindrical structure
Be12Ti block and Li2TiO3 pebbles
First wall
Li2TiO3 pebbles
Cylindrical structure
Be12Ti block
• Be12Ti Block
• Li2TiO3 (LTZO)
Design
• Honeycomb structure with loading Mixed pebble of Be12Ti and Li2TiO3
• Cylindrical structure with loading block shape of Be12Ti and Li2TiO3 pebble
Objective
As new design proposed, it is necessary to establish a database of beryllide blocks for the DEMO reactor design. Therefore, the purpose of this study is to construct a database of mechanical properties of beryllides fabricated by the plasma sintering method.
Y.Someya, et al. Fusion Engineering and Design 136 1306–1312 (2018)
Y.Someya et al., IAEA-CN-286/TECH/2-1
4
QST

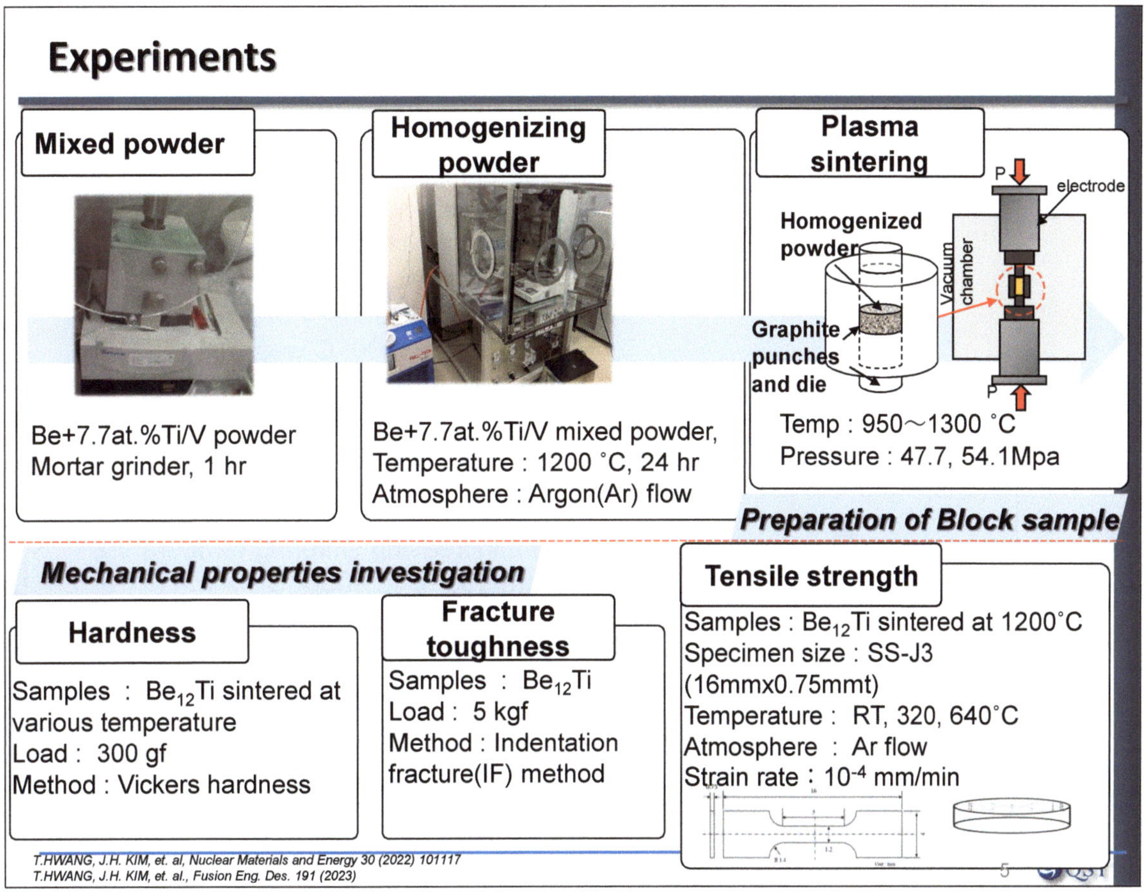
Experiments
Mixed powder
Be+7.7at.%Ti/V powder
Mortar grinder, 1 hr
Homogenizing powder
Be+7.7at.%Ti/V mixed powder,
Temperature : 1200 °C, 24 hr
Atmosphere : Argon(Ar) flow
Plasma sintering
P
electrode
Homogenized powder
Vacuum chamber
Graphite punches and die
P
Temp : 950~1300 °C
Pressure : 47.7, 54.1Mpa
Preparation of Block sample
Mechanical properties investigation
Hardness
Samples : Be12Ti sintered at various temperature
Load : 300 gf
Method : Vickers hardness
Fracture toughness
Samples : Be12Ti
Load : 5 kgf
Method : Indentation fracture(IF) method
Tensile strength
Samples : Be12Ti sintered at 1200°C
Specimen size : SS-J3 (16mmx0.75mmt)
Temperature : RT, 320, 640°C
Atmosphere : Ar flow
Strain rate : 10-4 mm/min
T.HWANG, J.H. KIM, et. al, Nuclear Materials and Energy 30 (2022) 101117
T.HWANG, J.H. KIM, et. al., Fusion Eng. Des. 191 (2023)
5

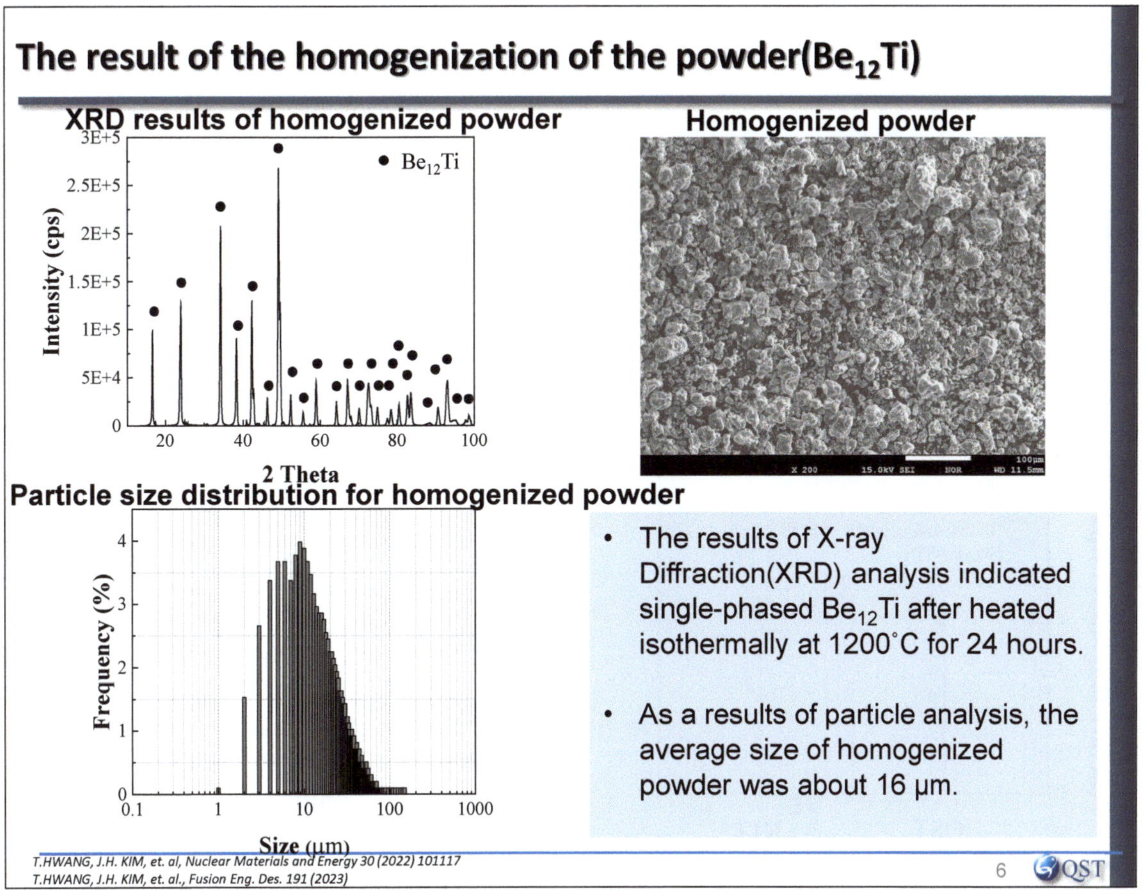
The result of the homogenization of the powder(Be12Ti)
XRD results of homogenized powder
Intensity (cps)
3E+5
2.5E+5
2E+5
1.5E+5
1E+5
5E+4
0
20
40
60
80
100
2 Theta
● Be12Ti
Homogenized powder
X 200
15.0kV SEI
100μm
Particle size distribution for homogenized powder
Frequency (%)
0
1
2
3
4
0.1
1
10
100
1000
Size (μm)
• The results of X-ray Diffraction(XRD) analysis indicated single-phased Be12Ti after heated isothermally at 1200°C for 24 hours.
• As a results of particle analysis, the average size of homogenized powder was about 16 μm.
T.HWANG, J.H. KIM, et. al, Nuclear Materials and Energy 30 (2022) 101117
T.HWANG, J.H. KIM, et. al., Fusion Eng. Des. 191 (2023)
6

Effect of sintering condition on density and hardness

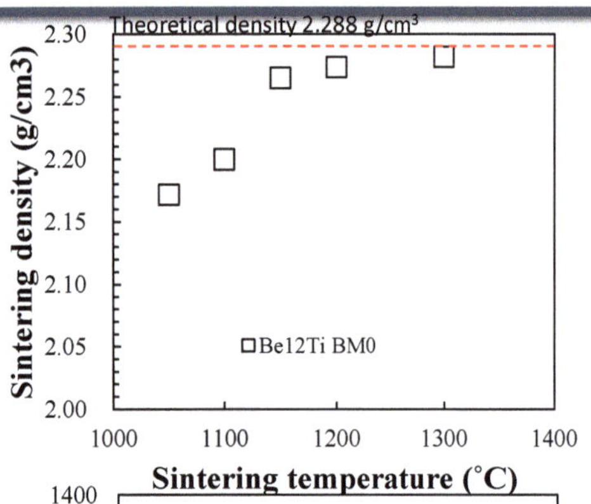

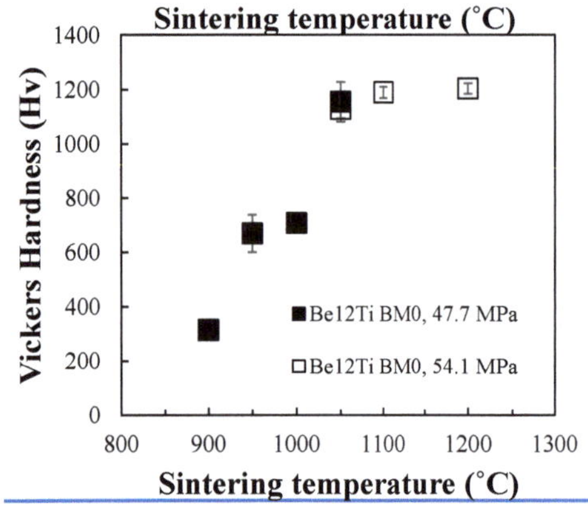

Sintering density

- The density increased sharply as increasing sintering temperature from 1050 ˚C to 1150 ˚C. Then, it showed slightly increase.
- The sintered density of the sample sintered at 1300 ˚C was close to the theoretical density[1] (2.288 g/cm^3).

Vickers hardness

- The Vickers hardness sharply increased with increasing sintering temperature from 900 ˚C to 1050 ˚C.
- The Vickers hardness showed saturation over 1050 ˚C.
- Even though the sintering pressure was different, the Vickers hardness was almost the same at 1050 ˚C

T.HWANG, J.H. KIM, et. al., Fusion Eng. Des. 191 (2023)

G.V. Samsonov, I.M., Handbook of Refractory compounds, Ifi/Plenum, New York, 1980

QST

Results of Fracture toughness calculation by IF method

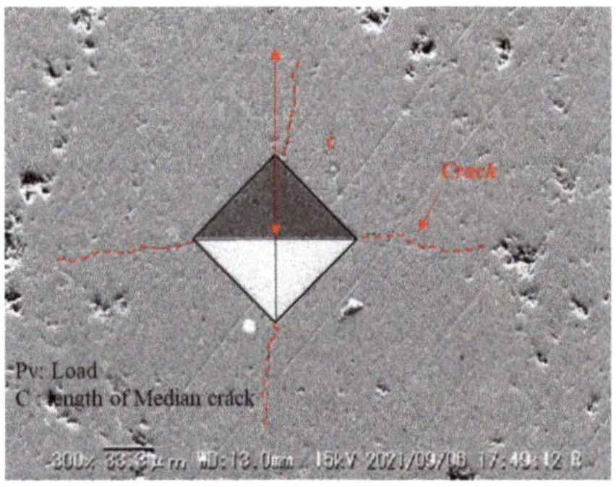

$$K_C = 0.016 \times \frac{P_v}{c^{1.5}} \times \left(\frac{E}{HV}\right)^{0.5} \quad (1)\ \text{Antis}$$

$$K_C = 0.0752 \times \frac{P_v}{c^{1.5}} \quad (2)\ \text{Evans and Charles}$$

$$K_C = 0.018 \times \frac{P_v}{c^{1.5}} \times \left(\frac{E}{HV}\right)^{0.5} \quad (3)\ \text{JIS R1607}$$

Pv : load during the Vickers test (N),
c : the crack length from the center,
E : Young's modulus (GPa),
HV : Vickers hardness (GPa)

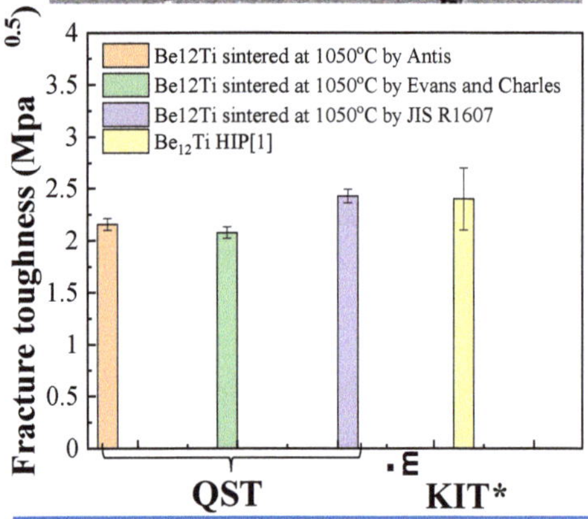

- The results of calculation for $Be_{12}Ti$ sintered at 1050˚C showed that the fracture toughness values were 2.07, 2.00 $MPa{\cdot}m^{0.5}$, and 2.44 $MPa{\cdot}m^{0.5}$ according to Evans & Charles, Anstis, and JIS equations, respectively.
- The fracture toughness value was similar with that of $Be_{12}Ti$ prepared by Hot Isostatic Pressing(HIP) (2.4 $MPa{\cdot}m^{0.5}$).

[1] R. Gaisin et al., J. Alloys Compd. 818 (2020),
T.Hwang et al., Nuclear Materials and Energy 30 (2022) 101117

*Karlsruhe institute of technology

8 QST

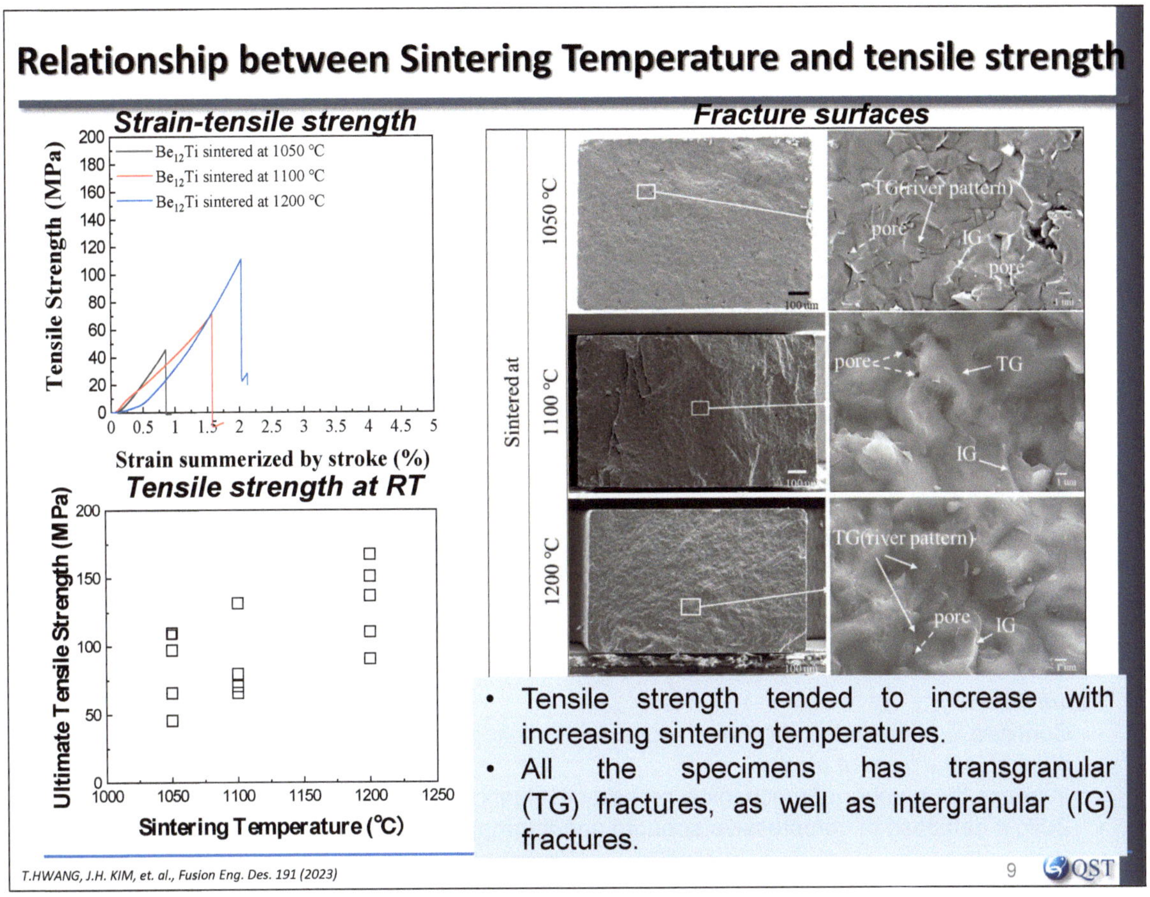
Relationship between Sintering Temperature and tensile strength
Strain-tensile strength
Be12Ti sintered at 1050 °C
Be12Ti sintered at 1100 °C
Be12Ti sintered at 1200 °C
Tensile Strength (MPa)
Strain summerized by stroke (%)
Tensile strength at RT
Ultimate Tensile Strength (MPa)
Sintering Temperature (°C)
Fracture surfaces
Sintered at
1050 °C
1100 °C
1200 °C
TG(river pattern)
pore
IG
TG
• Tensile strength tended to increase with increasing sintering temperatures.
• All the specimens has transgranular (TG) fractures, as well as intergranular (IG) fractures.
T.HWANG, J.H. KIM, et. al., Fusion Eng. Des. 191 (2023)
9
QST

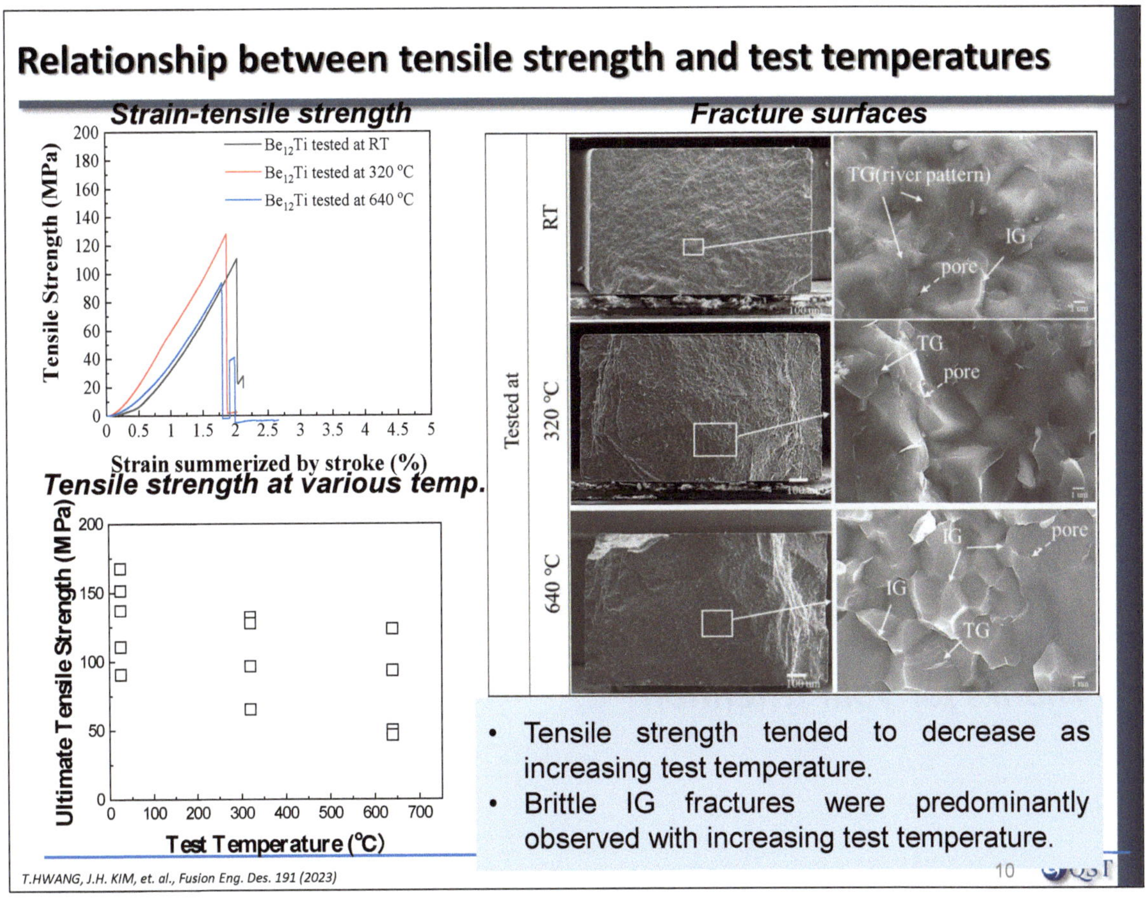
Relationship between tensile strength and test temperatures
Strain-tensile strength
Be12Ti tested at RT
Be12Ti tested at 320 °C
Be12Ti tested at 640 °C
Tensile Strength (MPa)
Strain summerized by stroke (%)
Tensile strength at various temp.
Ultimate Tensile Strength (MPa)
Test Temperature (°C)
Fracture surfaces
Tested at
RT
320 °C
640 °C
TG(river pattern)
IG
pore
TG
• Tensile strength tended to decrease as increasing test temperature.
• Brittle IG fractures were predominantly observed with increasing test temperature.
T.HWANG, J.H. KIM, et. al., Fusion Eng. Des. 191 (2023)
10

Summary

The mechanical properties of plasma sintered beryllide blocks have been investigated and the database is under establishing.

- The sintered density increased with increasing sintering temperature from 1050 °C to 1300 °C and ***reached almost the theoretical density of $Be_{12}Ti$ HIP.*** The Vickers hardness increased sharply with increasing sintering temperature from 900 °C to 1000 °C. The Vickers hardness showed saturation above 1000 °C.
- The fracture toughness of the sample sintered at 1050 °C showed the value between 2.00 to 2.43 MPa·$m^{0.5}$. This value was similar with that of $Be_{12}Ti$ prepared by HIP (2.4 MPa·$m^{0.5}$).
- QST has obtained data on the tensile strength of $Be_{12}Ti$ produced by plasma sintering for the first time in the world.
- Tensile test results showed that tensile strength increased with increasing sintering temperature and decreased with increasing test temperature.
- The results of SEM fracture surface observations indicated the presence of both TG and IG fractures. However, the IG fracture was predominantly observed with increasing temperature.
- ***According to the result of mechanical properties, the most suitable sintering temperature for $Be_{12}Ti$ seemed to 1200℃.***

Future Works

- Continue to improve the database necessary for loading neutron multipliers into the blanket in the form of blocks.
- Increase the number of mechanical property evaluations for more accuracy
- Build a database of compressive strength and thermal properties.

11 QST

Thanks for your attention

12

SESSION 6

Irradiation damage & He-implantation

6.1 Investigation of radiation damage effects in beryllium: updates on microstructure and properties of proton, neutron and He-ions irradiated samples

Slava Kuksenko

Beryllium and beryllium-based alloys are under extensive investigation by different nuclear facilities communities. The paper gives an overview of the recent results obtained in the UK Atomic Energy Authority on beryllium samples, particularly during collaborative studies with the Karlsruhe Institute of Technology (Germany), the University of Huddersfield (UK) and the international Radiation Damage In Accelerator Target Environments (RaDIATE) collaboration and include:

- Detailed SEM investigation of beryllium after neutron irradiation at different DEMO relevant conditions during the HIDOBE-2 campaign in the HFR reactor;
- Atom probe tomography mapping of the fine-scale distribution of impurities and transmutants in beryllium after neutron irradiation at different DEMO relevant conditions during the HIDOBE-2 campaign in the HFR reactor;
- Updates on the in-situ and ex-situ TEM studies of He implanted beryllium, TiBe12 and CrBe12 alloys and micromechanical tests.

The paper will also give an overview of the future plans.

Presenting Author: Slava Kuksenko
Presenting Author's Affiliation: Culham Centre for Fusion Energy, UK Atomic Energy Authority, UK
Email: slava.kuksenko@ukaea.uk

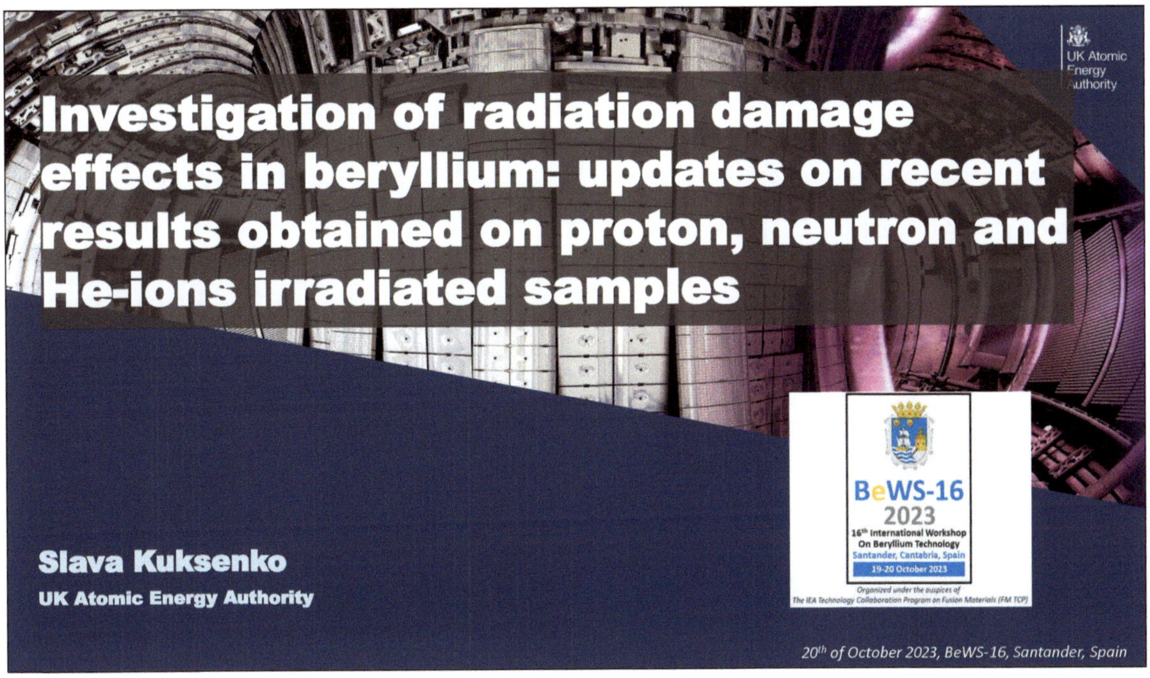
UK Atomic Energy Authority
Investigation of radiation damage effects in beryllium: updates on recent results obtained on proton, neutron and He-ions irradiated samples
BeWS-16
2023
16th International Workshop On Beryllium Technology
Santander, Cantabria, Spain
19-20 October 2023
Slava Kuksenko
UK Atomic Energy Authority
20th of October 2023, BeWS-16, Santander, Spain

Acknowledgments
MRF (UKAEA) operations team
Pavel Vladimirov, Vladimir Chakin, Mikhail Klimenkov and the KIT team
Prof Steve Roberts and MFFP, Micromechanics and Atom Probe groups of the University of Oxford
Patrick Hurh and the Fermilab team
Chris Densham and the RAL team
This work was part funded by the Fermi Research Alliance, LLC under Contract No. DE-AC02-07CH11359 with the United States Department of Energy, and by the UK Science and Technology Facilities Council.
This work was part funded by the RCUK Energy Programme [grant number EP/T012250/1]"
UK Atomic Energy Authority
KIT
Karlsruher Institut für Technologie
MATERIALS RESEARCH FACILITY
UNIVERSITY OF OXFORD
RaDIATE
Fermilab
Science & Technology Facilities Council
Rutherford Appleton Laboratory
2
Slava Kuksenko, 20th of October 2023, BeWS-16, Santander, Spain

4
Be
Beryllium
9.012

Beryllium has a unique combination of mechanical and physical and of particular importance for nuclear application:

- First wall material in JET experiment
- Neutron multiplier candidate material for tritium breeding blanket of the future DEMO fusion reactors
- Reflector and moderation in research reactors and spallation neutron sources
- Beam windows and target components material in neutrino sources

		Approximate He production in, appm/dpa
W	Fission LW reactor	0.0008
Fe	Fission LW reactor	0.35
	DEMO fusion reactor	10
Be	HFR, (Petten, Netherlands)	160
	SM-3 high-flux reactor	330
	DEMO fusion reactor	700
	Neutrino target NuMI (FNAL, USA)	4000

3 |

Slava Kuksenko, 20th of October 2023, BeWS-16, Santander, Spain

Outline

- Hardening under the high energy proton irradiation at 50°C (in the NuMI beamline) and helium implantation at 50°C and 200°C;
- He behavior in high-temperature neutron irradiated beryllium (HIDOBE-2 campaign, HFR).
- Updates on $TiBe_{12}$ and $CrBe_{12}$ obtained with micromechanical tests
- Thermal shock effects is beryllium
- Investigation of correlation between neutronic performance of beryllium and its microstructure and chemical composition
- Erosion and tritium retention in JET plasma facing beryllium
- Combined Effects of Light Gas and Damage Accumulation in Beryllium (presentation of J. Sharp at BeWS-16)

UK Atomic Energy Authority
Fermilab
RaDIATE
University of Huddersfield
KIT
Institute of High Energy Physics Chinese Academy of Sciences
CERN
ess

4 |

Slava Kuksenko, 20th of October 2023, BeWS-16, Santander, Spain

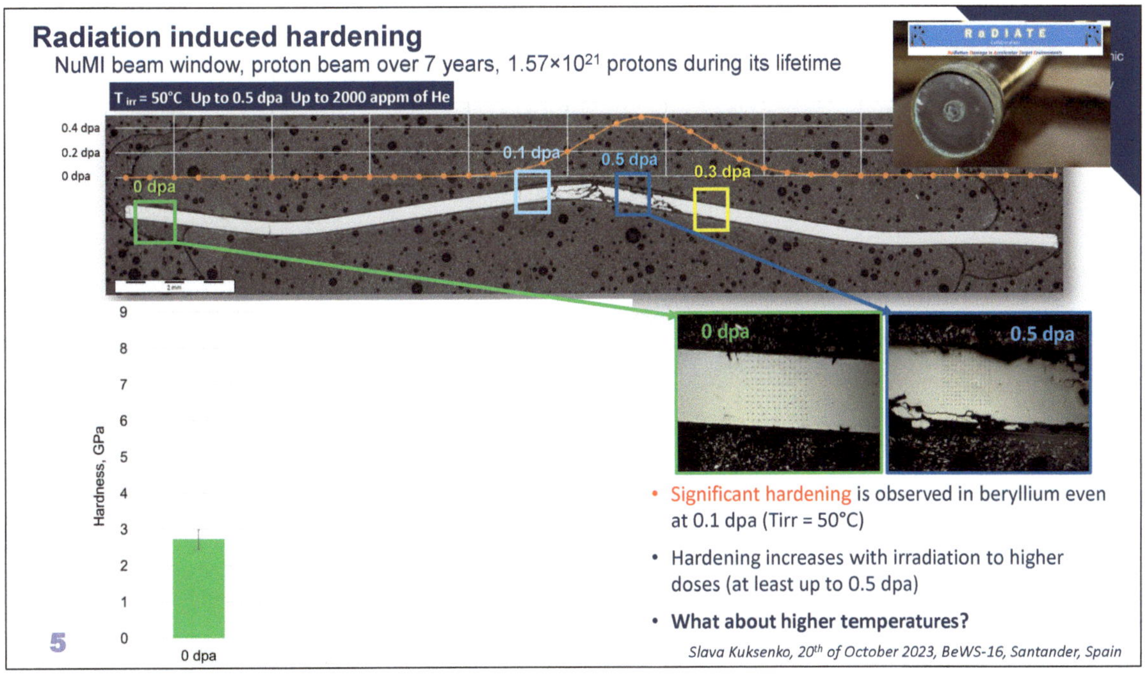
Radiation induced hardening
NuMI beam window, proton beam over 7 years, 1.57×10²¹ protons during its lifetime
T irr = 50°C Up to 0.5 dpa Up to 2000 appm of He
RaDIATE
0.4 dpa
0.2 dpa
0 dpa
0 dpa
0.1 dpa
0.5 dpa
0.3 dpa
0 dpa
0.5 dpa
Hardness, GPa
9
8
7
6
5
4
3
2
1
0
0 dpa
Significant hardening is observed in beryllium even at 0.1 dpa (Tirr = 50°C)
Hardening increases with irradiation to higher doses (at least up to 0.5 dpa)
What about higher temperatures?
5
Slava Kuksenko, 20th of October 2023, BeWS-16, Santander, Spain

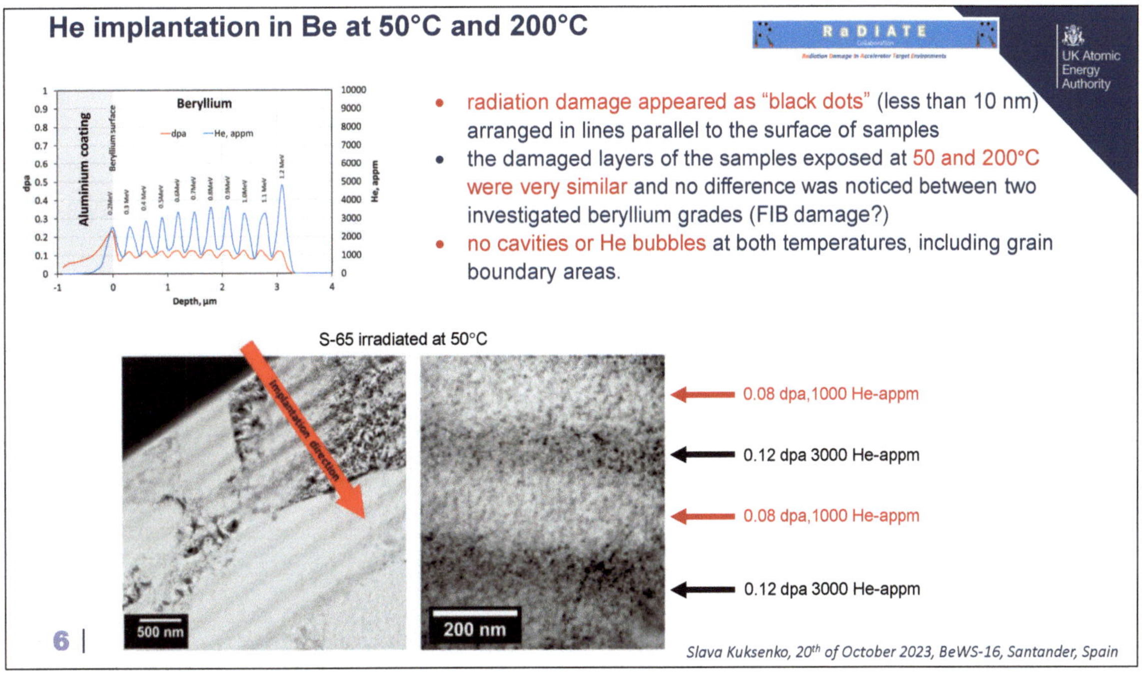
He implantation in Be at 50°C and 200°C
RaDIATE
UK Atomic Energy Authority
Beryllium
dpa
He, appm
Aluminium coating
Beryllium surface
Depth, µm
radiation damage appeared as "black dots" (less than 10 nm) arranged in lines parallel to the surface of samples
the damaged layers of the samples exposed at 50 and 200°C were very similar and no difference was noticed between two investigated beryllium grades (FIB damage?)
no cavities or He bubbles at both temperatures, including grain boundary areas.
S-65 irradiated at 50°C
implantation direction
500 nm
200 nm
0.08 dpa,1000 He-appm
0.12 dpa 3000 He-appm
0.08 dpa,1000 He-appm
0.12 dpa 3000 He-appm
6
Slava Kuksenko, 20th of October 2023, BeWS-16, Santander, Spain

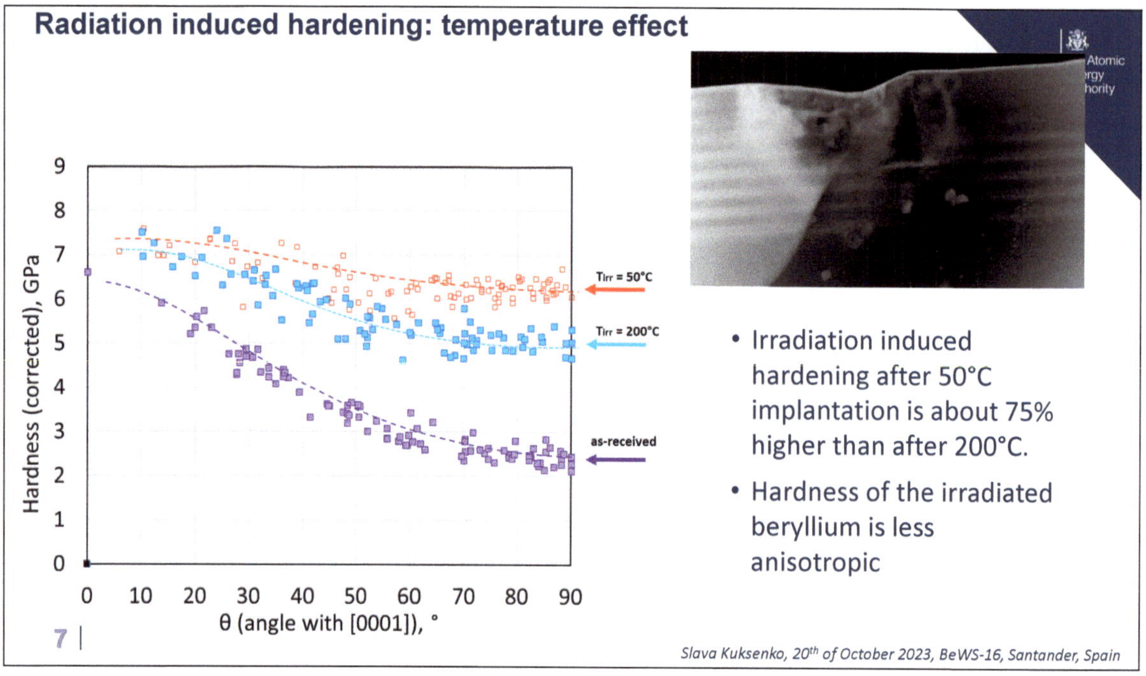
Radiation induced hardening: temperature effect
Hardness (corrected), GPa
θ (angle with [0001]), °
Tirr = 50°C
Tirr = 200°C
as-received
• Irradiation induced hardening after 50°C implantation is about 75% higher than after 200°C.
• Hardness of the irradiated beryllium is less anisotropic
7
Slava Kuksenko, 20th of October 2023, BeWS-16, Santander, Spain

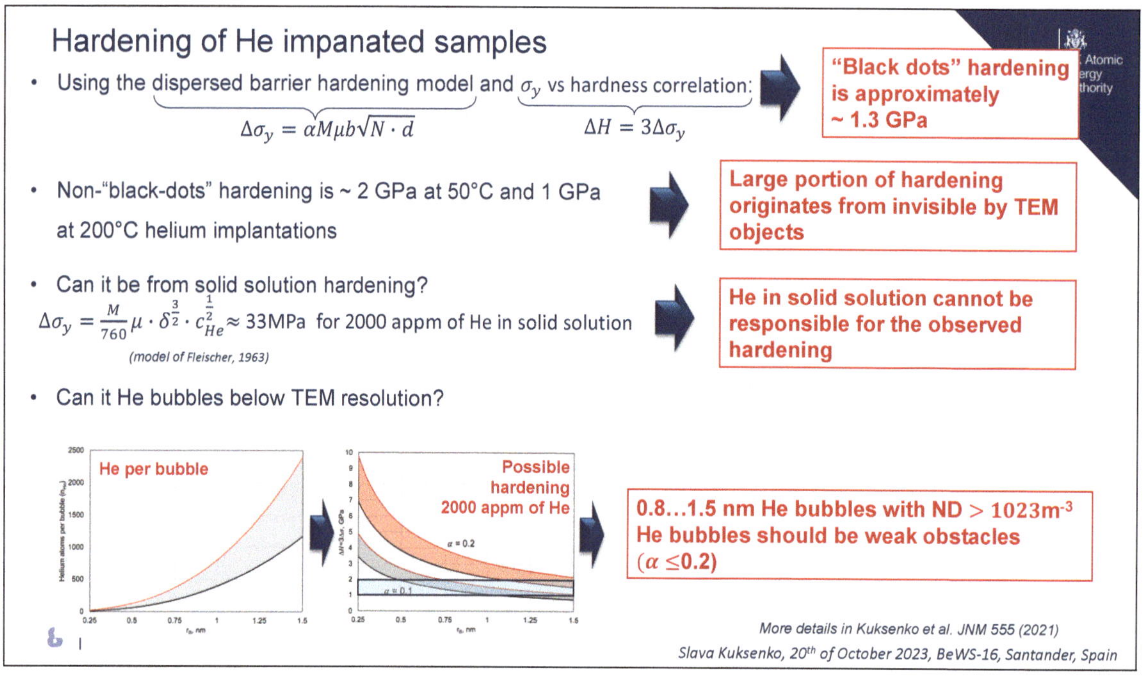
Hardening of He impanated samples
• Using the dispersed barrier hardening model and σ_y vs hardness correlation:
$\Delta\sigma_y = \alpha M \mu b \sqrt{N \cdot d}$
$\Delta H = 3\Delta\sigma_y$
"Black dots" hardening is approximately ~ 1.3 GPa
• Non-"black-dots" hardening is ~ 2 GPa at 50°C and 1 GPa at 200°C helium implantations
Large portion of hardening originates from invisible by TEM objects
• Can it be from solid solution hardening?
$\Delta\sigma_y = \frac{M}{760}\mu \cdot \delta^{\frac{3}{2}} \cdot c_{He}^{\frac{1}{2}} \approx$ 33MPa for 2000 appm of He in solid solution
(model of Fleischer, 1963)
He in solid solution cannot be responsible for the observed hardening
• Can it He bubbles below TEM resolution?
He per bubble
Possible hardening 2000 appm of He
0.8…1.5 nm He bubbles with ND > 1023m-3 He bubbles should be weak obstacles ($\alpha \leq 0.2$)
More details in Kuksenko et al. JNM 555 (2021)
Slava Kuksenko, 20th of October 2023, BeWS-16, Santander, Spain

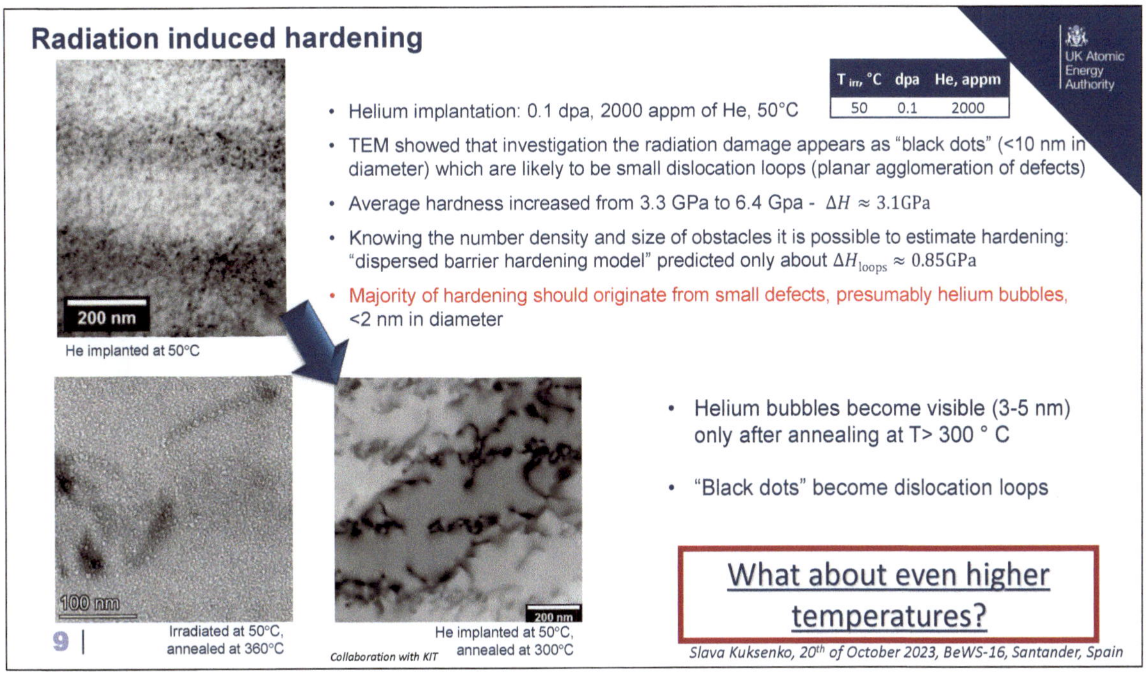
Radiation induced hardening
UK Atomic Energy Authority
T irr, °C | dpa | He, appm
50 | 0.1 | 2000
• Helium implantation: 0.1 dpa, 2000 appm of He, 50°C
• TEM showed that investigation the radiation damage appears as "black dots" (<10 nm in diameter) which are likely to be small dislocation loops (planar agglomeration of defects)
• Average hardness increased from 3.3 GPa to 6.4 Gpa - $\Delta H \approx 3.1\text{GPa}$
• Knowing the number density and size of obstacles it is possible to estimate hardening: "dispersed barrier hardening model" predicted only about $\Delta H_{\text{loops}} \approx 0.85\text{GPa}$
• Majority of hardening should originate from small defects, presumably helium bubbles, <2 nm in diameter
200 nm
He implanted at 50°C
100 nm
Irradiated at 50°C, annealed at 360°C
200 nm
He implanted at 50°C, annealed at 300°C
Collaboration with KIT
• Helium bubbles become visible (3-5 nm) only after annealing at T> 300 ° C
• "Black dots" become dislocation loops
What about even higher temperatures?
9 |
Slava Kuksenko, 20th of October 2023, BeWS-16, Santander, Spain

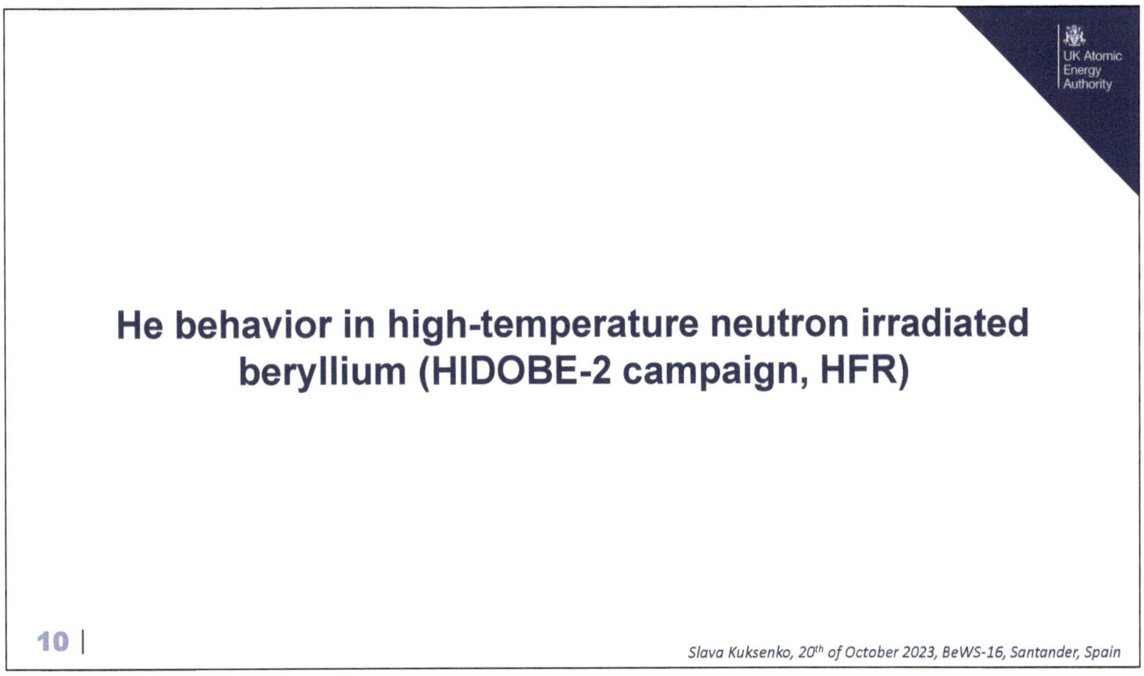
UK Atomic Energy Authority
He behavior in high-temperature neutron irradiated beryllium (HIDOBE-2 campaign, HFR)
10 |
Slava Kuksenko, 20th of October 2023, BeWS-16, Santander, Spain

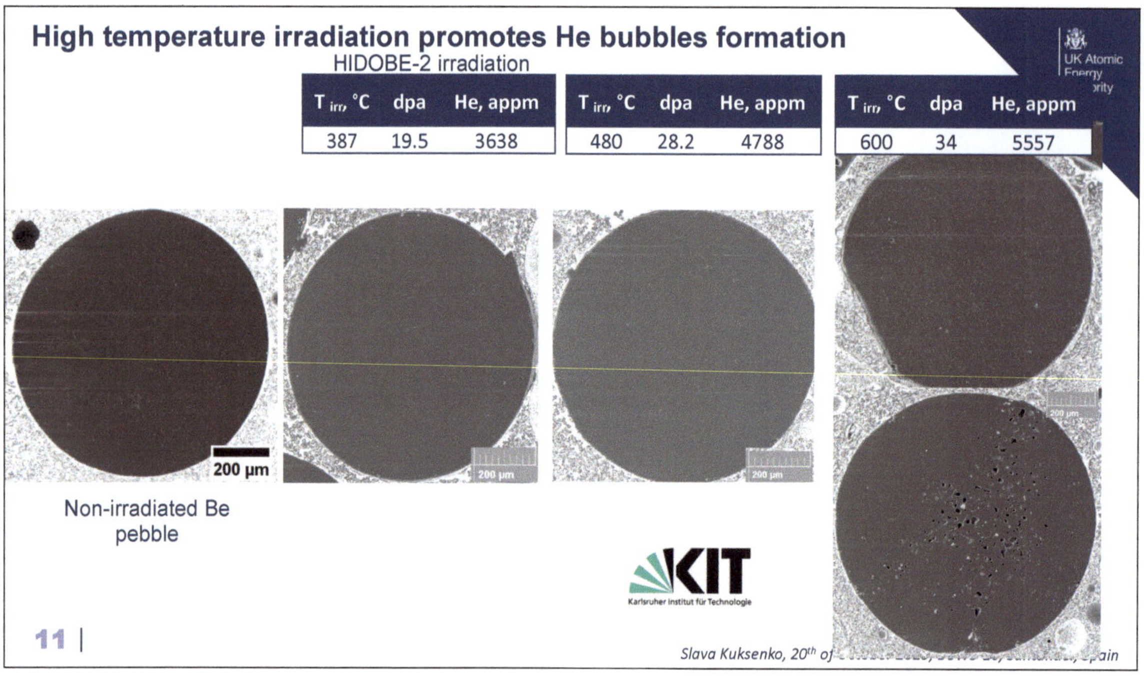
High temperature irradiation promotes He bubbles formation
HIDOBE-2 irradiation
T irr, °C dpa He, appm
387 19.5 3638
T irr, °C dpa He, appm
480 28.2 4788
T irr, °C dpa He, appm
600 34 5557
200 µm
Non-irradiated Be pebble
KIT
Karlsruher Institut für Technologie
UK Atomic Energy Authority
11 |
Slava Kuksenko, 20th of

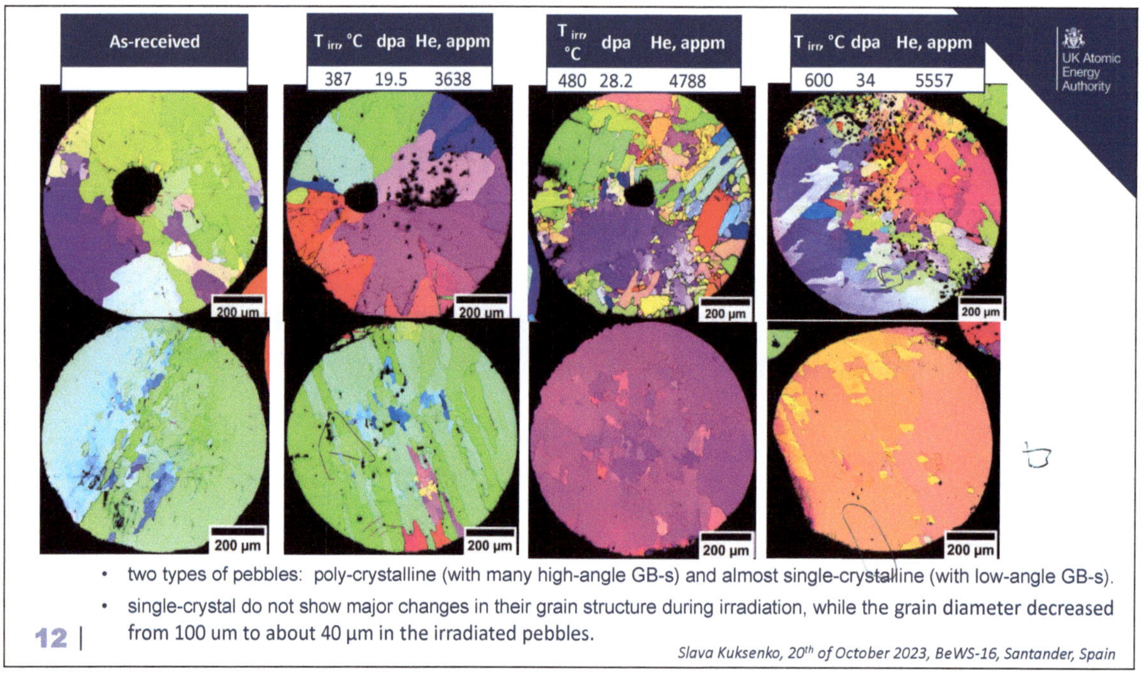
As-received
T irr, °C dpa He, appm
387 19.5 3638
T irr, °C dpa He, appm
480 28.2 4788
T irr, °C dpa He, appm
600 34 5557
UK Atomic Energy Authority
200 µm
• two types of pebbles: poly-crystalline (with many high-angle GB-s) and almost single-crystalline (with low-angle GB-s).
• single-crystal do not show major changes in their grain structure during irradiation, while the grain diameter decreased from 100 um to about 40 µm in the irradiated pebbles.
12 |
Slava Kuksenko, 20th of October 2023, BeWS-16, Santander, Spain

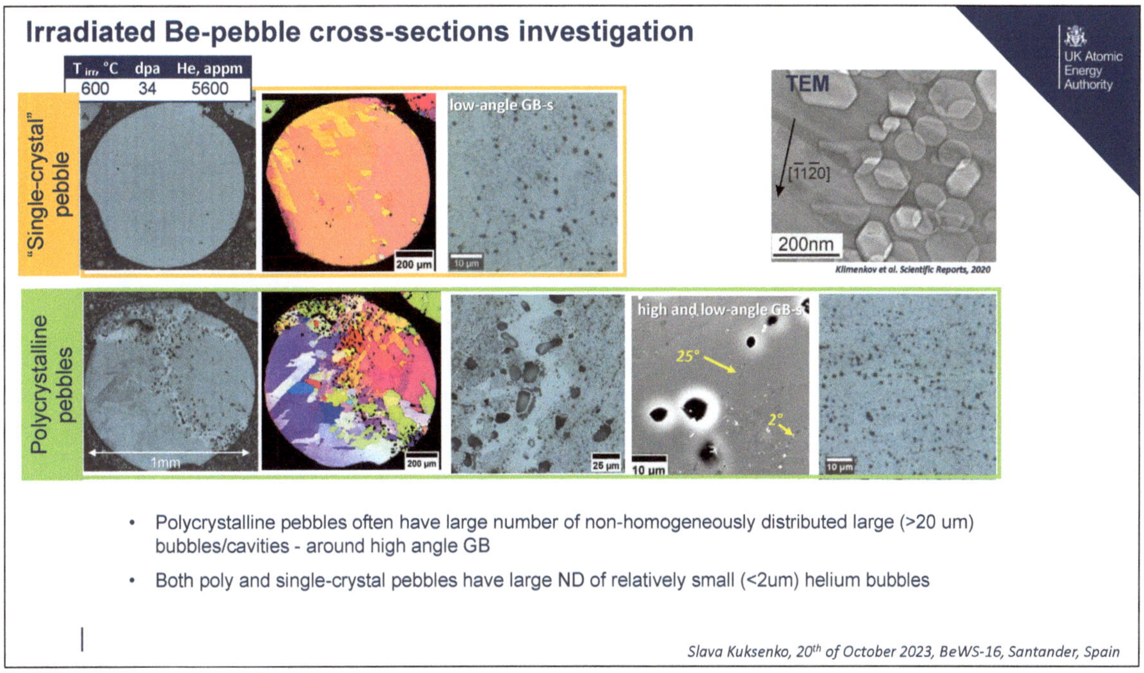
Irradiated Be-pebble cross-sections investigation
UK Atomic Energy Authority
T irr, °C dpa He, appm
600 34 5600
"Single-crystal" pebble
200 µm
10 µm
low-angle GB-s
TEM
[1120]
200nm
Klimenkov et al. Scientific Reports, 2020
Polycrystalline pebbles
1mm
200 µm
25 µm
high and low-angle GB-s
25°
2°
10 µm
10 µm
• Polycrystalline pebbles often have large number of non-homogeneously distributed large (>20 um) bubbles/cavities - around high angle GB
• Both poly and single-crystal pebbles have large ND of relatively small (<2um) helium bubbles
Slava Kuksenko, 20th of October 2023, BeWS-16, Santander, Spain

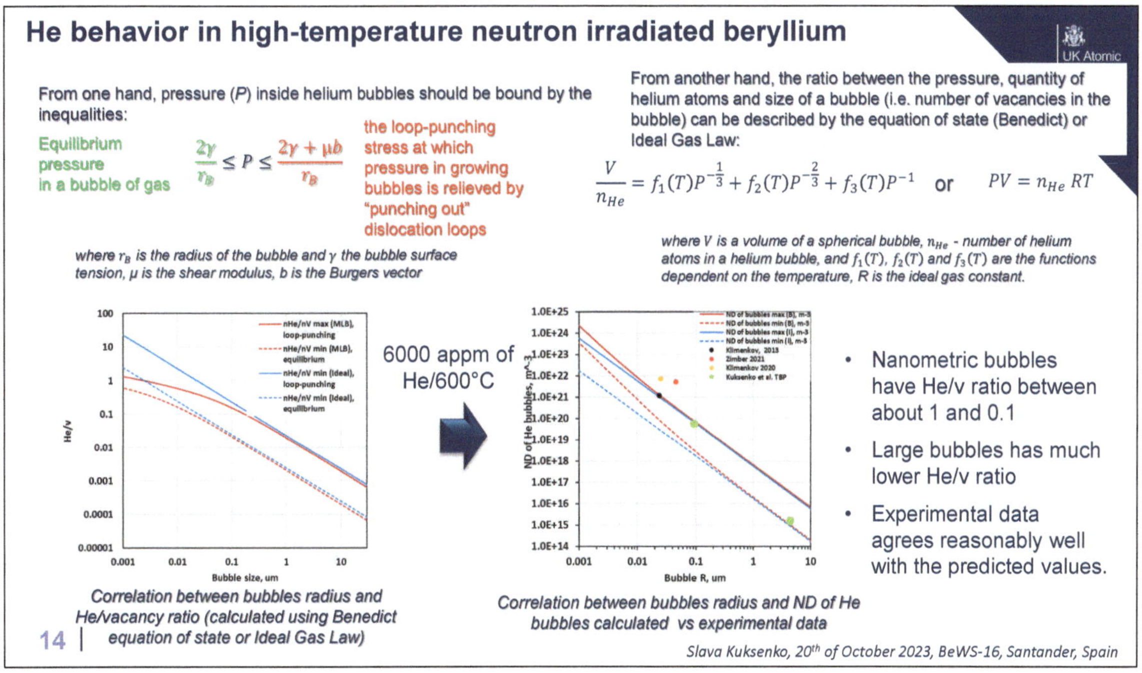
He behavior in high-temperature neutron irradiated beryllium
UK Atomic
From one hand, pressure (P) inside helium bubbles should be bound by the inequalities:
Equilibrium pressure in a bubble of gas
$\frac{2\gamma}{r_B} \leq P \leq \frac{2\gamma + \mu b}{r_B}$
the loop-punching stress at which pressure in growing bubbles is relieved by "punching out" dislocation loops
where r_B is the radius of the bubble and γ the bubble surface tension, μ is the shear modulus, b is the Burgers vector
From another hand, the ratio between the pressure, quantity of helium atoms and size of a bubble (i.e. number of vacancies in the bubble) can be described by the equation of state (Benedict) or Ideal Gas Law:
$\frac{V}{n_{He}} = f_1(T)P^{-\frac{1}{3}} + f_2(T)P^{-\frac{2}{3}} + f_3(T)P^{-1}$ or $PV = n_{He}RT$
where V is a volume of a spherical bubble, n_{He} - number of helium atoms in a helium bubble, and $f_1(T)$, $f_2(T)$ and $f_3(T)$ are the functions dependent on the temperature, R is the ideal gas constant.
He/v
Bubble size, um
6000 appm of He/600°C
ND of He bubbles, m^-3
Bubble R, um
• Nanometric bubbles have He/v ratio between about 1 and 0.1
• Large bubbles has much lower He/v ratio
• Experimental data agrees reasonably well with the predicted values.
Correlation between bubbles radius and He/vacancy ratio (calculated using Benedict equation of state or Ideal Gas Law)
Correlation between bubbles radius and ND of He bubbles calculated vs experimental data
14
Slava Kuksenko, 20th of October 2023, BeWS-16, Santander, Spain

He behavior in high-temperature neutron irradiated beryllium

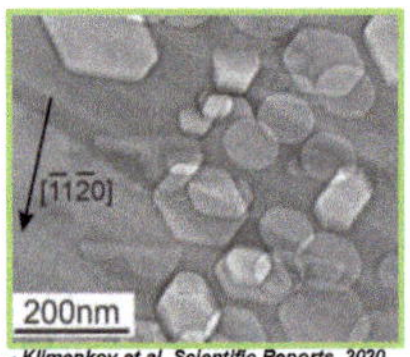

Klimenkov et al. Scientific Reports, 2020

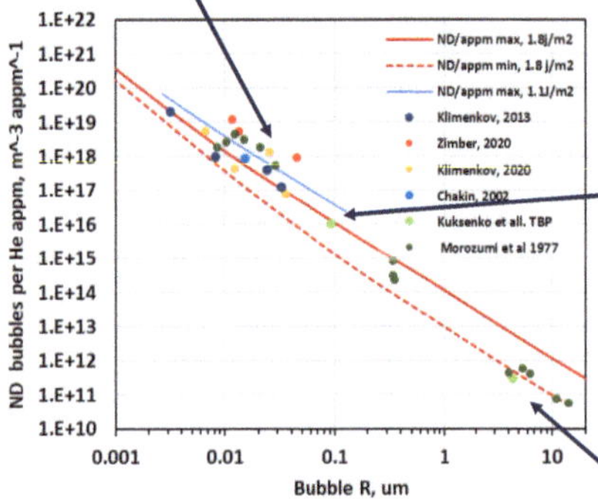

Calculated and measured ND of He bubbles (appm normalized) vs sizes appm range 2000-6000 appm, T range 380 to 650°C

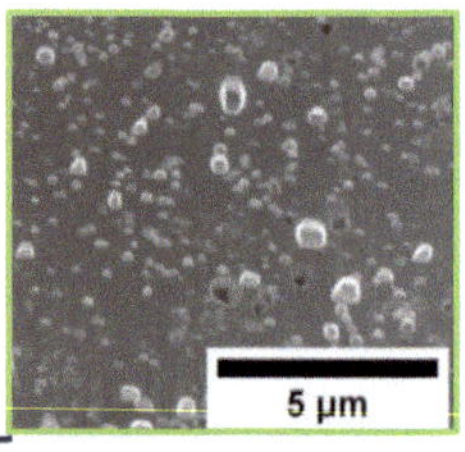

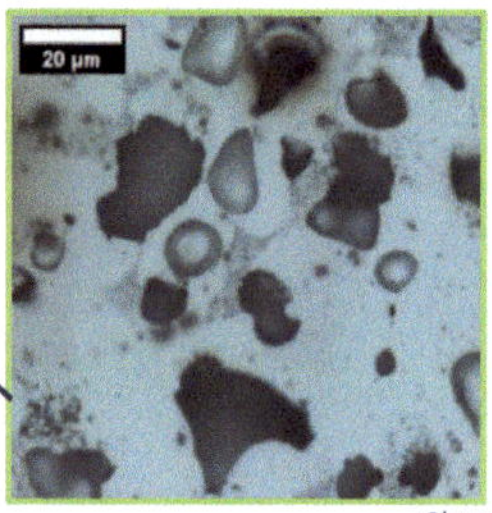

- In the size range between about 5 to 100 nm the equation of state underpredict ND of He bubbles.
- However, presence of tritium is expected to decrease surface energy of Be from 1.8 to 1.1 J/m^2 (Bachurin & Vladimirov, 2017) and cylindrical shape of bubbles should be considered.
- Correction for shape and presence of T increase agreement between calculations and experiments.
- In the same sample local swelling of 4.8% was observed inside grains (small bubbles) and 36% in the areas of high density of high-angle grain boundaries.
- Non-homogeneous swelling may lead to build-up of significant internal stresses.

15 |

Slava Kuksenko, 20th of October 2023, BeWS-16, Santander, Spain

Micromechanical tests of novel neutron multiplier materials

16 |

Slava Kuksenko, 20th of October 2023, BeWS-16, Santander, Spain

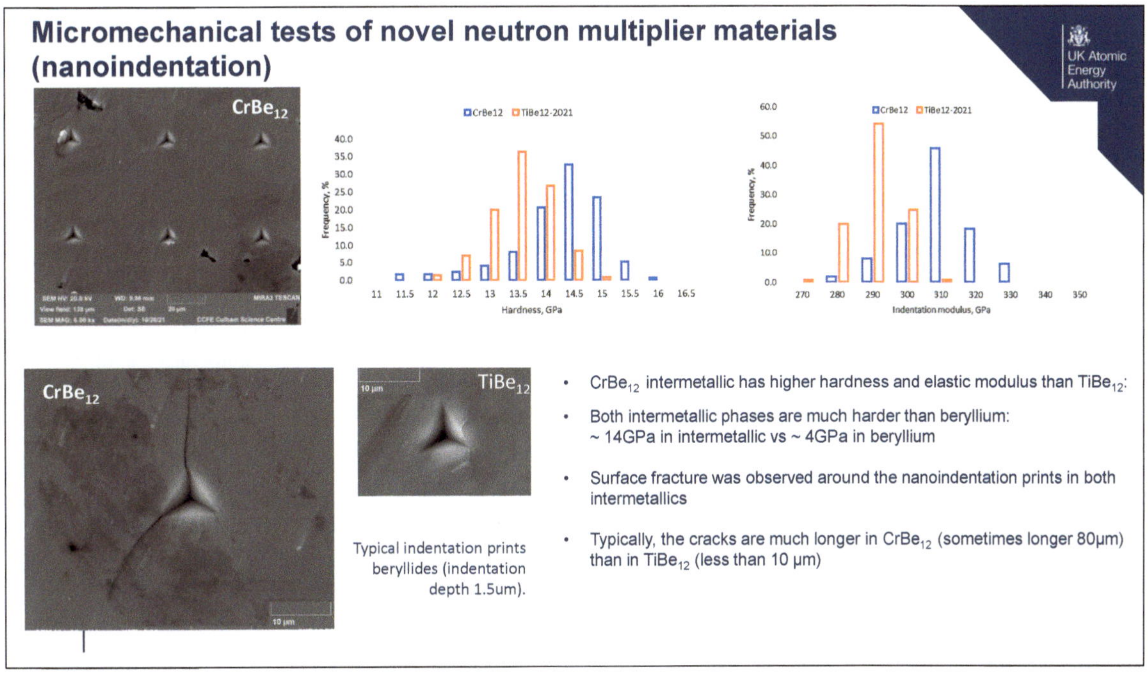
Micromechanical tests of novel neutron multiplier materials (nanoindentation)
UK Atomic Energy Authority
CrBe12
CrBe12 TiBe12-2021
Frequency, %
Hardness, GPa
Indentation modulus, GPa
CrBe12
TiBe12
Typical indentation prints beryllides (indentation depth 1.5um).
CrBe12 intermetallic has higher hardness and elastic modulus than TiBe12;
Both intermetallic phases are much harder than beryllium: ~ 14GPa in intermetallic vs ~ 4GPa in beryllium
Surface fracture was observed around the nanoindentation prints in both intermetallics
Typically, the cracks are much longer in CrBe12 (sometimes longer 80μm) than in TiBe12 (less than 10 μm)

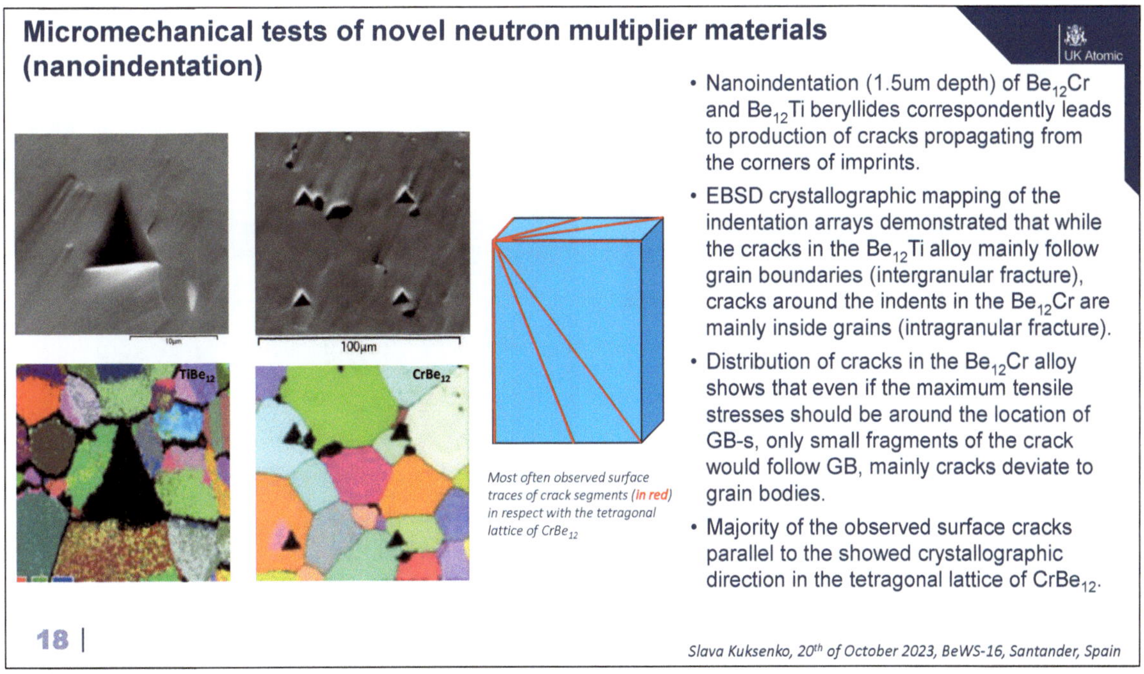
Micromechanical tests of novel neutron multiplier materials (nanoindentation)
UK Atomic
100μm
TiBe12
CrBe12
Most often observed surface traces of crack segments (in red) in respect with the tetragonal lattice of CrBe12
Nanoindentation (1.5um depth) of Be12Cr and Be12Ti beryllides correspondently leads to production of cracks propagating from the corners of imprints.
EBSD crystallographic mapping of the indentation arrays demonstrated that while the cracks in the Be12Ti alloy mainly follow grain boundaries (intergranular fracture), cracks around the indents in the Be12Cr are mainly inside grains (intragranular fracture).
Distribution of cracks in the Be12Cr alloy shows that even if the maximum tensile stresses should be around the location of GB-s, only small fragments of the crack would follow GB, mainly cracks deviate to grain bodies.
Majority of the observed surface cracks parallel to the showed crystallographic direction in the tetragonal lattice of CrBe12.
18
Slava Kuksenko, 20th of October 2023, BeWS-16, Santander, Spain

Micromechanical tests of novel neutron multiplier materials (nanoindentation)

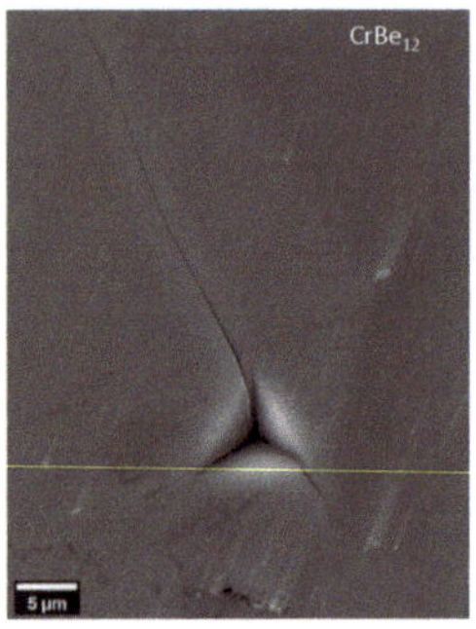

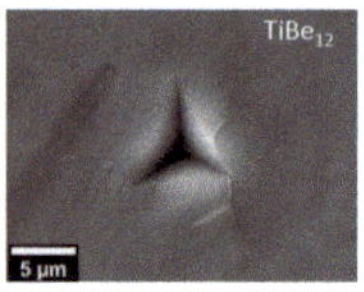

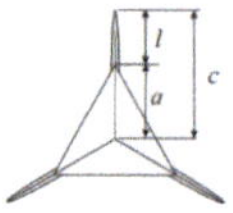

Fracture toughness from the Berkovich indentation induced cracks:

$$K_c = 1.073 x_v \left(a/l\right)^{1/2} \left(\frac{E}{H}\right)^{2/3} \frac{P}{c^{3/2}}$$

E is the elasticity modulus in GPa, H is the hardness in GPa, P is the indentation load in N; a, l and c are measured as shown by the scatch; $x_v = 0.015$

- The cracks in $Be_{12}Ti$ were up to 6um whereas the cracks up to 30…50 um long were observed in $Be_{12}Cr$; This **implies lower fracture toughness of the $CrBe_{12}$ sample**
- **Fracture toughness** estimated from the indentation induced cracks are 6.9±3.5 MPa·m$^{1/2}$ and 1.6±0.9 MPa·m$^{1/2}$ for the $Be_{12}Ti$ and $Be_{12}Cr$ correspondently.
- The fracture toughness value of $Be_{12}Cr$ agrees well with the previously reported values in other beryllides (0.8-4.0 MPa·m$^{1/2}$)
- The fracture toughness value of $Be_{12}Ti$ is significantly higher than the previously reported values in other beryllides (0.8-4.0 MPa·m$^{1/2}$ including $Be_{12}Ti$ (1.7-2.4 MPa·m$^{1/2}$).
- It is too early to collude on the superior fracture resistance of the investigated $Be_{12}Ti$ alloy, this will need confirmation by other fracture toughness measurement techniques.

19 | Official

Slava Kuksenko, 20th of October 2023, BeWS-16, Santander, Spain

Micromechanical tests of novel neutron multiplier materials (microcantilever fracture)

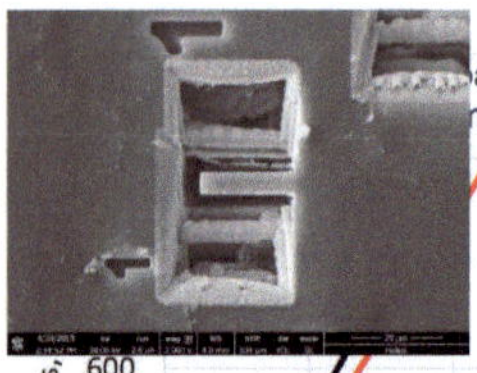

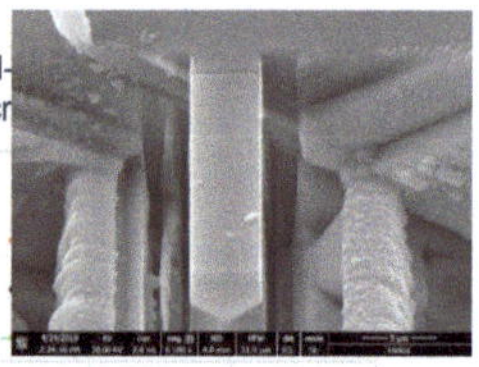

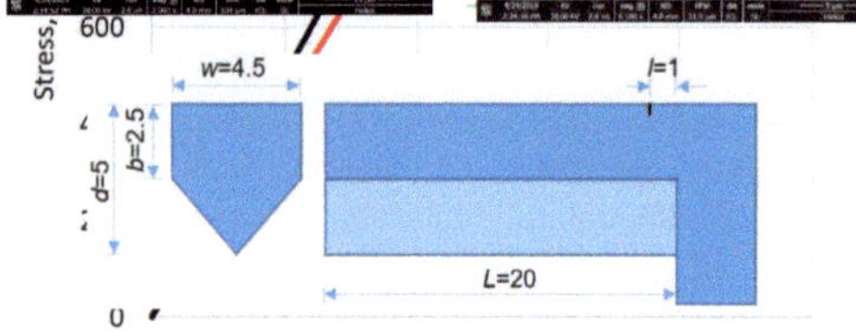

General geometry of the cantilevers prepared with FIB in the TiBe12 sample and their typical dimensions in µm.

Gaisin, Kuksenko et al. to be submitted

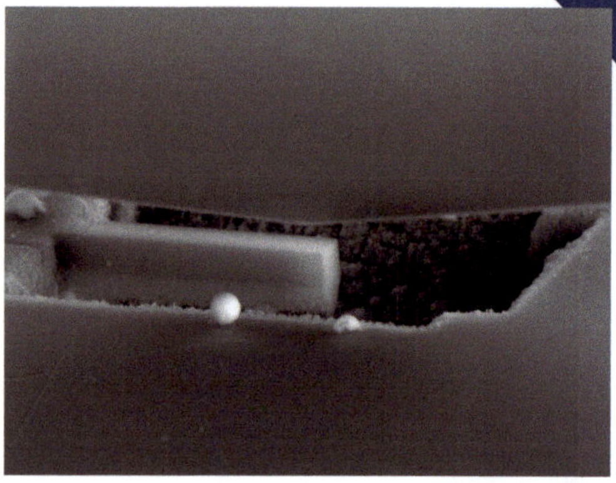

- In $TiBe_{12}$ and $CrBe_{12}$ phases load changes linearly until a sudden failure. Negligible plasticity in $TiBe_{12}$ and $CrBe_{12}$ cantilevers (see **black** and **red** load displacement curves)
- Be phase in the Ti-Be alloy is less hard and deforms plastically during cantilever bending (see green load-displacement curve)
- Room temperature fracture toughness of $TiBe_{12}$ and $CrBe_{12}$ is lower than 2 MPa·m$^{1/2}$

20 |

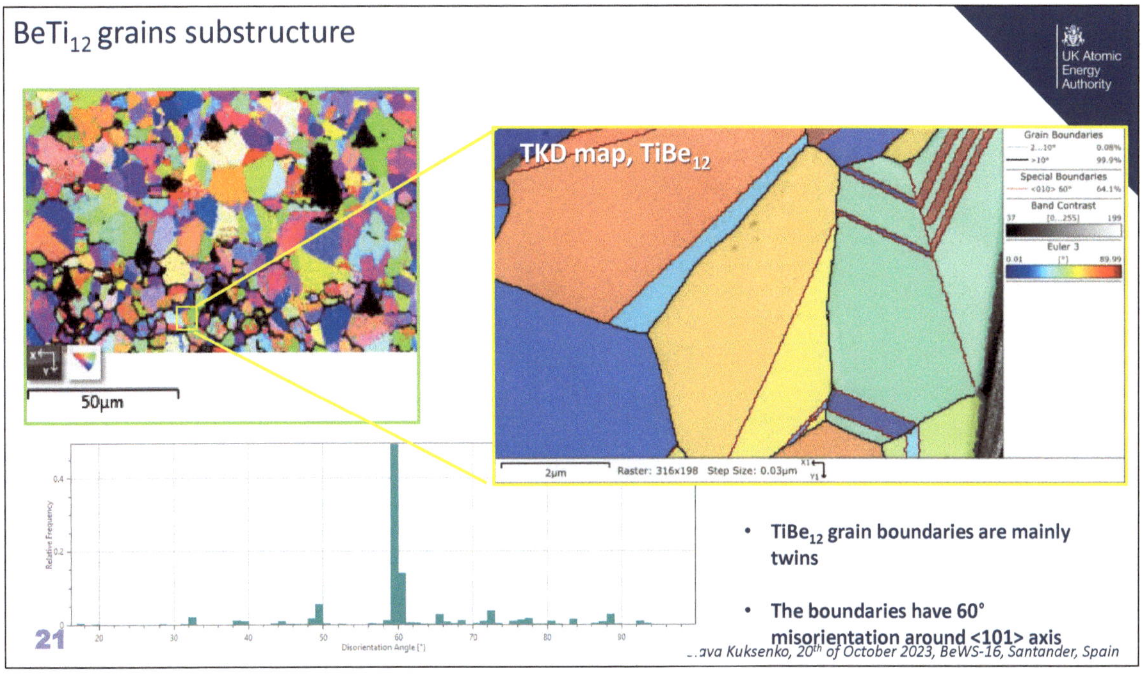
BeTi12 grains substructure
UK Atomic Energy Authority
50μm
TKD map, TiBe12
Grain Boundaries
Special Boundaries
Band Contrast
Euler 3
2μm
Raster: 316x198 Step Size: 0.03μm
Disorientation Angle [°]
Relative Frequency
TiBe12 grain boundaries are mainly twins
The boundaries have 60° misorientation around <101> axis
21
...ava Kuksenko, 20th of October 2023, BeWS-16, Santander, Spain

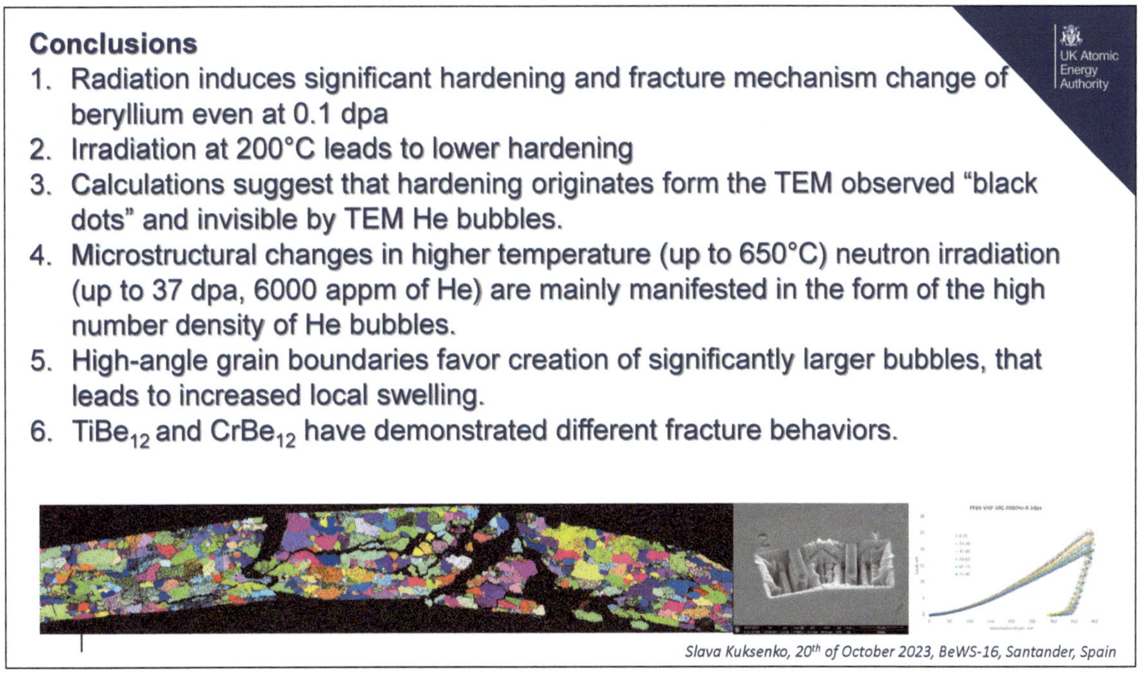
Conclusions
1. Radiation induces significant hardening and fracture mechanism change of beryllium even at 0.1 dpa
2. Irradiation at 200°C leads to lower hardening
3. Calculations suggest that hardening originates form the TEM observed “black dots” and invisible by TEM He bubbles.
4. Microstructural changes in higher temperature (up to 650°C) neutron irradiation (up to 37 dpa, 6000 appm of He) are mainly manifested in the form of the high number density of He bubbles.
5. High-angle grain boundaries favor creation of significantly larger bubbles, that leads to increased local swelling.
6. TiBe12 and CrBe12 have demonstrated different fracture behaviors.
UK Atomic Energy Authority
Slava Kuksenko, 20th of October 2023, BeWS-16, Santander, Spain

UK Atomic Energy Authority

- This work was part funded by the Fermi Research Alliance, LLC under Contract No. DE-AC02-07CH11359 with the United States Department of Energy, and by the UK Science and Technology Facilities Council.
- This work was part funded by the RCUK Energy Programme [grant number EP/T012250/1]"

Thank you!

23 |

Slava Kuksenko, 20th of October 2023, BeWS-16, Santander, Spain

6.2 The swelling property of neutron multipliers after irradiation with He ions under high temperature

Yutaka Sugimoto, Jae-Hwan Kim, Taehyun Hwang, Mitsutaka Miyamoto, Yoshiaki Akatsu, Suguru Nakano, Shota Yokohama, Masaru Nakamichi

Beryllium and beryllium intermetallic compounds (beryllides), such as $Be_{12}Ti$ and $Be_{12}V$, are neutron multipliers in blanket systems of Demonstration (DEMO) fusion reactors. Neutron multipliers are one of the tritium breeding functional materials for fusion reactors. However, they are exposed to high neutron fluence. Swelling of irradiated neutron multipliers can occur as a result of the formation and growth of bubble, and can load on structural materials. In order to ensure the safe operation of fusion reactors, it is necessary to evaluate the growth behavior of bubbles in order to understand the swelling properties of beryllium and beryllides. Therefore, direct observation experiments of the growth behavior are necessary.

This study aimed to investigate the growth behaviors of helium bubbles in beryllium and beryllide under irradiation with 3 keV-He^{+} ions at high temperatures, using an in-situ transmission electron microscope equipped with an ion gun. Bubble densities and sizes were measured in the beryllium and beryllides under irradiations, and the temperture-dependency of the swelling for the irradiated beryllium and beryllides were evaluated. The results indicated that the irradiated beryllium exhibited the increasing of bubble density and the decreasign of bubble size with increasing temperature.

Presenting Author: Yutaka Sugimoto
Presenting Author's Affiliation: National Institutes for Quantum and Radiological Science and Technology, Japan
Email: sugimoto.yutaka@qst.go.jp

The 16th International Workshop on Beryllium Technology (BeWS-16)
@Santander, Spain 19-20.10.2023

The swelling property of Neutron multipliers after irradiation with He ions under high temperature

Yutaka Sugimoto[1*], Mitsutaka Miyamoto[2] Kazutomo Hara[2], Jae-Hwan Kim[1], Taehyun Hwang[1], Mitsutaka Haruta[3], Hiroki Kurata[3], Suguru Nakano[1], Yoshiaki Akatsu[1],Yokohama Shota[1], Masaru Nakamichi[1]

[1] *Fusion Energy Research and Development Directorate, National Institutes for Quantum Science and Technology, Rokkasho, Aomori 2-166, Japan*
[2] *Department of Material Science, Shimane University, Matsue, Shimane 690-8504, Japan*
[3] *Institute for Chemical Research, Kyoto University, Gokasho Uji-city, Kyoto 611-0011, Japan*

1

Presentation outline

1. Introduction
- The gas retention affects the material property for beryllium as a PFM
- The bubble behaviors affects the gas retention property for beryllium

2. Research method
- In-situ TEM equipped with ion-gun and QMS at Shimane University
- STEM-EELS with high energy and high spatial resolution at Kyoto university

3. Results
- The relationship between the morphology and the internal pressure of bubble
- The difference of the internal pressure affects the bubble growth and the bubble mobility

4. Discussion

5. Current research

swelling properties of beryllium and beryllides using in-situ TEM equipped with ion gun

2

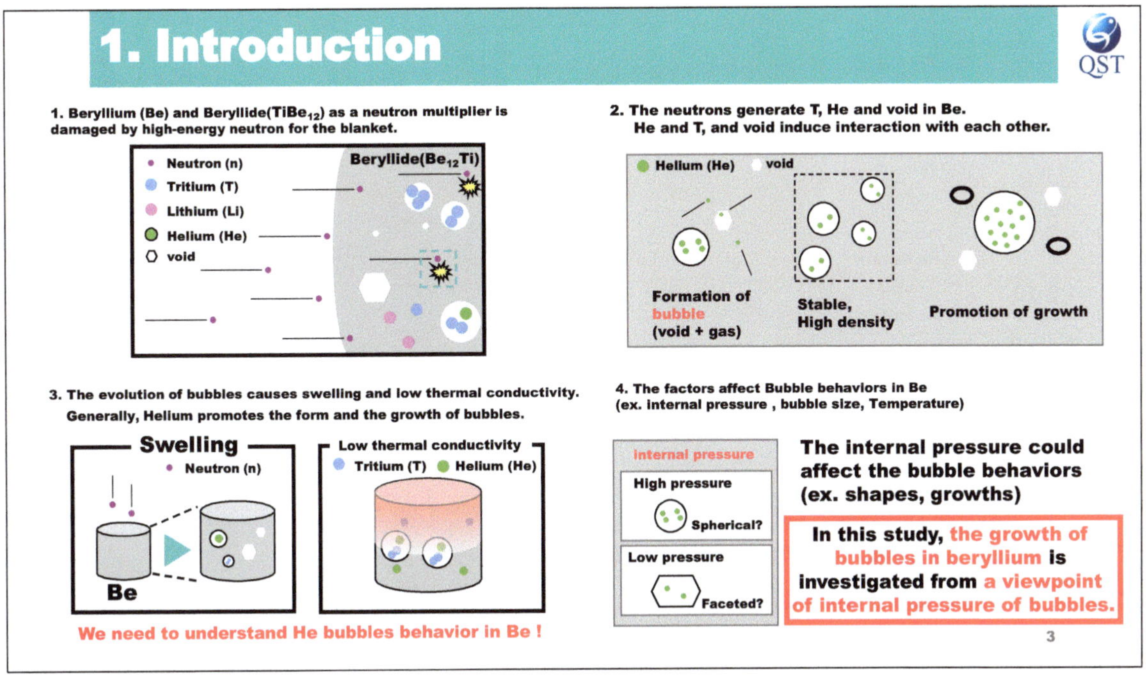
1. Introduction
QST
1. Beryllium (Be) and Beryllide(TiBe$_{12}$) as a neutron multiplier is damaged by high-energy neutron for the blanket.
Neutron (n)
Tritium (T)
Lithium (Li)
Helium (He)
void
Beryllide(Be$_{12}$Ti)
2. The neutrons generate T, He and void in Be.
He and T, and void induce interaction with each other.
Helium (He)
void
Formation of bubble (void + gas)
Stable, High density
Promotion of growth
3. The evolution of bubbles causes swelling and low thermal conductivity.
Generally, Helium promotes the form and the growth of bubbles.
Swelling
Neutron (n)
Be
Low thermal conductivity
Tritium (T)
Helium (He)
We need to understand He bubbles behavior in Be !
4. The factors affect Bubble behaviors in Be
(ex. internal pressure , bubble size, Temperature)
internal pressure
High pressure
Spherical?
Low pressure
Faceted?
The internal pressure could affect the bubble behaviors (ex. shapes, growths)
In this study, the growth of bubbles in beryllium is investigated from a viewpoint of internal pressure of bubbles.
3

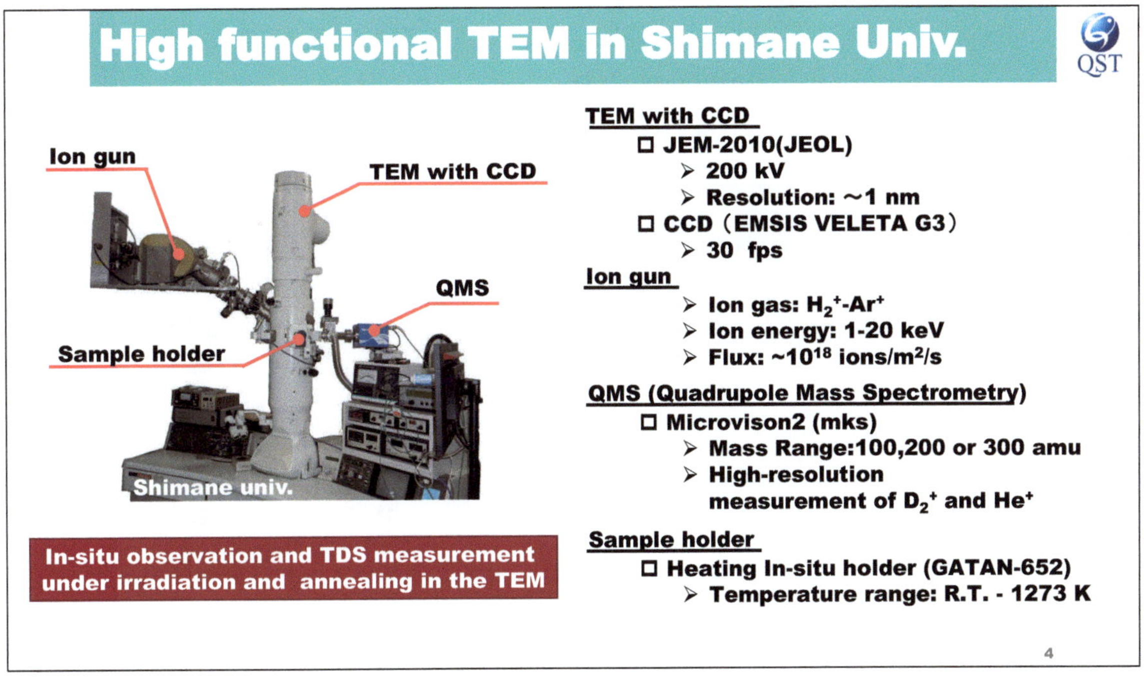
High functional TEM in Shimane Univ.
QST
Ion gun
TEM with CCD
QMS
Sample holder
Shimane univ.
In-situ observation and TDS measurement under irradiation and annealing in the TEM
TEM with CCD
JEM-2010(JEOL)
200 kV
Resolution: ~1 nm
CCD (EMSIS VELETA G3)
30 fps
Ion gun
Ion gas: H_2^+-Ar^+
Ion energy: 1-20 keV
Flux: ~10^{18} ions/m^2/s
QMS (Quadrupole Mass Spectrometry)
Microvison2 (mks)
Mass Range:100,200 or 300 amu
High-resolution measurement of D_2^+ and He^+
Sample holder
Heating In-situ holder (GATAN-652)
Temperature range: R.T. - 1273 K
4

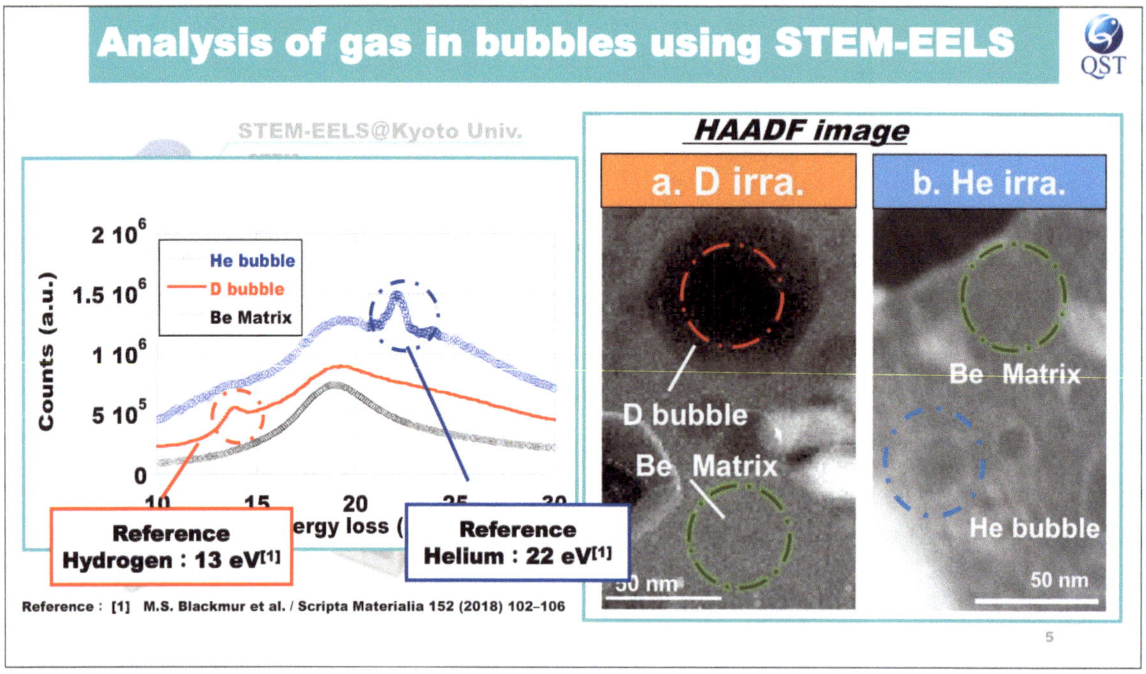
Analysis of gas in bubbles using STEM-EELS
QST
STEM-EELS@Kyoto Univ.
Counts (a.u.)
2 10^6
1.5 10^6
1 10^6
5 10^5
0
He bubble
D bubble
Be Matrix
20
Reference
Hydrogen : 13 eV[1]
Reference
Helium : 22 eV[1]
HAADF image
a. D irra.
b. He irra.
D bubble
Be Matrix
Be Matrix
He bubble
50 nm
Reference : [1] M.S. Blackmur et al. / Scripta Materialia 152 (2018) 102–106
5

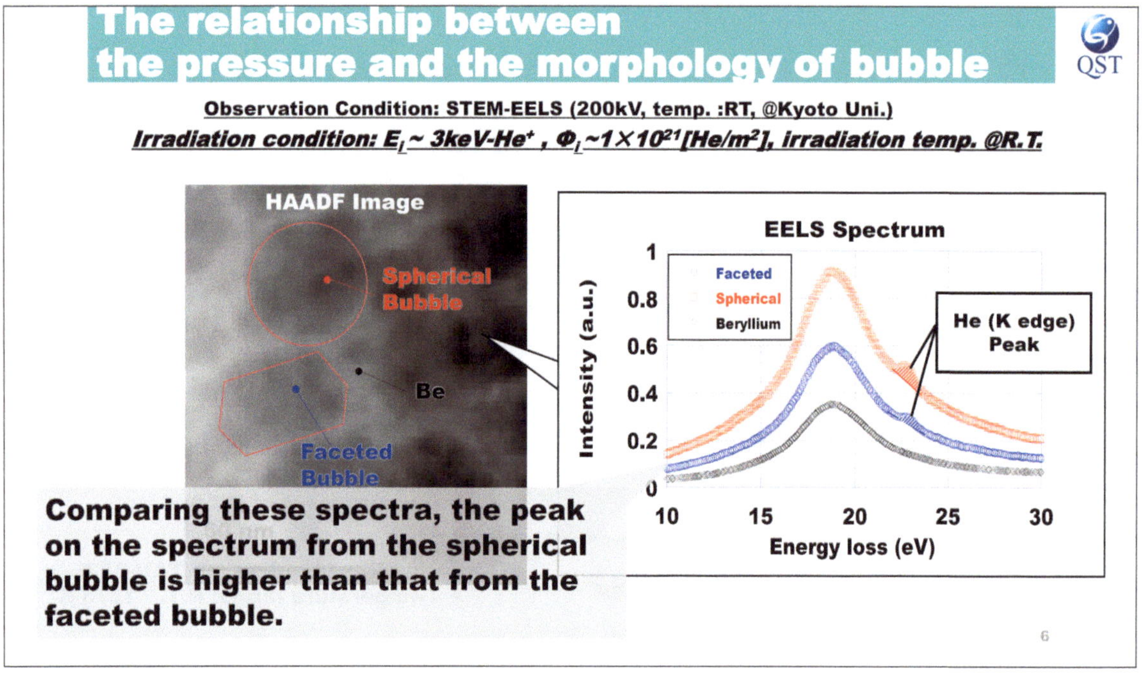
The relationship between
the pressure and the morphology of bubble
QST
Observation Condition: STEM-EELS (200kV, temp. :RT, @Kyoto Uni.)
Irradiation condition: E_i~ 3keV-He⁺ , $Φ_i$~1×10²¹[He/m²], irradiation temp. @R.T.
HAADF Image
Spherical Bubble
Be
Faceted Bubble
EELS Spectrum
Intensity (a.u.)
1
0.8
0.6
0.4
0.2
0
10
15
20
25
30
Energy loss (eV)
Faceted
Spherical
Beryllium
He (K edge) Peak
Comparing these spectra, the peak on the spectrum from the spherical bubble is higher than that from the faceted bubble.
6

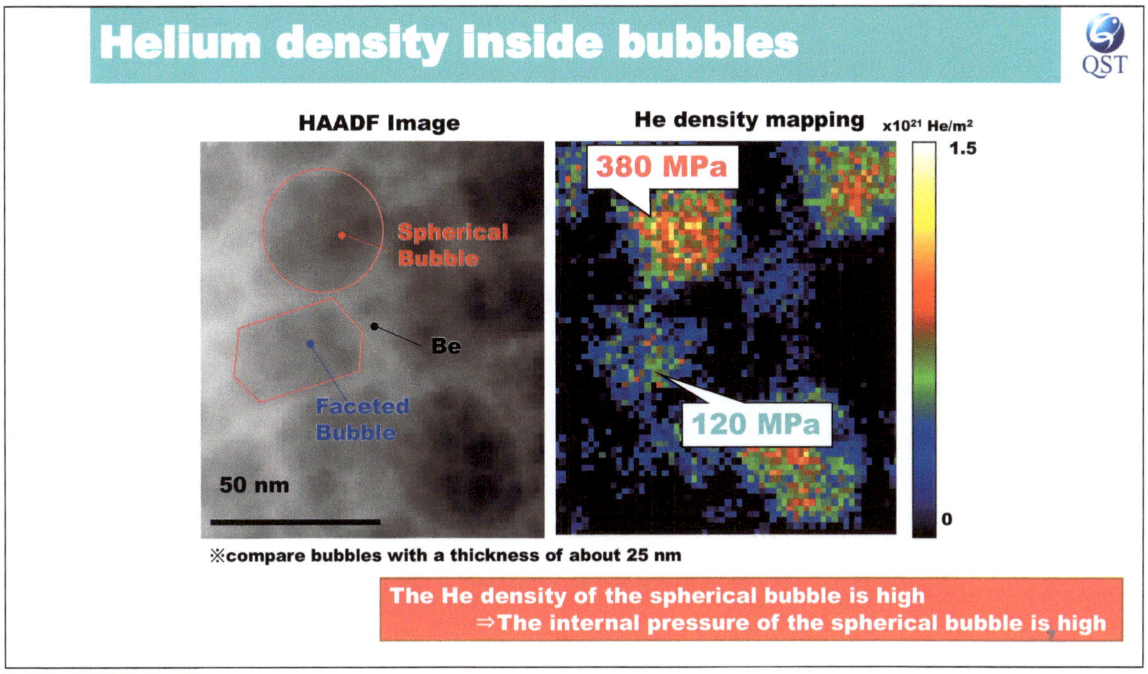
Helium density inside bubbles
QST
HAADF Image
He density mapping
x10²¹ He/m²
1.5
0
Spherical Bubble
Be
Faceted Bubble
50 nm
380 MPa
120 MPa
※compare bubbles with a thickness of about 25 nm
The He density of the spherical bubble is high
⇒The internal pressure of the spherical bubble is high

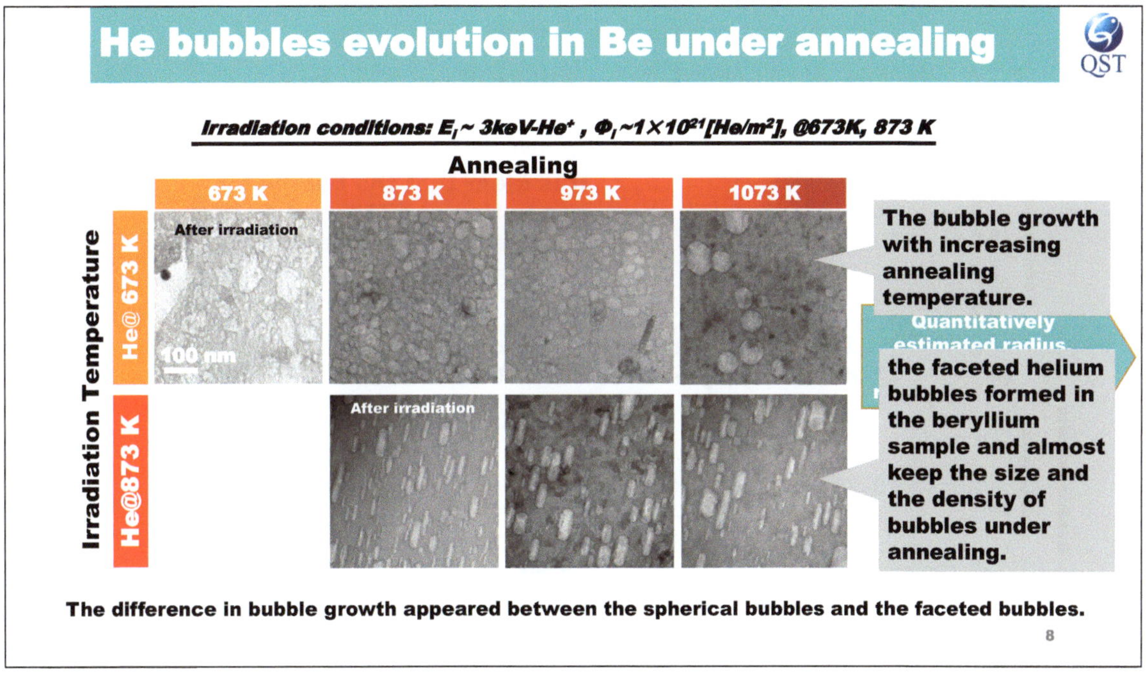
He bubbles evolution in Be under annealing
QST
Irradiation conditions: E_i~ 3keV-He⁺ , Φ_i~1×10²¹[He/m²], @673K, 873 K
Annealing
673 K
873 K
973 K
1073 K
Irradiation Temperature
He@ 673 K
He@873 K
After irradiation
100 nm
After irradiation
The bubble growth with increasing annealing temperature.
Quantitatively
the faceted helium bubbles formed in the beryllium sample and almost keep the size and the density of bubbles under annealing.
The difference in bubble growth appeared between the spherical bubbles and the faceted bubbles.
8

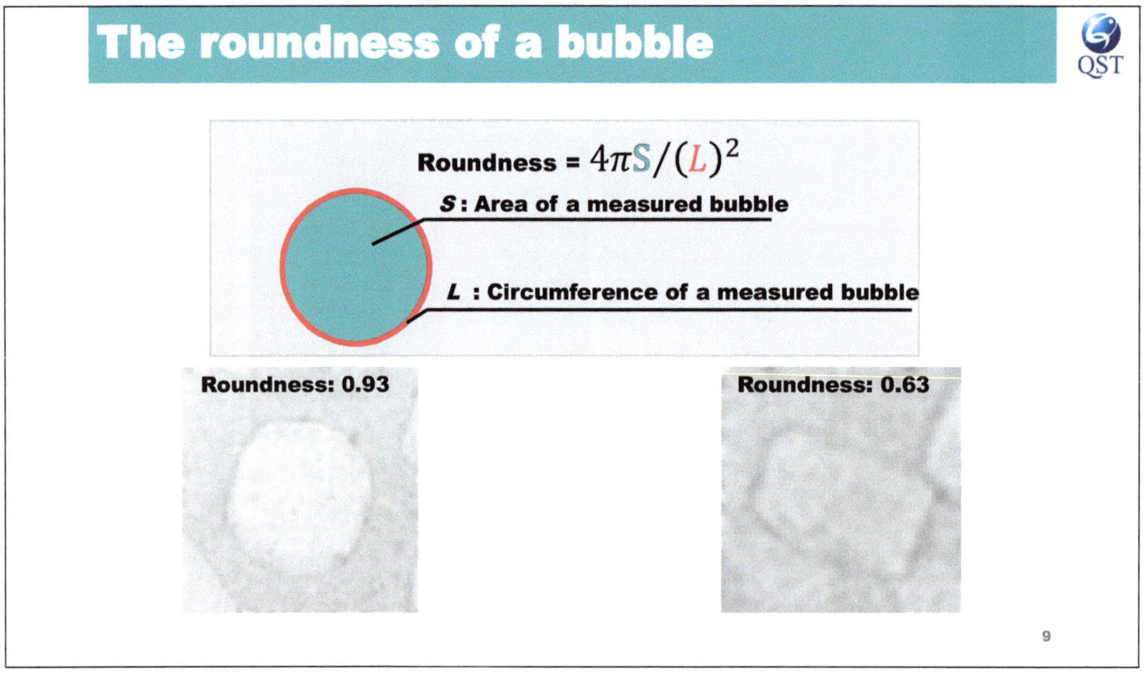
The roundness of a bubble
QST
Roundness = $4\pi S/(L)^2$
S : Area of a measured bubble
L : Circumference of a measured bubble
Roundness: 0.93
Roundness: 0.63
9

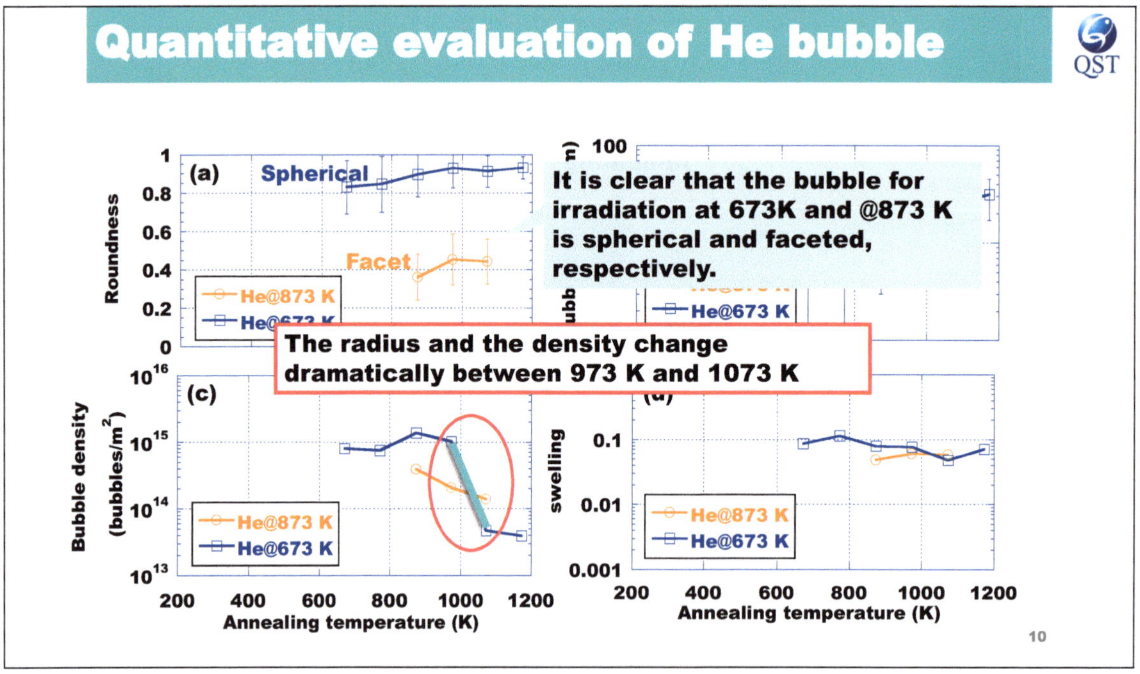
Quantitative evaluation of He bubble
QST
(a)
Spherical
Facet
Roundness
He@873 K
It is clear that the bubble for irradiation at 673K and @873 K is spherical and faceted, respectively.
He@673 K
The radius and the density change dramatically between 973 K and 1073 K
(c)
Bubble density (bubbles/m^2)
swelling
He@873 K
He@673 K
Annealing temperature (K)
10

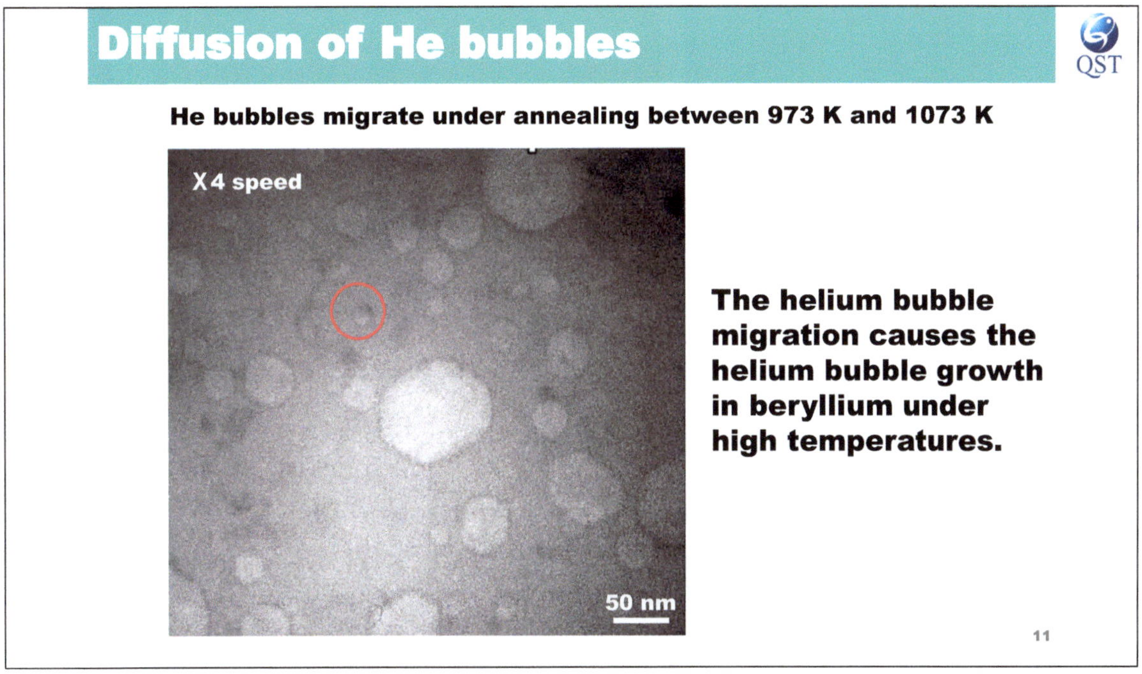
Diffusion of He bubbles
QST
He bubbles migrate under annealing between 973 K and 1073 K
X4 speed
50 nm
The helium bubble migration causes the helium bubble growth in beryllium under high temperatures.
11

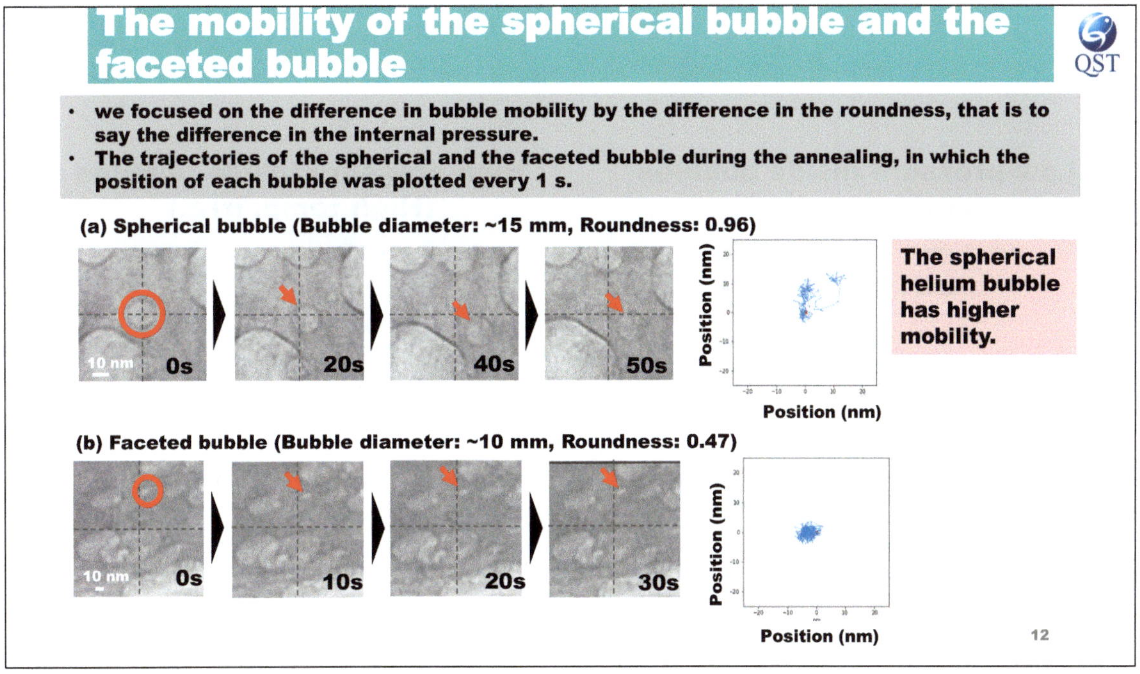
The mobility of the spherical bubble and the faceted bubble
QST
• we focused on the difference in bubble mobility by the difference in the roundness, that is to say the difference in the internal pressure.
• The trajectories of the spherical and the faceted bubble during the annealing, in which the position of each bubble was plotted every 1 s.
(a) Spherical bubble (Bubble diameter: ~15 mm, Roundness: 0.96)
10 nm
0s
20s
40s
50s
Position (nm)
Position (nm)
The spherical helium bubble has higher mobility.
(b) Faceted bubble (Bubble diameter: ~10 mm, Roundness: 0.47)
10 nm
0s
10s
20s
30s
Position (nm)
Position (nm)
12

The mobility of the spherical bubble and the faceted bubble

QST

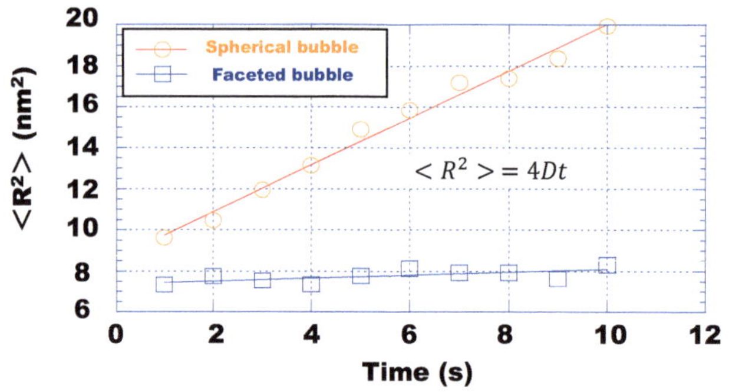

Bubble diffusion coefficient

Spherical Helium bubble: 2.8×10^{-19} m²/s
Faceted Helium bubble: 3.3×10^{-20} m²/s

13

The growth process of He bubble

QST

Low-pressure inside bubbles (Low temp. irra.)

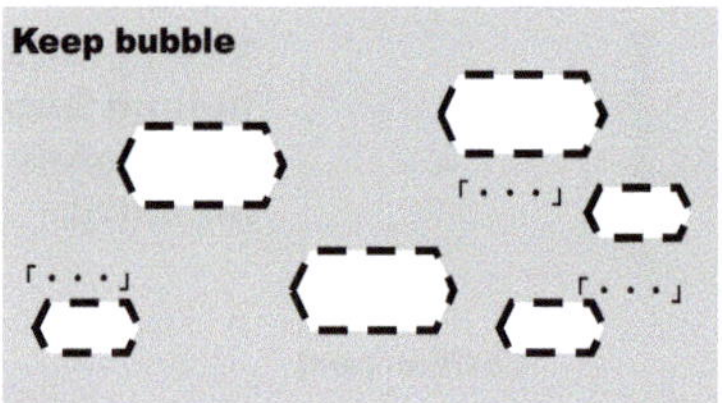

Diffusion coefficient is low

High-pressure inside bubbles (High temp. irra.)

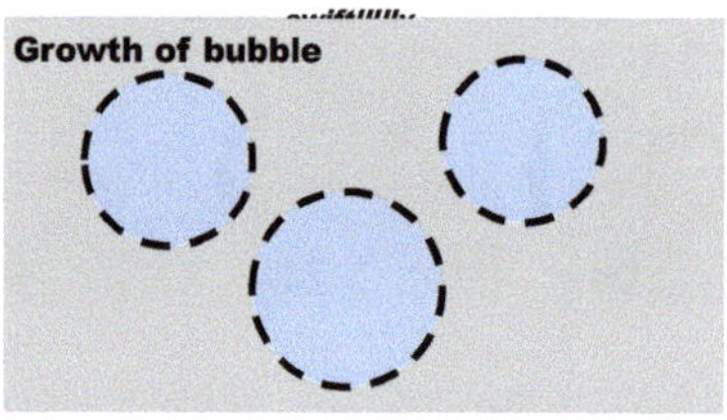

Diffusion coefficient is high

The internal pressure of bubbles would affect the growth of bubble

14

Conclusion

QST

We investigated the relationship between the internal pressure and the growth of bubbles in Be under annealing

Mophology of bubbles and internal pressure of bubbles

- ✓ **STEM-EELS analysis revealed the internal pressure of the spherical bubble and that of the faceted bubble was 293 MPa and 163 MPa, respectively.**

In-situ TEM observation showed the growth of He bubbles under annealing

- ✓ **The bubbles in the Be irradiated with He ions at 673 K is higher roundness than that at 873 K.**
- ✓ **In the Be irradiated with He ions at 673 K, the bubble radius increased sufficiently with decreasing the bubble density at 1073K.**

Diffusion coefficients of the spherical He bubble and the faceted He bubble

- ✓ $D_{spherical}$ = 2.8×10^{-19} m^2/s **for the spherical bubble**
- ✓ D_{facted} = 3.3×10^{-20} m^2/s **for the faceted bubble**

Consequently, the spherical He bubbles with high pressure in Be facilitated bubble growth owing to their high mobility.

QST

Thank you for your attention!

16

6.3 High temperature helium ion irradiation of Be as a rapid-throughput proxy for neutron irradiation experiments

Joanne Sharp, Slava Kuksenko, Graeme Greaves, Jonathan Hinks, Stephen Donnelly

In this study we aim to qualitatively compare the effects of irradiation of Be with 300 keV He ions to the effects of neutron irradiation at 380-600 °C in the 4-year HIgh DOse irradiation of Beryllium (HIDOBE) experiment. Ion irradiation can achieve He implantation loadings in 1-2 hours comparable to those seen after years of transmutation He production in a neutron irradiation environment, and may therefore serve as a rapid material screening method for fusion applications. Ion irradiation, however, causes less displacement damage per He than neutrons. The objective therefore was to find which aspects of the HIDOBE samples would be replicated by He irradiation and which would not.

Microsections of S-65 grade Be 2-3 μm in thickness and $\approx$ 30×15 μm in extent were prepared by focused ion beam (FIB) at UKAEA and subsequently irradiated in the MIAMI-2 (Microscope and Ion Accelerator for Materials Investigation) combined transmission electron microscope (TEM) and ion accelerator at the University of Huddersfield. He^+ ions at 300 kV were used to irradiate samples at three temperatures corresponding to the temperatures in the HIDOBE study: 387 °C, 480 °C and 600 °C (+/- 2 °C) to a fluence of 10^{17} ions/cm^2. After irradiation, samples were cut by FIB parallel to the ion beam to produce sections containing the whole irradiated depth, with the irradiated surface at the top and a band of He at the implantation Bragg peak approximately halfway down the sample. The re-sections were examined using TEM for BF/DF and diffraction investigation of defects, defocus analysis of bubbles, and EDS analysis of precipitates and grain boundary segregation.

Bubbles occurred at the He Bragg peak in grain interiors at all temperatures; bubbles nucleated preferentially on grain boundaries and precipitates, and larger bubbles were seen at these locations. Fine (single nm) spheroidal grain interior bubbles were found in the sample irradiated at 387 °C; 20-30 nm faceted bubbles at 480 °C; and 50-100 nm faceted bubbles with bubbles of >200 nm on grain boundaries in the sample irradiated at 600 °C. These results correspond to the results of the HIDOBE-02 neutron experiment. Dislocation analysis by conventional dark field TEM was complicated by surface Ga contamination from FIB, which is particularly a problem with Be. However, decoration of larger dislocation loops with reprecipitated particles containing Fe and Al found in HIDOBE-02 samples (e.g. Zimber and Vladmirov, J. Nucl. Mat. 2022) would have been observable but was not seen in the He-irradiated samples. This is possibly because He-irradiated samples spent less time at high temperature for Fe and Al to diffuse to the defects (1-2 hours compared with 4 years).

To compare with the re-sectioning study, FIB samples of thickness <200 nm were irradiated in situ at the same temperatures with 37 kV He^+ (to ensure the He Bragg peak fell within the sample thickness). The samples were pre-characterised using Transmission Kikuchi Diffraction so that grain boundary characters were known. Bubbles were found to nucleate more on disordered high-angle grain boundaries than on boundaries with a coincident site lattice-like misorientation, as found in the HIDOBE-02 Be pebbles. In the thin samples, the irradiation effects of the electron beam become significant and cannot be left out of the analysis of high temperature in situ irradiation studies of Be.

We conclude that TEM in situ He ion irradiation of Be can produce some of the main features of neutron irradiation: bubbles with the same shape and size change with temperature, and the same preferential bubble nucleation behaviour on microstructural features. Re-precipitation on dislocations was not reproduced and FIB artefacts complicate dislocation analysis. The effects of the electron beam on this light metal at high temperatures are significant and analysis of this is ongoing.

Presenting Author: Dr Joanne Sharp
Presenting Author's Affiliation: University of Huddersfield, UK

University of HUDDERSFIELD
Inspiring global professionals

Helium ion irradiation studies of beryllium: comparison with neutron irradiation effects

Jo Sharp[1], Graeme Greaves[1], Jonathan Hinks[1], Slava Kuksenko[2], Ramil Gaisin[3], Pavel Vladimirov[3]

1: MIAMI facility, University of Huddersfield, UK
2: UKAEA
3: Karlsruhe Institute of Technology

j.h.sharp@hud.ac.uk

HEA Global Teaching Excellence Award

Why Be?

University of HUDDERSFIELD
Inspiring global professionals

Nuclear fusion:
tritium + deuterium → helium + energy
(clean energy, explore space, etc)

Not a whole lot of tritium around.
We need to use the neutrons from the fusion reaction to "breed" more tritium in breeder modules...

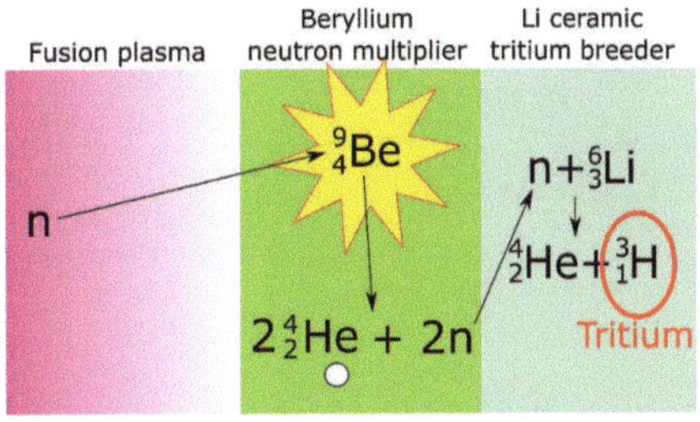

1: https://www.iter.org/sci/FusionFuels

HEA Global Teaching Excellence Award

(Semi-)Ex situ experiments

University of HUDDERSFIELD
Inspiring global professionals

Irradiation with He^+ ions:

- Sample not active
- Fast – good for screening materials
- But less dpa per He

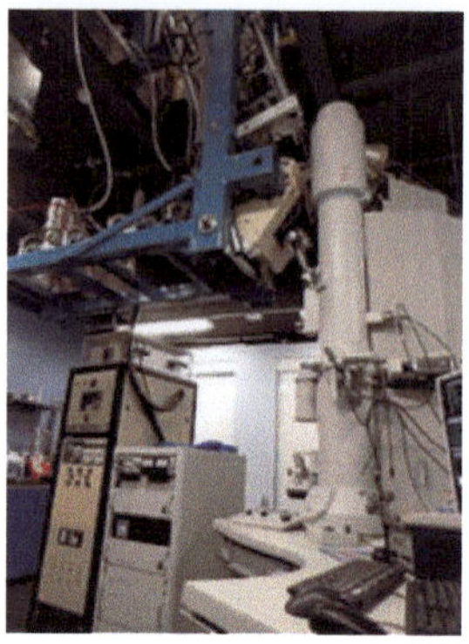

My goal: find out the implications for He^+ as an analogue for neutron irradiation

HEA Global Teaching Excellence Award

Re-lift experiments

University of HUDDERSFIELD
Inspiring global professionals

This is "semi-ex-situ" – it is done in the TEM of MIAMI-2 but when it is irradiated the sample is thick enough to contain the whole irradiated volume.

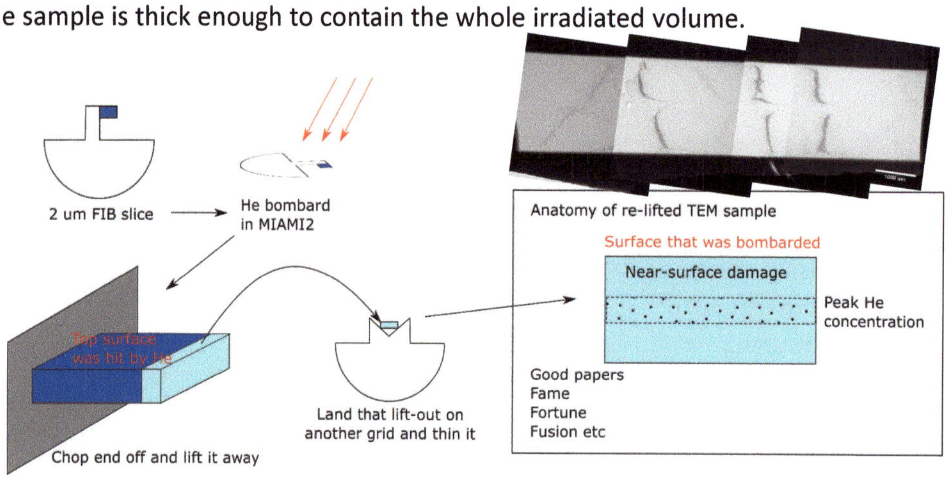

Have to use UKAEA's Be-safe FIB – Royal Mail is an important stakeholder in this project!

HEA Global Teaching Excellence Award

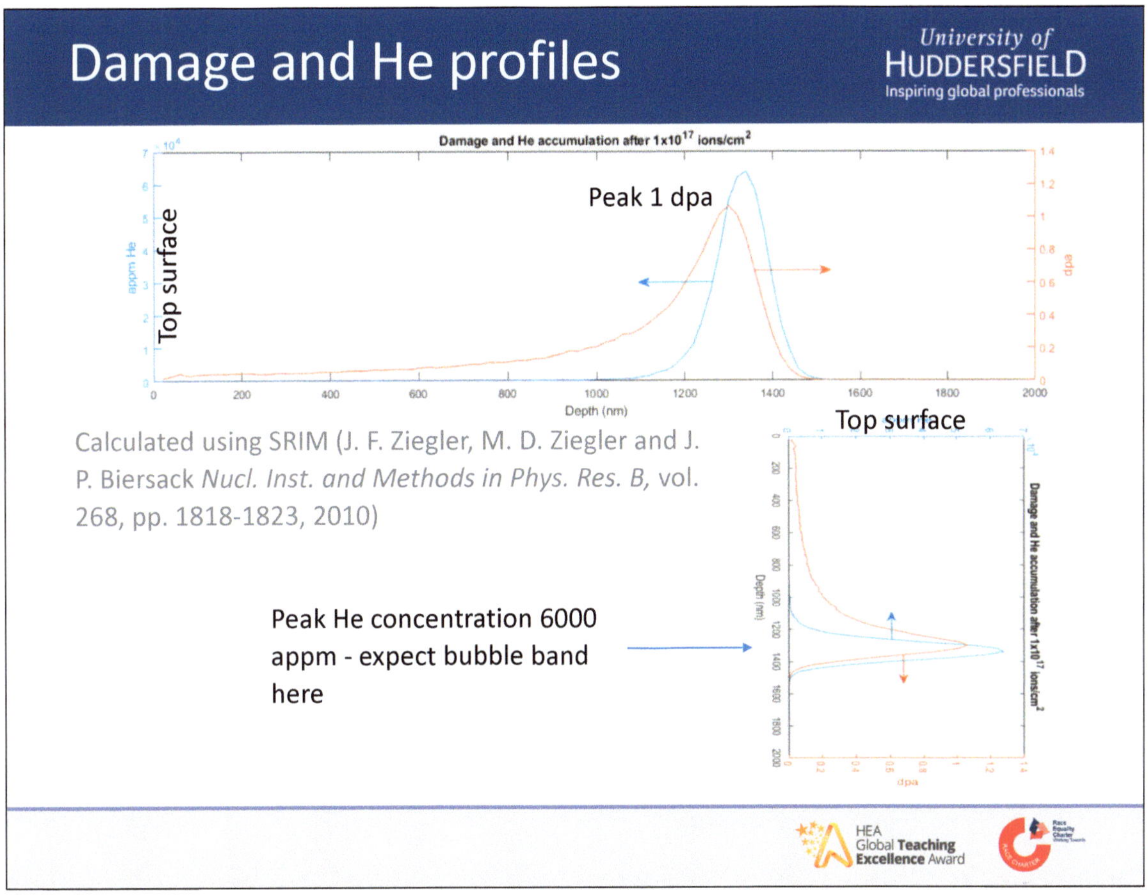
Damage and He profiles
University of HUDDERSFIELD
Inspiring global professionals
Damage and He accumulation after 1x10^17 ions/cm^2
Peak 1 dpa
Top surface
appm He
dpa
Depth (nm)
Calculated using SRIM (J. F. Ziegler, M. D. Ziegler and J. P. Biersack Nucl. Inst. and Methods in Phys. Res. B, vol. 268, pp. 1818-1823, 2010)
Top surface
Peak He concentration 6000 appm - expect bubble band here
HEA Global Teaching Excellence Award

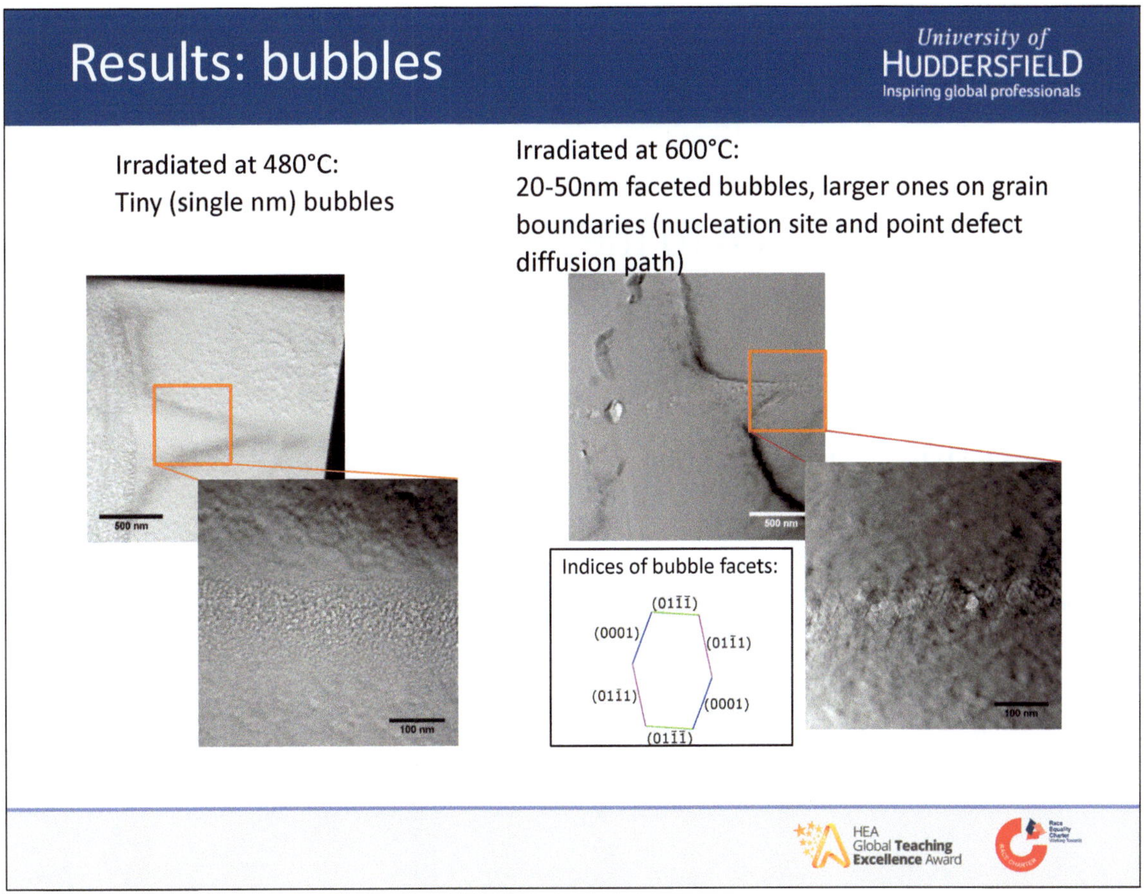
Results: bubbles
University of HUDDERSFIELD
Inspiring global professionals
Irradiated at 480°C:
Tiny (single nm) bubbles
Irradiated at 600°C:
20-50nm faceted bubbles, larger ones on grain boundaries (nucleation site and point defect diffusion path)
500 nm
100 nm
500 nm
100 nm
Indices of bubble facets:
(0001)
(0001)
HEA Global Teaching Excellence Award

Comparison with neutron results

University of HUDDERSFIELD
Inspiring global professionals

480°C: faceted bubbles, 10s of nm

Bubbles are less developed in He ion experiments (critical size for spherical → faceted)

600°C: faceted bubbles, mode ~100nm

Similar shapes in He ion experiments but less growth

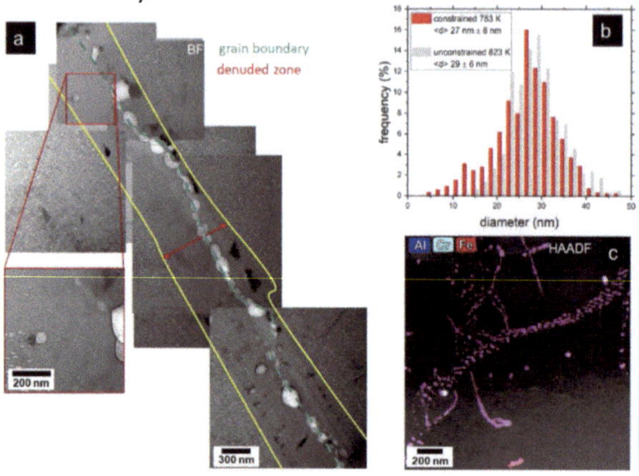

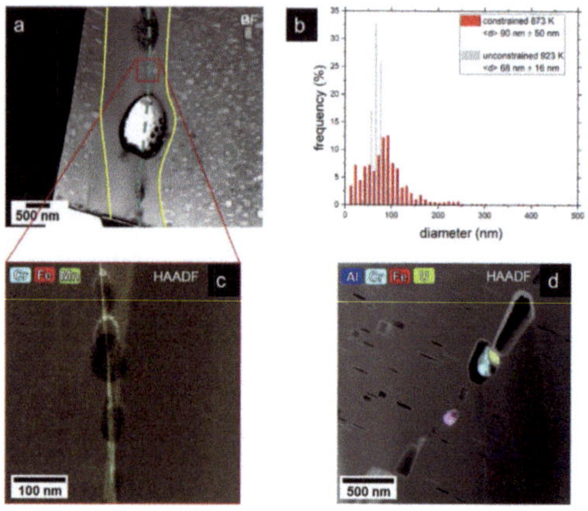

N. Zimber, P. Vladimirov, M. Klimenkov, V. Kuksenko, *J. Nucl. Mat.* **540** (2020) 152374 figs 4 & 6

HEA Global Teaching Excellence Award

Why ions → smaller bubbles

University of HUDDERSFIELD
Inspiring global professionals

- Vacancies help He diffuse to bubbles, and themselves add to the size of the bubble (vacancy = unit of hole-ness)
- Less damage per He from ion irradiation compared to neutron irradiation → fewer point defects → smaller bubbles
- Less time at high temperature → less time to diffuse too

HEA Global Teaching Excellence Award

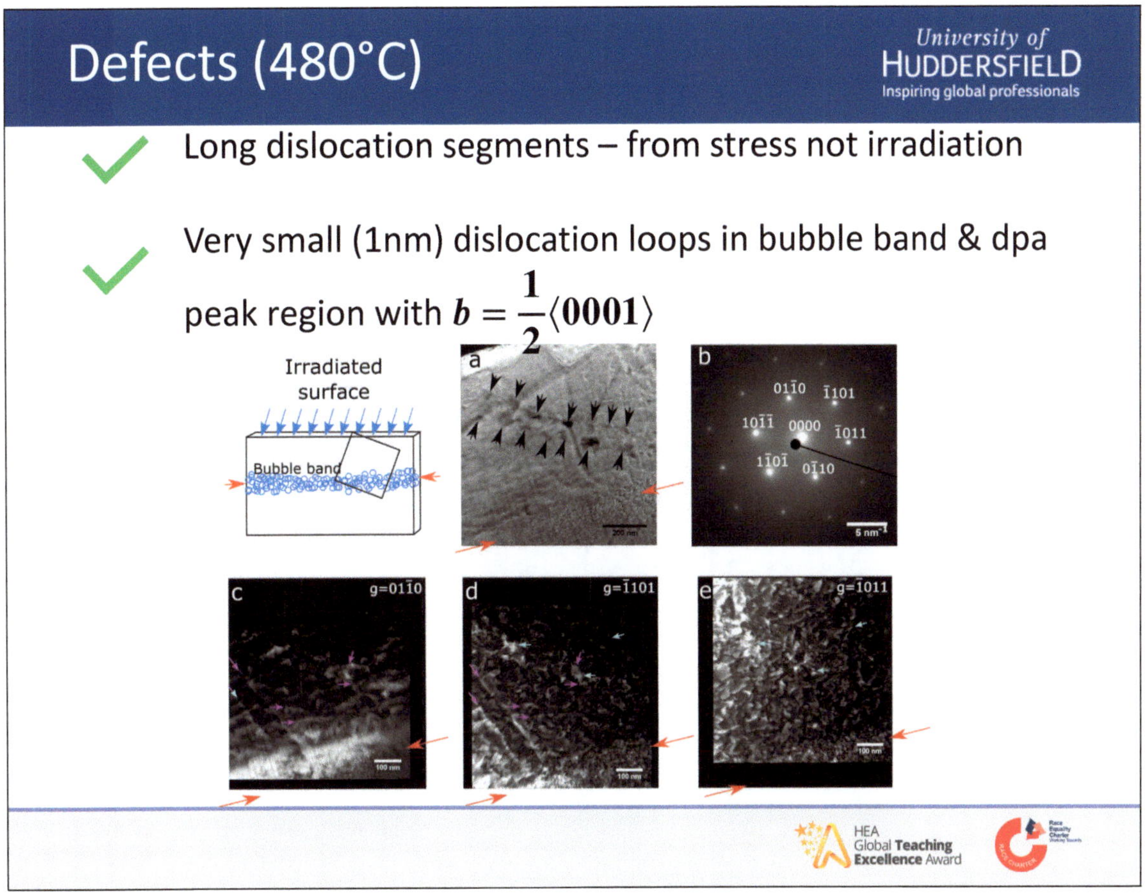
Defects (480°C)
University of HUDDERSFIELD
Inspiring global professionals
Long dislocation segments – from stress not irradiation
Very small (1nm) dislocation loops in bubble band & dpa peak region with $b = \frac{1}{2}\langle 0001 \rangle$
Irradiated surface
Bubble band
a
b
0000
c
g=01$\bar{1}$0
d
g=$\bar{1}$101
e
g=$\bar{1}$011
HEA Global Teaching Excellence Award

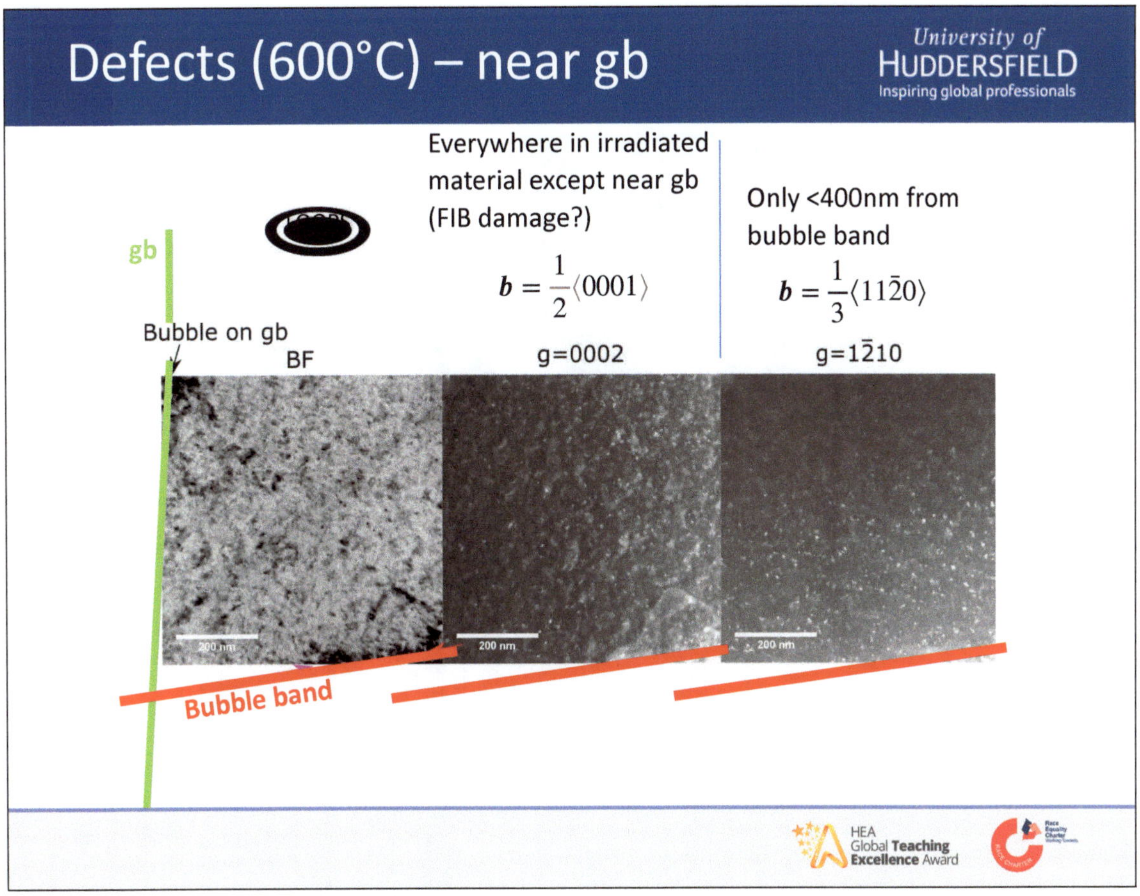
Defects (600°C) – near gb
University of HUDDERSFIELD
Inspiring global professionals
Everywhere in irradiated material except near gb (FIB damage?)
$b = \frac{1}{2}\langle 0001 \rangle$
Only <400nm from bubble band
$b = \frac{1}{3}\langle 11\bar{2}0 \rangle$
gb
Bubble on gb
BF
g=0002
g=1$\bar{2}$10
200 nm
200 nm
200 nm
Bubble band
HEA Global Teaching Excellence Award

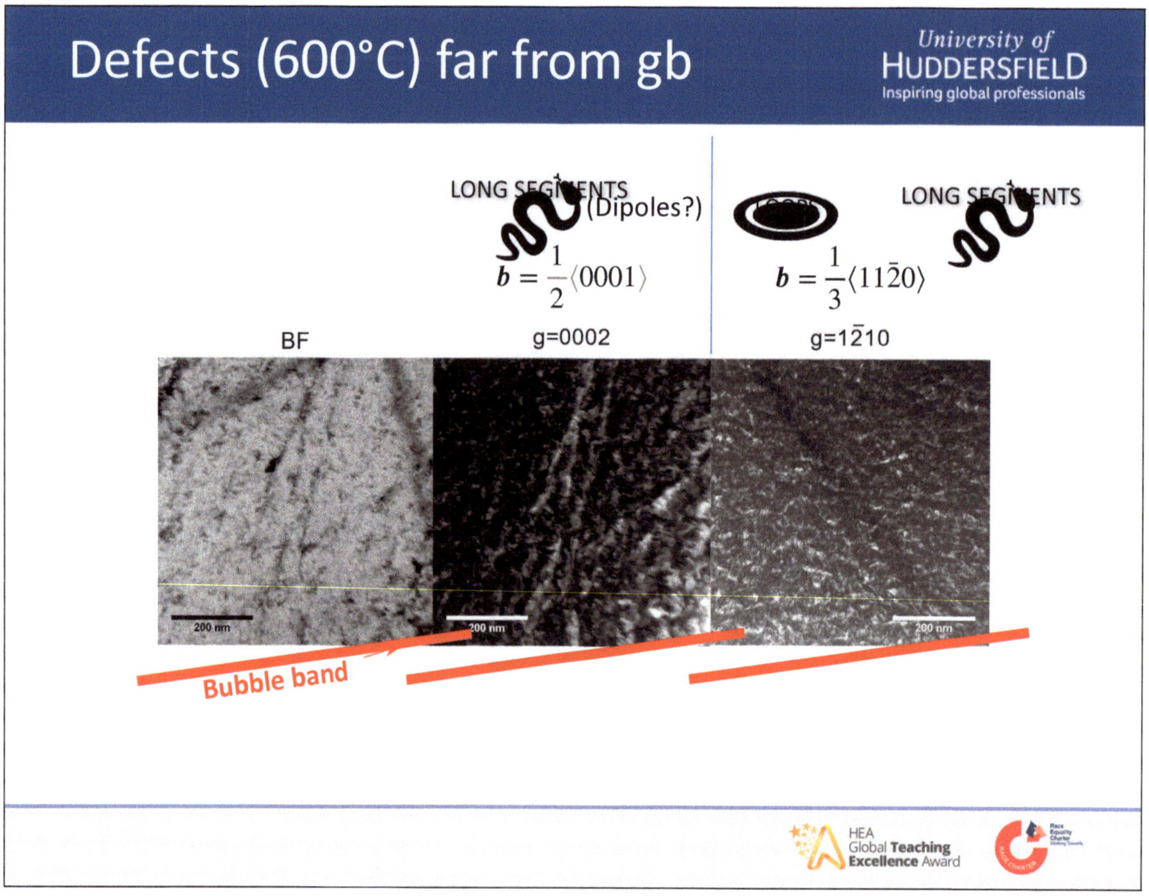
Defects (600°C) far from gb
University of
HUDDERSFIELD
Inspiring global professionals
LONG SEGMENTS
(Dipoles?)
$b = \frac{1}{2}\langle 0001 \rangle$
LONG SEGMENTS
$b = \frac{1}{3}\langle 11\bar{2}0 \rangle$
BF
g=0002
g=1$\bar{2}$10
200 nm
200 nm
200 nm
Bubble band
HEA
Global Teaching
Excellence Award

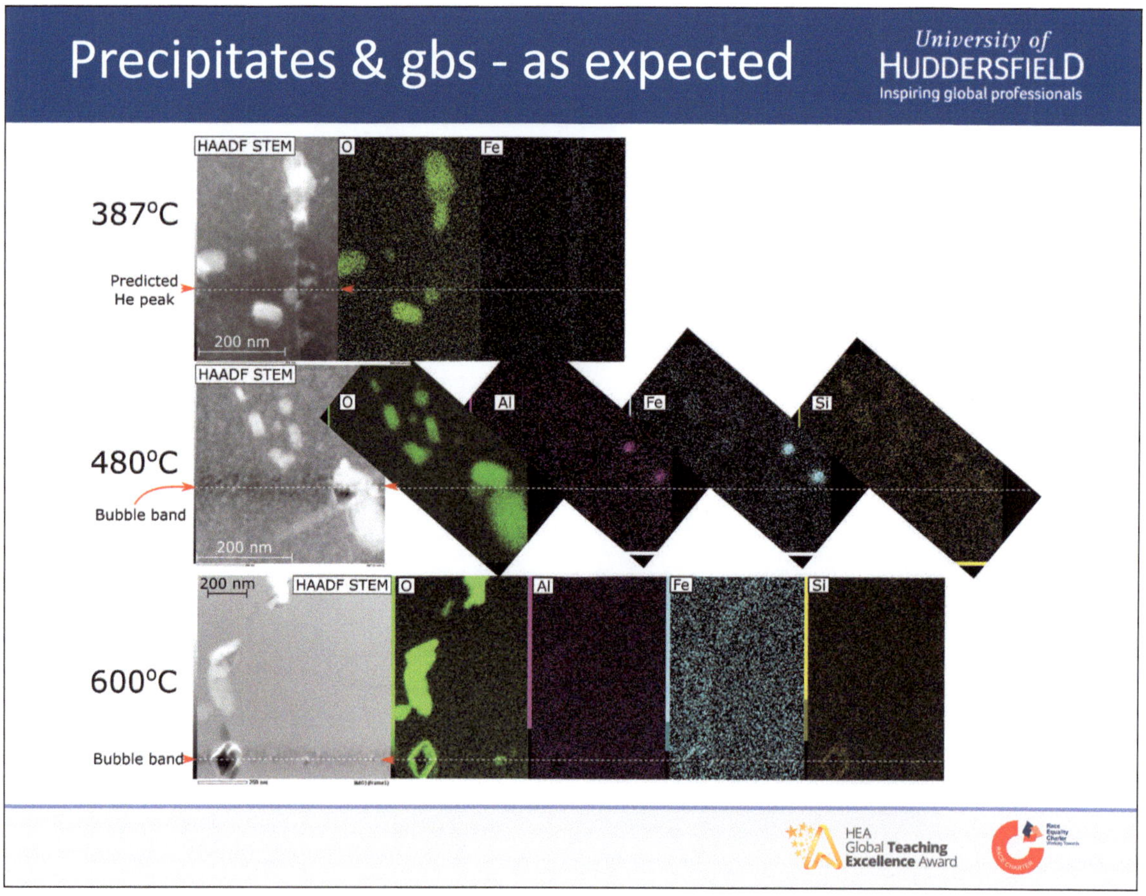
Precipitates & gbs - as expected
University of
HUDDERSFIELD
Inspiring global professionals
HAADF STEM
O
Fe
387°C
Predicted
He peak
200 nm
HAADF STEM
O
Al
Fe
Si
480°C
Bubble band
200 nm
200 nm
HAADF STEM
O
Al
Fe
Si
600°C
Bubble band
HEA
Global Teaching
Excellence Award

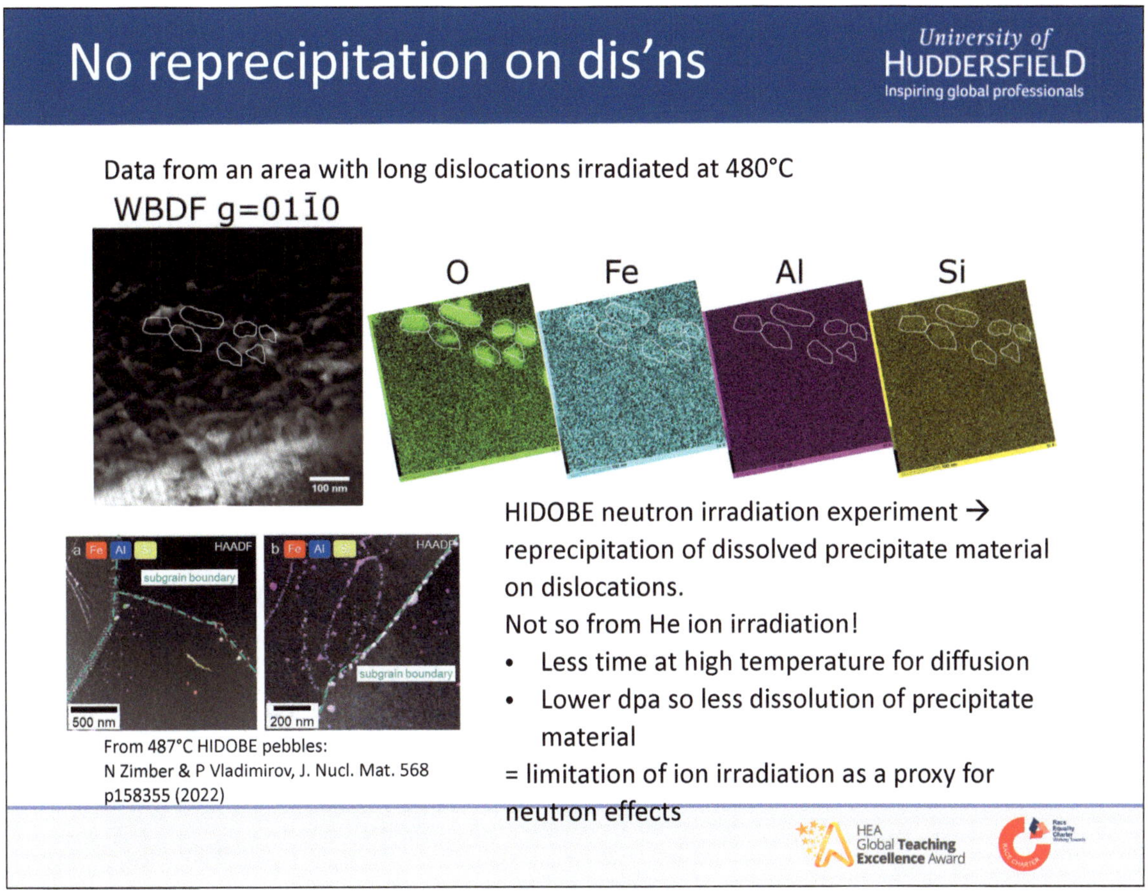
No reprecipitation on dis'ns
University of HUDDERSFIELD
Inspiring global professionals
Data from an area with long dislocations irradiated at 480°C
WBDF g=01$\bar{1}$0
100 nm
O
Fe
Al
Si
a Fe Al Si HAADF
subgrain boundary
500 nm
b Fe Al Si HAADF
subgrain boundary
200 nm
From 487°C HIDOBE pebbles:
N Zimber & P Vladimirov, J. Nucl. Mat. 568 p158355 (2022)
HIDOBE neutron irradiation experiment → reprecipitation of dissolved precipitate material on dislocations.
Not so from He ion irradiation!
• Less time at high temperature for diffusion
• Lower dpa so less dissolution of precipitate material
= limitation of ion irradiation as a proxy for neutron effects
HEA Global Teaching Excellence Award

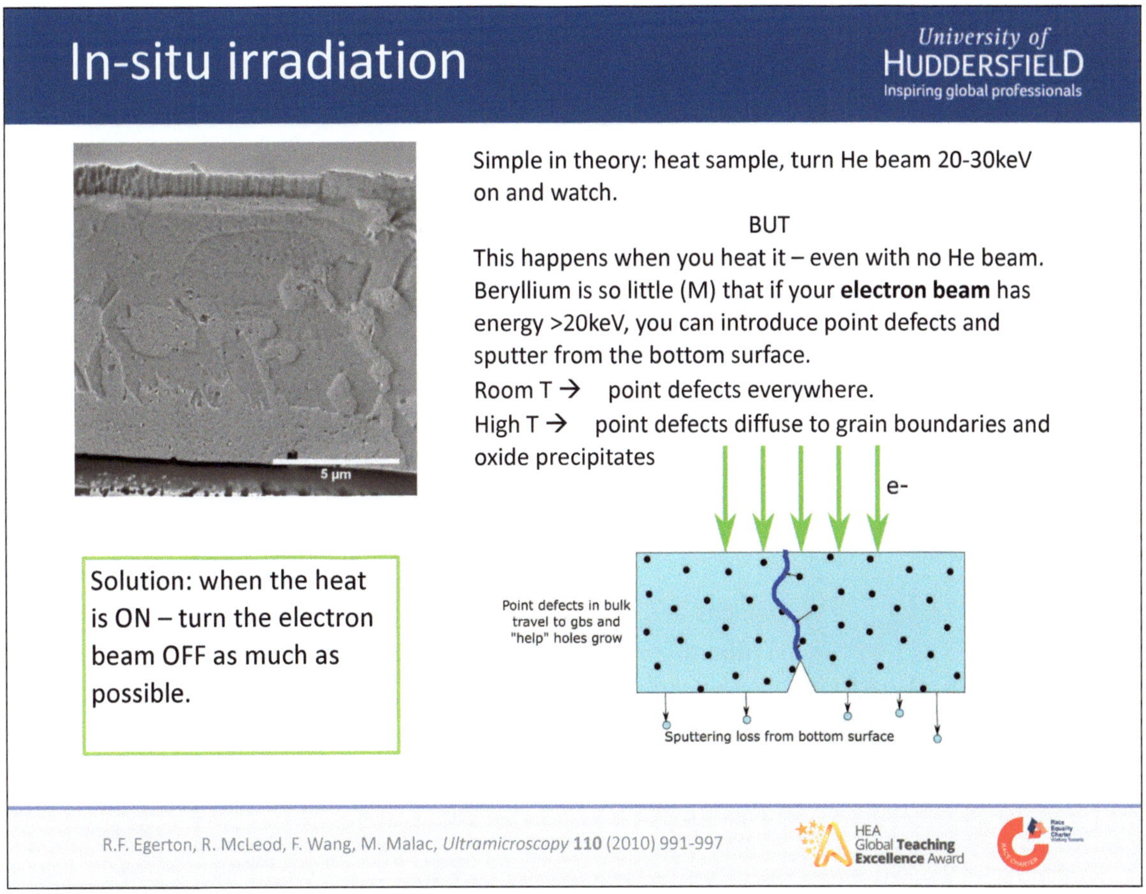
In-situ irradiation
University of HUDDERSFIELD
Inspiring global professionals
5 µm
Simple in theory: heat sample, turn He beam 20-30keV on and watch.
BUT
This happens when you heat it – even with no He beam.
Beryllium is so little (M) that if your electron beam has energy >20keV, you can introduce point defects and sputter from the bottom surface.
Room T → point defects everywhere.
High T → point defects diffuse to grain boundaries and oxide precipitates
e-
Point defects in bulk travel to gbs and "help" holes grow
Sputtering loss from bottom surface
Solution: when the heat is ON – turn the electron beam OFF as much as possible.
R.F. Egerton, R. McLeod, F. Wang, M. Malac, Ultramicroscopy 110 (2010) 991-997
HEA Global Teaching Excellence Award

Holing – still useful data

University of HUDDERSFIELD
Inspiring global professionals

Electron beam damage IS irradiation damage – could we use this to get closer appm/dpa to neutron irradiation for Be?!

But have to balance with mass loss (nothing of TEM specimen left to see!)

For now – stick with He – maybe use e-beam effect in later experiments (semi-ex-situ??)

HEA Global Teaching Excellence Award

Bubbles especially at gbs

University of HUDDERSFIELD
Inspiring global professionals

In situ irradiation at 400°C, 20kV He
Single nm bubbles everywhere, larger arranged on grain boundaries

Defocus +1200nm

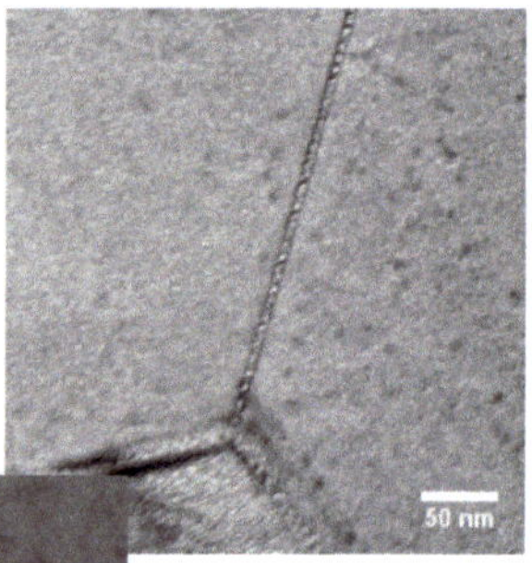

Pre-annealing

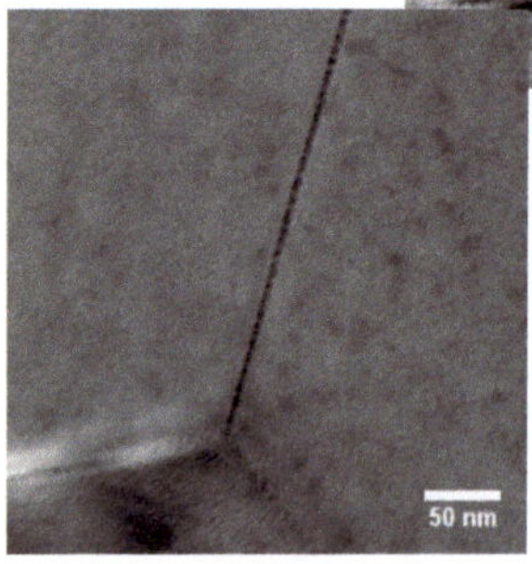

Defocus -1700nm

Ga EDS map – Ga is present (see later) but wider than the bubble contrast and continuous.
Bubbles are still visible through Ga after annealing.

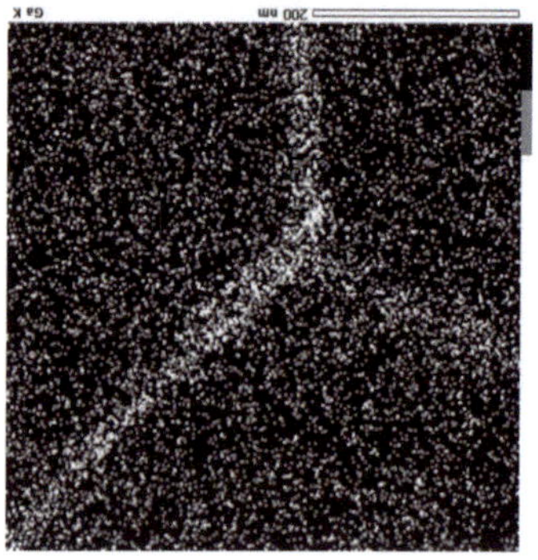

HEA Global Teaching Excellence Award

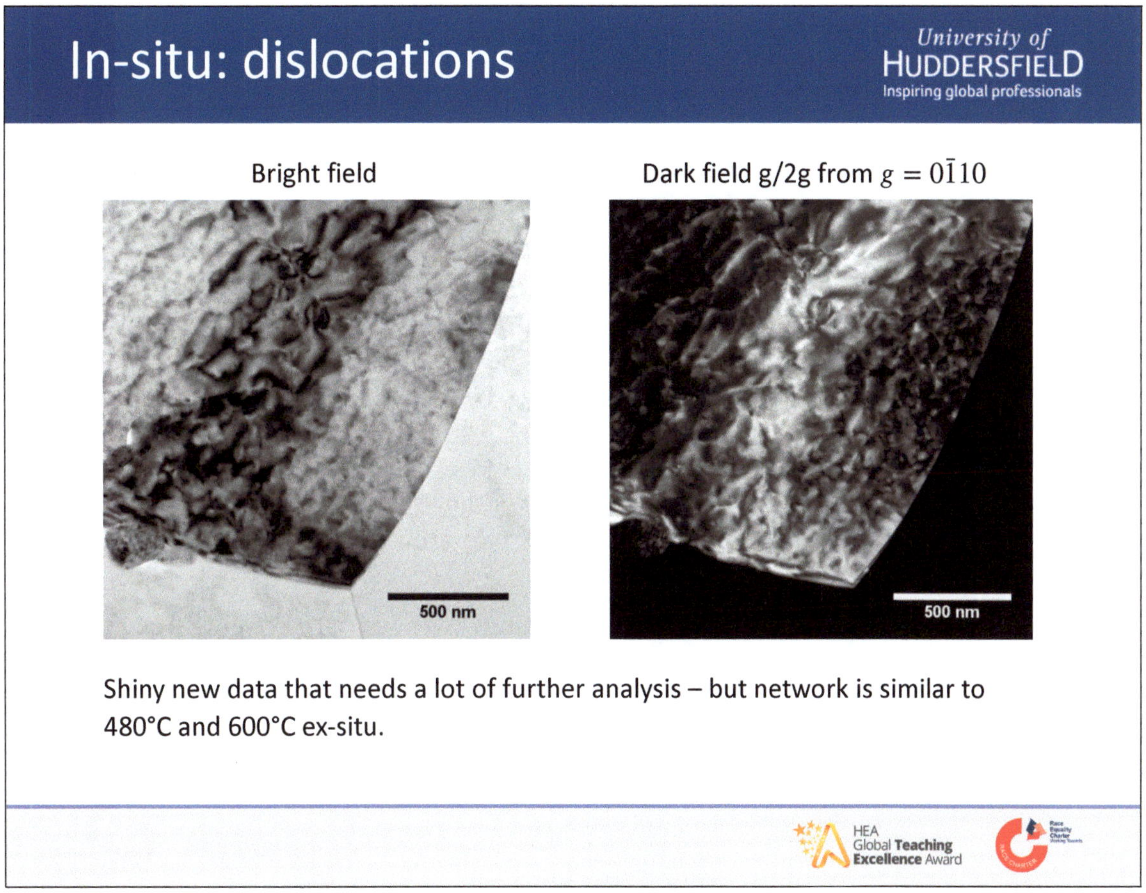
In-situ: dislocations
University of HUDDERSFIELD
Inspiring global professionals
Bright field
Dark field g/2g from $g = 0\bar{1}10$
500 nm
500 nm
Shiny new data that needs a lot of further analysis – but network is similar to 480°C and 600°C ex-situ.
HEA Global Teaching Excellence Award

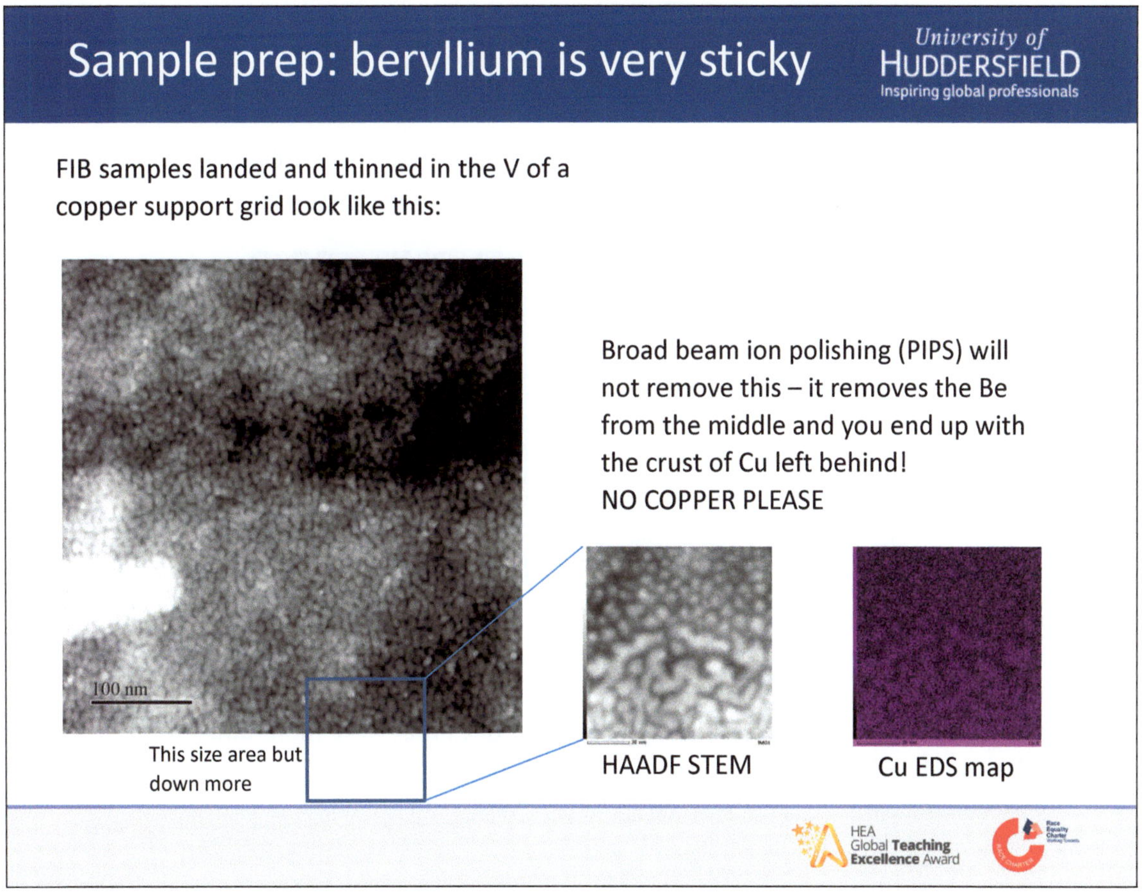
Sample prep: beryllium is very sticky
University of HUDDERSFIELD
Inspiring global professionals
FIB samples landed and thinned in the V of a copper support grid look like this:
100 nm
This size area but down more
Broad beam ion polishing (PIPS) will not remove this – it removes the Be from the middle and you end up with the crust of Cu left behind!
NO COPPER PLEASE
HAADF STEM
Cu EDS map
HEA Global Teaching Excellence Award

Ga also sticks

HAADF STEM | Higher mag | Ga EDS map

Oxide precipitate

Hole

What are these??

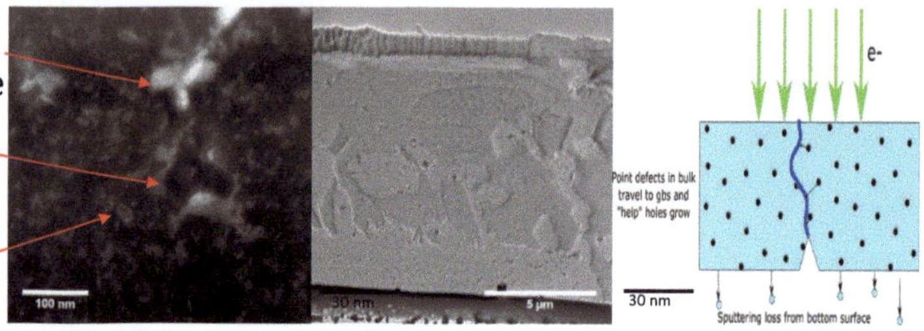

Adherent Ga looks like dislocations – this is not helpful for imaging dislocations!

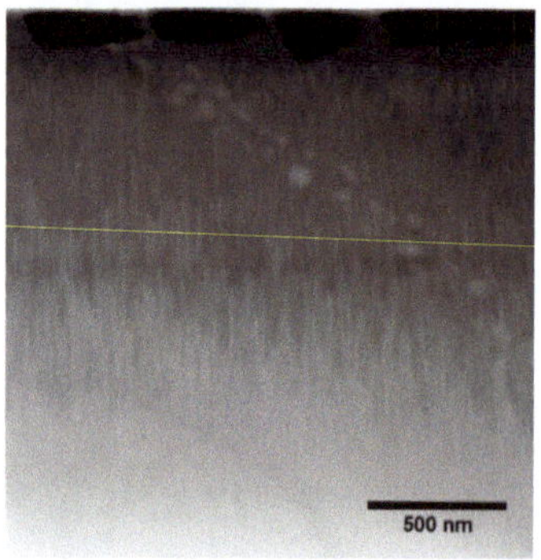

This is not a Ga/Be phase (phase diagram shows they don't mix) – it is Ga running around on the surface.

Broad beam ion milling removes Ga but also causes severe curtaining artefacts

Removing Ga

Annealing at 300°C for 1 hour in vacuum (e.g. in-situ holder) makes Ga migrate to grain boundaries. This obscures the grain boundaries so do work on those and **then** anneal.

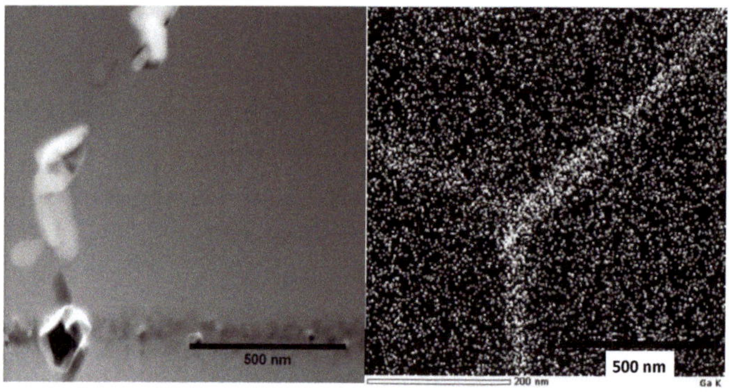

Prismatic slip systems activate at around 200°C – so some dislocations can cross-slip and escape from the TEM sample surfaces during annealing. Need to bear this in mind when observing dislocations after annealing – we are seeing fewer than there were originally, and the ones left were the ones that were more tangled. BUT – can still observe trends between samples.

Conclusions

University of HUDDERSFIELD Inspiring global professionals

- Bubbles are nm-size at 480°C, larger and more faceted at 600°C – facets are slip planes and terraces thereof. Bubbles grow larger on grain boundaries and precipitates.
- Bubbles are smaller and less developed in He ion experiments than in neutron experiments – less time at high temperature, and less point defects
- The dislocation network is also less developed in He ion than neutron experiments – loops have not had time and energy to grow.
- Reprecipitation on dislocations is not seen in He ion work – less displacement from ppts and less diffusion
- Sample prep: keep away from Cu grids and from the PIPS.
- Treat as a very beam-sensitive specimen when hot!

HEA Global Teaching Excellence Award

Acknowledgements

University of HUDDERSFIELD Inspiring global professionals

- Slava Kuksenko and Max Rigby-Bell for FIB at UKAEA which has been funded by and is part of the UK's National Nuclear User Facility and Henry Royce Institute for Advanced Materials and the RCUK Energy Programme (Grant No. EP/T012250/1
- The team at Karlsruhe Institute of Technology for collaboration and letting us use their figures: Nikolai Zimber, Pavel Vladimirov, many others – "This work has been carried out within the framework of the EUROfusion Consortium, funded by the European Union via the Euratom Research and Training Programme (Grant Agreement No 101052200 — EUROfusion). Views and opinions expressed are however those of the author(s) only and do not necessarily reflect those of the European Union or the European Commission. Neither the European Union nor the European Commission can be held responsible for them"
- Everybody at MIAMI for helping keep the ions coming
- EPSRC for MONEY: EP/T027193/1 for my project, EP/M028283/1 for building MIAMI-2 in the first place

HEA Global Teaching Excellence Award

6.4 Tritium content in partially oxidized neutron-irradiated beryllium pebbles

Rudolfs Janis Zabolockis and Elina Pajuste

Beryllium is a metal with a broad scope of applications in science and industry. Particularly it finds use in nuclear fusion reactors. There beryllium can be used as a neutron multiplier, and multiple studies have been published regarding its use as a plasma-facing material. However, reactions with neutrons yield gas inclusions (such as the radioactive isotope of hydrogen–tritium) in the metal and pose health risks in case of reactor malfunction.

In this study, neutron-irradiated beryllium pebbles ($\approx$1 mm diameter, mass $\approx$1 mg, irradiation for 294 days with a neutron (E > 0,1 MeV) fluence $(3-4) \cdot 10^{25}$ m^{-2}, at a temperature range from 423 to 823 K) have been investigated. The pebbles were thermally treated in thermogravimetric/ differential thermal analysis furnace SEIKO 6300 ExSTAR, up to 1548 K, heating rate 10 K/min in airflow (12 L/h).

Tritium in the air flow was assessed using tritium edelgas monitor TEM2100 (DDH32 detector, P10 counting gas, 4.3 kV). The thermally treated pebbles were dissolved in concentrated H_2SO_4 and distilled. The distillate was measured using a liquid scintillation counter (LSC) PerkinElmer Tri-Carb 2900TR. The complete tritium inventory was assessed by combining data from tritium monitoring as well as LSC.

Thermal treatment of beryllium pebbles yielded a mass increase of the pebbles due to oxidation and tritiated gas release. Beryllium pebbles after thermal treatment were coated with an oxide layer, however, by mass it was determined that they have not been fully oxidized. Dissolution in concentrated H_2SO_4 and subsequent distilling yielded tritiated water. Therefore, it must be emphasized that neutron-irradiated beryllium pebbles (as described before) still contains tritium even after thermal treatment up to 1548K and partial oxidation.

Presenting Author: Rudolfs Janis Zabolockis
Presenting Author's Affiliation: University of Latvia, Riga, Latvia
Email: rudolfs_janis.zabolockis@lu.lv

UNIVERSITY OF LATVIA
INSTITUTE OF CHEMICAL PHYSICS
BeWS-16
2023
UNIVERSITY OF LATVIA
FACULTY OF CHEMISTRY
Tritium content in partially oxidized neutron-irradiated beryllium pebbles
Rudolfs Janis Zabolockis1,2*, Elina Pajuste1,2
1University of Latvia, Institute of Chemical Physics
2 University of Latvia, Faculty of Chemistry
*rudolfs_janis.zabolockis@lu.lv
16th International Workshop on Beryllium Technology
19-20 October 2023, Santander, Spain

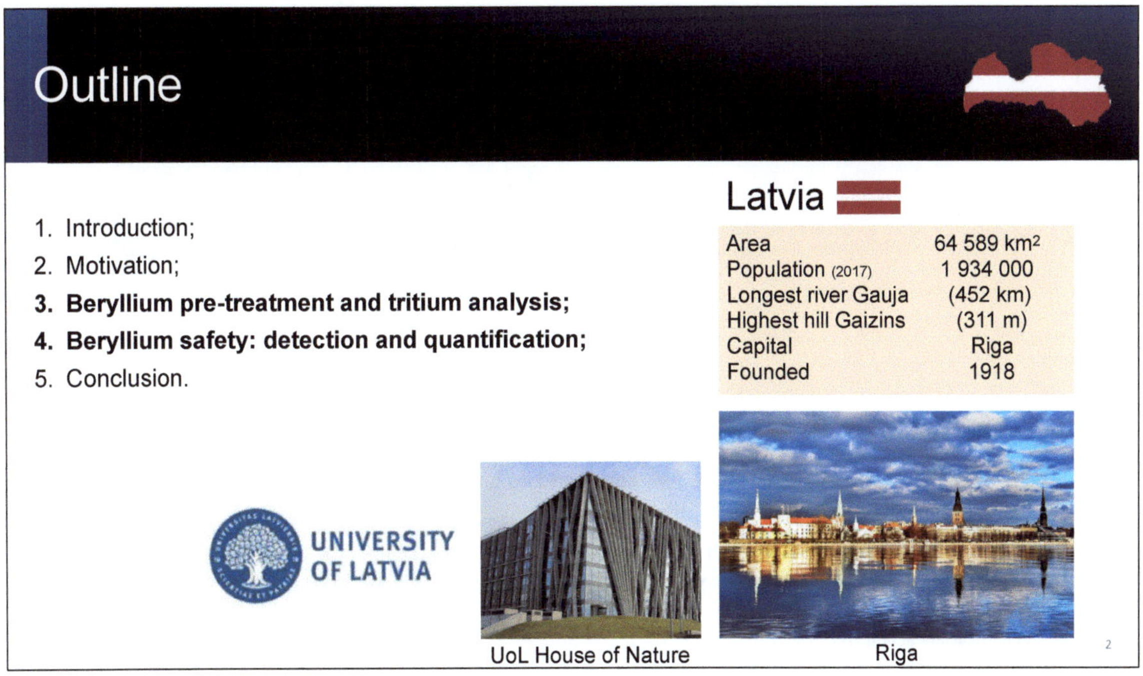
Outline
1. Introduction;
2. Motivation;
3. Beryllium pre-treatment and tritium analysis;
4. Beryllium safety: detection and quantification;
5. Conclusion.
Latvia
Area 64 589 km2
Population (2017) 1 934 000
Longest river Gauja (452 km)
Highest hill Gaizins (311 m)
Capital Riga
Founded 1918
UNIVERSITY OF LATVIA
UoL House of Nature
Riga
2

Introduction

- **Beryllium** can be used as a **plasma facing material** and as a **neutron multiplier**;
- After temperature excursions in an oxidizing environment Be can form BeO microparticles. If Be has been under neutron irradiation, **it contains tritium**;
- Tritium is a **radioactive isotope of hydrogen,** usable as a fusion fuel (**waste product** of nuclear **fission** reactors);

Fig. 2. **ITER water cooling system [2]**

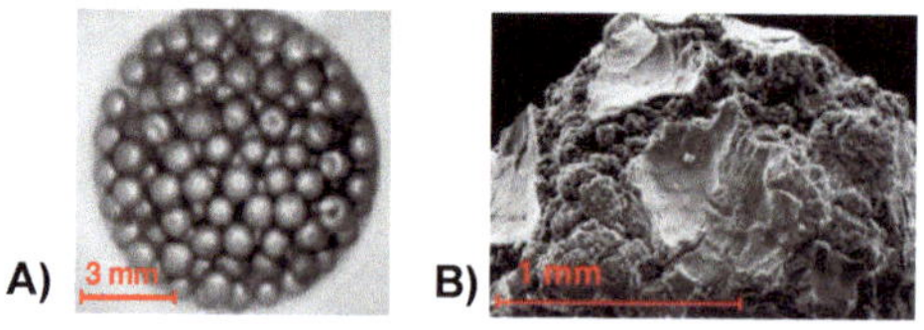

Fig. 3. **A) Be 1mm pebbles, B) thermally treated Be SEM [3]**

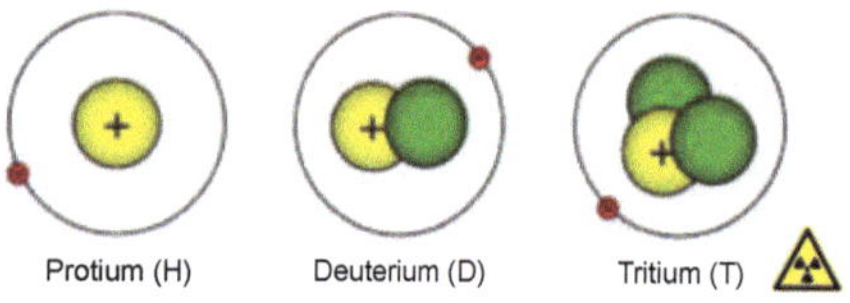

Fig. 1. **Hydrogen isotopes [1]**

[1] Accessable: https://www.giss.nasa.gov/research/briefs/2006_schmidt_03/ (reviewed 19.10.2023.)
[2] ITER - the new way to energy https://www.iter.or
[3] Zabolockis et al.. 9th International Conference on Radiation in Various Fields of Research. 14-18.06.2021., Montenegro

3

Motivation I

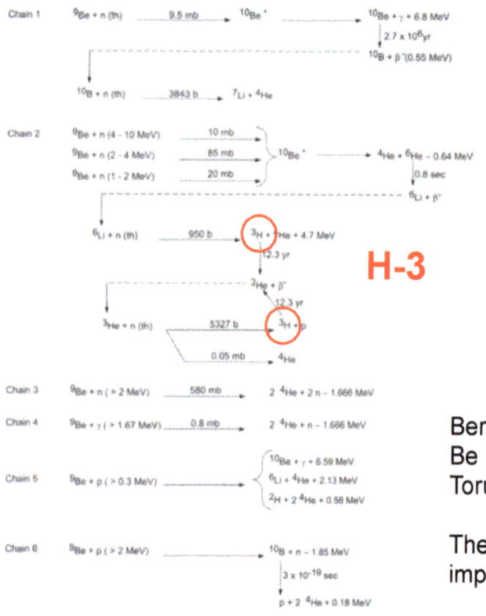

Fig. 1. **Be and particle interaction [1]**

Fig. 2. **Be in the Joint European Torus [2]**

Beryllium has been proposed as neutron multiplier.
Be has been used as plasma facing material in the Joint European Torus [2].

Therefore, the work with Be and Be containing materials is important and some aspects still need to be characterized in detail.

[1] [6]. J. E. Evans, "Reaction Products in High nvt Irradiated Beryllium", 1956, United States Atomic Energy Commission, IDO-16364.
[2] S.Brezinsek et al., Nucl. Fusion 54 (2014) 103001 (11pp) doi:10.1088/0029-5515/54/10/103001

4

Motivation II

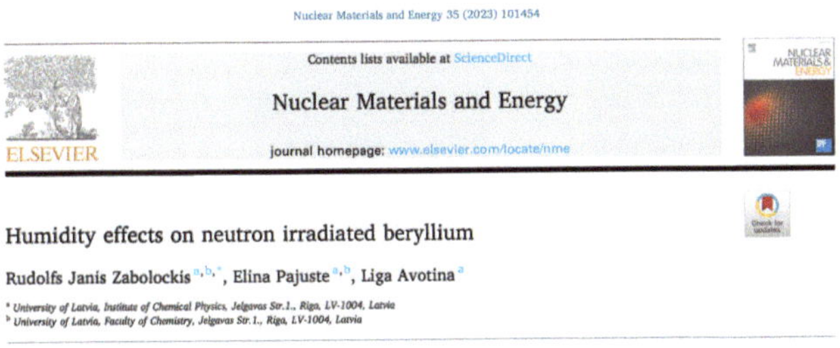
Nuclear Materials and Energy 35 (2023) 101454

Contents lists available at ScienceDirect

Nuclear Materials and Energy

journal homepage: www.elsevier.com/locate/nme

Humidity effects on neutron irradiated beryllium

Rudolfs Janis Zabolockis[a,b,*], Elina Pajuste[a,b], Liga Avotina[a]

[a] University of Latvia, Institute of Chemical Physics, Jelgavas Str.1., Riga, LV-1004, Latvia
[b] University of Latvia, Faculty of Chemistry, Jelgavas Str.1., Riga, LV-1004, Latvia

H-3 (T) can be removed from Be by thermal treatment. Assuming LOVA/ LOCA conditions (as reviewed in previous research),

Is there any tritium left?

In addition – while working with beryllium, it is important to monitor the area. Therefore, the Be contamination monitoring is ***performed prior and after*** the experimental work.

Beryllium contamination testing methods – to assess occupational and environmental toxic risks

5

Experimental

Beryllium pre-treatment and tritium analysis

Analysis scheme

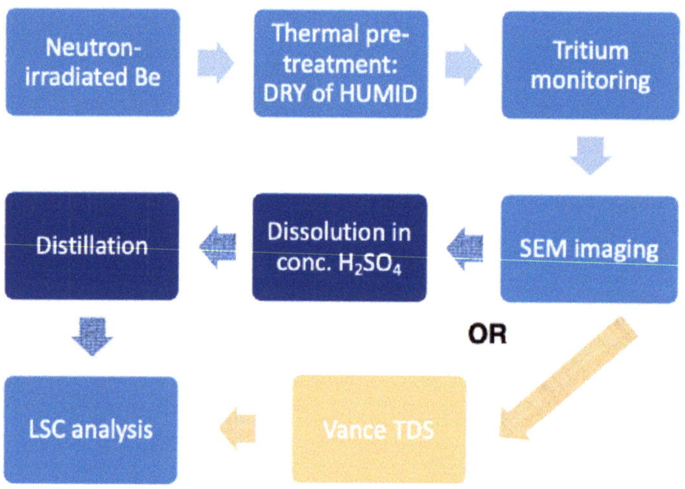

7

Be pre-treatment

Samples:

Be pebbles, **neutron-irradiated** (E > 0.1 MeV for 294 days, fluence 3-4 · 10^{25} m^{-2}, at a temperature range from 423 to 823 K, 2004 PBA) Ø ~ 1mm produced by Rotating Electrode Process (REP) [1].

Methods:

1. Thermogravimetric/Differential Thermal Analysis (TG/DTA) with SEIKO Exstar 6300; air flow 12 L/h, heating rate 10K/min, up to 1548K.
2. Scanning electron microscopy with Hitachi S-4800 equipped with EDX system Bruker XFlash Quad 5040 123eV.

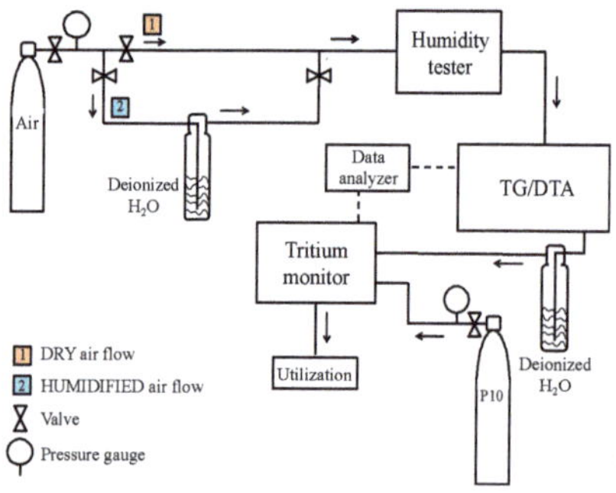

Sample pre-treatment parameters:

- DRY (RH<5%)
- HUMIDIFIED (RH>95%)

Fig. 1. **Beryllium thermal treatment system with humidity control [2]**

8

[1] Pajuste, E., Kizane G., Avotiņa L., Zariņš A. *Journal of Nuclear Materials*. 2015. 465. 293-300.
[2] Zabolockis, R. J., Pajuste, E., Avotina, L. *Journal of Nuclear Materials and Energy*. 2023. 35. 101454

Tritium analysis

Methods:

1. Dissolution in conc. H_2SO_4 and tritium monitoring with TEM2100A (Fig. 1.)
2. Vance device for total tritium release (TDS) (Fig. 3.)
3. Tritium measurement using liquid scintillation (LSC) PerkinElmer Tricarb Tr-2900 (Fig. 2.)

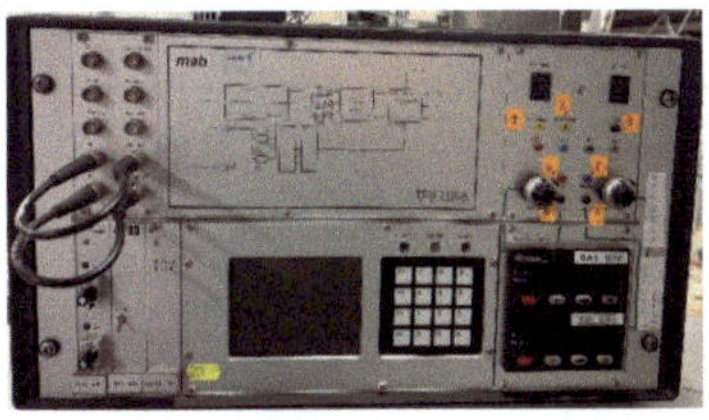

Fig. 1. **Tritium monitor TEM2100**

Fig.2. **Liquid scintillator TriCarb TR2900**

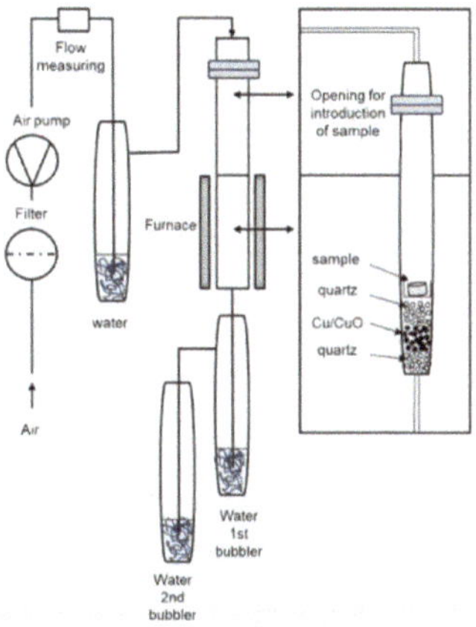

Fig. 3. **Vance TDS system for detritiation**

9

Results

Beryllium pre-treatment and tritium analysis

Co-60 dosimetry

Beryllium contains very low amounts of Co-59, therefore, **after neutron irradiation Co-60 is present in Be**.

This poses a **radiological risk due to β and γ radiation** (shielding – glass 0.4 mm, lead at least 16 mm to achieve 1/10 of initial dose rate).

Co-60, however, allows detection for tracing purposes.

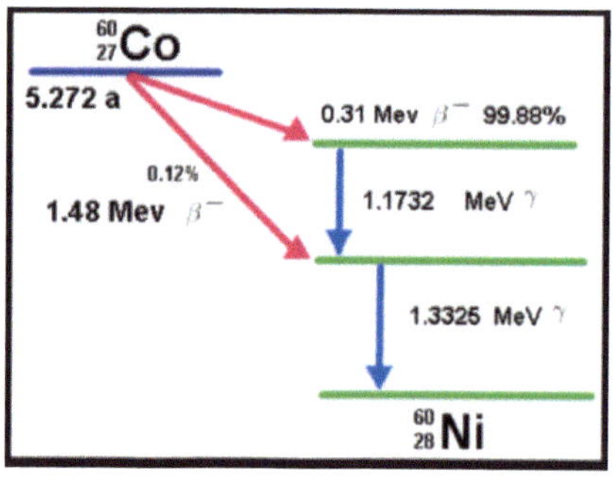

Fig. 1. **Co-60 decay chain [1]**

Co-60 was measured using gamma spectrometry (Identifinder 2) and Geiger-Müller counter (Inspector Alert) giving on average **0.401 μSv/h (d = 10 cm, 2023) per pebble.**

Recalculating to end of irradiation for **activity** (**2004**) gives **~ 38 kBq/ pebble** and thus $2.54 \cdot 10^9$ atoms of Co-60. In mass terms this gives **25.3 fg** ($2.53 \cdot 10^{-14}$ g) of Co-60/ 1mm Be pebble.

11

[1] Ioan, Mihail-Razvan & Cristian, Postolache & Fugaru, Viorel & Bercea, S. & Celarel, Aurelia & CENUSA, C.. (2018). Computational method for the determination of intense gamma-rays sources activity by using GEANT4. Romanian Reports in Physics.

Pre-treatment of Be

Pebble pre-treatment yielded partially oxidized pebbles:

$$Be(s) + 0.5O_2(g) \xrightarrow{temp.} \mathbf{BeO}(s)$$
$$Be(s) + H_2O(g) \xrightarrow{temp.} \mathbf{BeO}(s) + H_2(g)$$
$$3Be(s) + N_2(g) \xrightarrow{temp.} Be_3N_2(s)$$

$$Be_3N_2(g) + 1.5O_2(g) \xrightarrow{temp.} 3\mathbf{BeO}(s) + N_2(g)$$
$$0.5Be_3N_2(g) + 3H_2O(l) \xrightarrow{temp.} 1.5Be(OH)_2(s) + NH_3(g)$$
$$Be(OH)_2(s) \xrightarrow{temp.} \mathbf{BeO}(s) + H_2O(g)$$

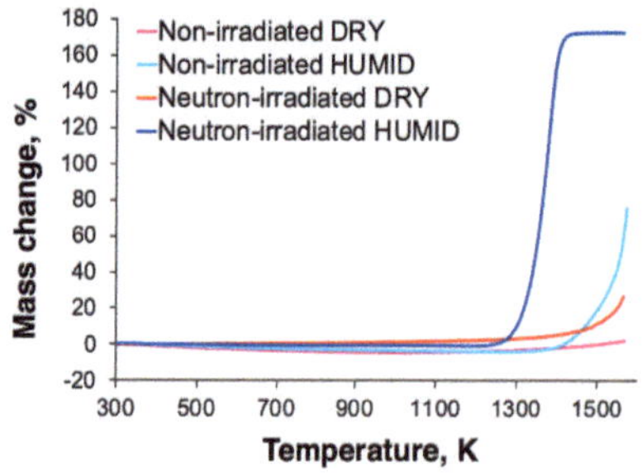

Fig. 1. **Mass change of Be pebbles depending on treatment conditions [1]**

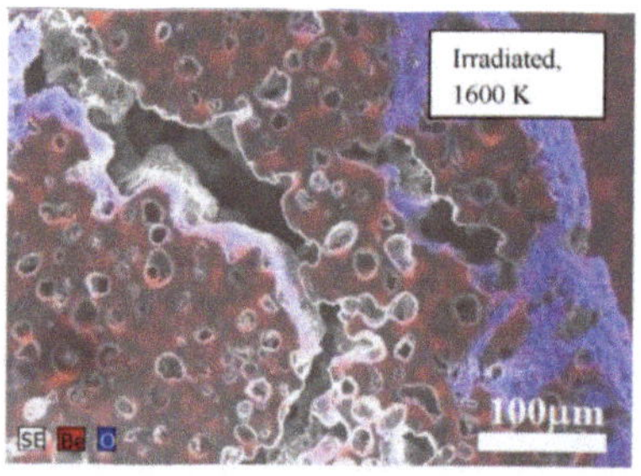

Fig. 2. **EDX of beryllium pebble after thermal treatment [2]**

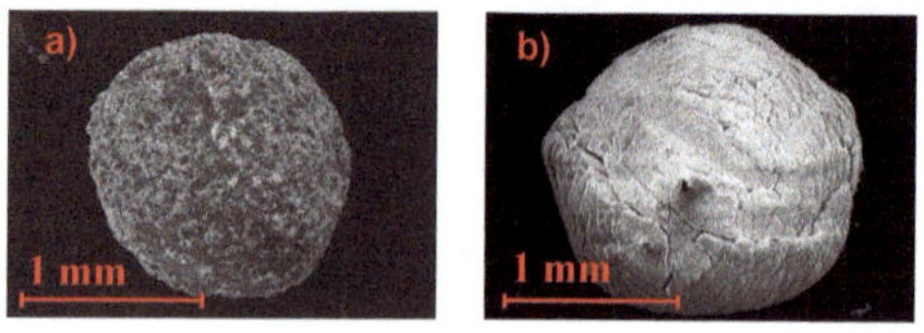

Fig. 3. **Beryllium pebbles after thermal treatment under SEM: a) neutron-irradiated DRY, b) neutron-irradiated HUMID**

12

[1] Zabolockis, R. J., Pajuste, E., Avotina, L. *Journal of Nuclear Materials and Energy*. 2023. 36. 101454
[2] Pajuste. E.. Kizane G.. Avotiņa L.. Zariņš A. *Journal of Nuclear Materials*. 2015. 465. 293-300.

Tritium release during pre-treatment

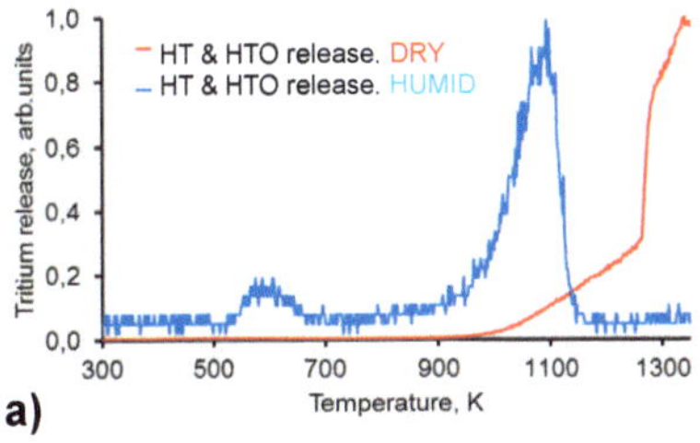

a)

Neutron-irradiated Be, which had accumulated **tritium (T)** releases it rapidly at temperatures above 900 K.

Heating in humid air additionally releases tritium above 500 K. Tritium in gas form is released very similarly in DRY and HUMID airflow.

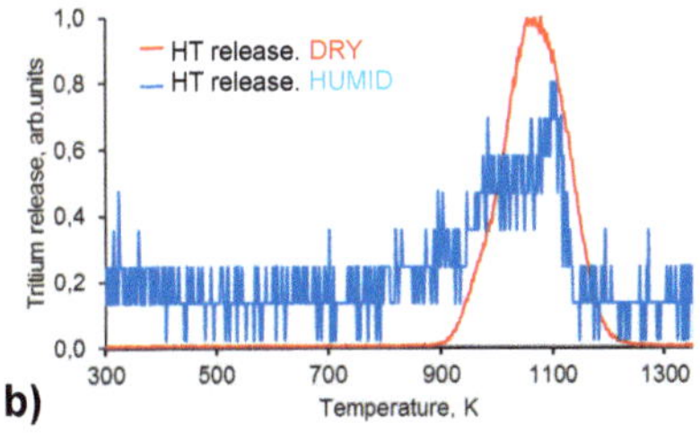

b)

The largest total amount of tritium released during pre-treatment for HUMID airflow was observed to be **92.6 ± 0.2 MBq** per pebble, thus, **~ 93 MBq/g** recalculating back to the end of irradiation.

Fig. 1. **a) HT and HTO, b) HT release from neutron irradiated beryllium in dry and humid airflow**

13

Tritium after pre-treatment: LSC analysis

The beryllium pebbles before and after heating

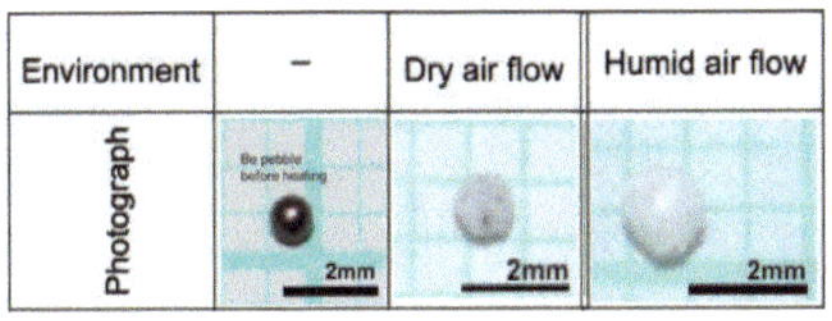

Fig. 1. **Photographs of initial and treated Be pebbles**

$$Be_{(s)} + H_2SO_{4_{(aq)}} \rightarrow BeSO_{4_{(aq)}} + H_{2_{(g)}}$$

$$BeO_{(s)} + H_2SO_{4_{(l)}} \xrightarrow{temp.} BeSO_{4_{(aq)}} + H_2O_{(l)}$$

Fig. 1. **Pre-treated Be pebble dissolution in H_2SO_4**

Neutron irradiated Be pebbles (~1 mg pebbles) after thermal treatment in **DRY airflow** still was found to contain 1899 ± 23 Bq by Vance TDS method, thus, it can be assumed that Be pebbles after dry pre-treatment **contain ~ 19 kBq/g of tritium.**

Therefore, **around 0.2 % of total tritium** is still left in the partially oxidized Be pebbles.

14

Experimental

Beryllium safety: detection and quantification

Method standardization and verification

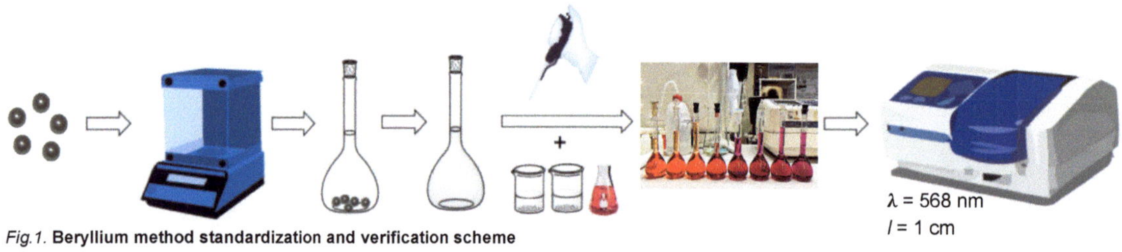

Fig.1. **Beryllium method standardization and verification scheme**

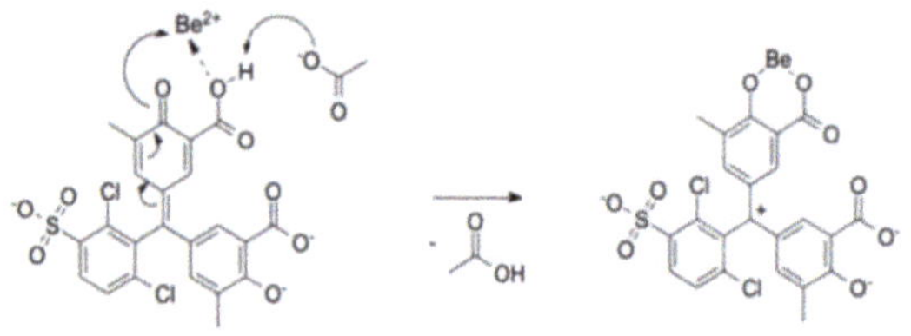

Fig. 2. **Beryllium-CAS reaction mechanism**

Photometric determination reactants:

- **1 mL 0.05% EDTA;**
- **10 mL acetate buffer system (pH = 4.6);**
- **2 mL 0.05% CAS.**

Photometric determination parameters:

- **λ = 568 nm; *l* = 1 cm;**
- **At least 3 paralel measurements**

16

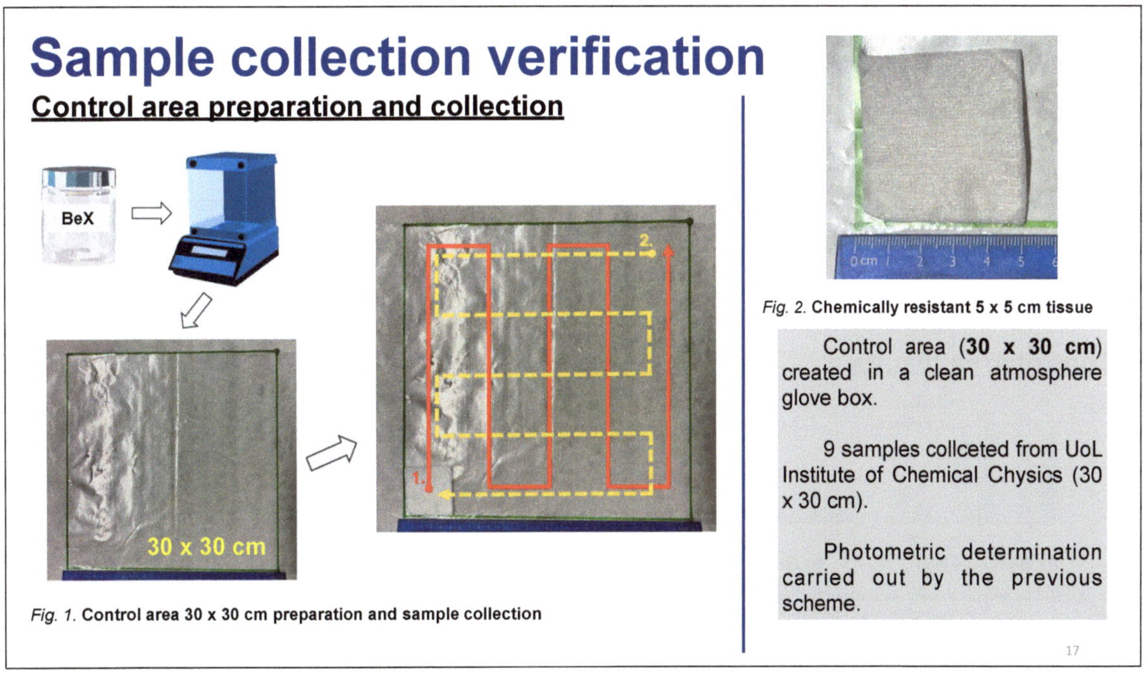
Sample collection verification
Control area preparation and collection
BeX
30 x 30 cm
1.
2.
Fig. 1. Control area 30 x 30 cm preparation and sample collection
Fig. 2. Chemically resistant 5 x 5 cm tissue
Control area (30 x 30 cm) created in a clean atmosphere glove box.
9 samples collceted from UoL Institute of Chemical Chysics (30 x 30 cm).
Photometric determination carried out by the previous scheme.
17

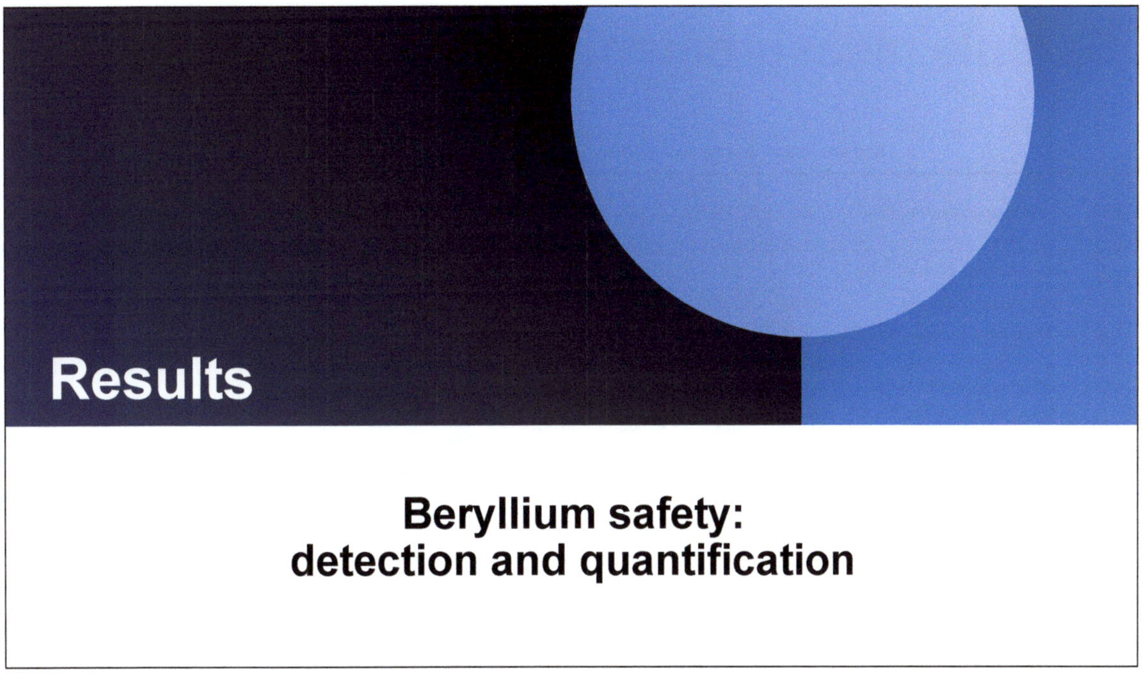
Results
Beryllium safety:
detection and quantification

Standardization results

Calibration curves

$$m_{Be^{2+}_{phot.}} = \frac{A-b}{a} \bullet V_{phot}$$

Photometric determination of Be^{2+} ion mass concentration resulted in a linear range of **0.050 mg/L** to **3.5 mg/L.**

Molar absorption coefficient ε:

- In literature **2200** 1/(cm•M) [1];
- Experimentally determined **(2207 ± 22)** 1/(cm•M).

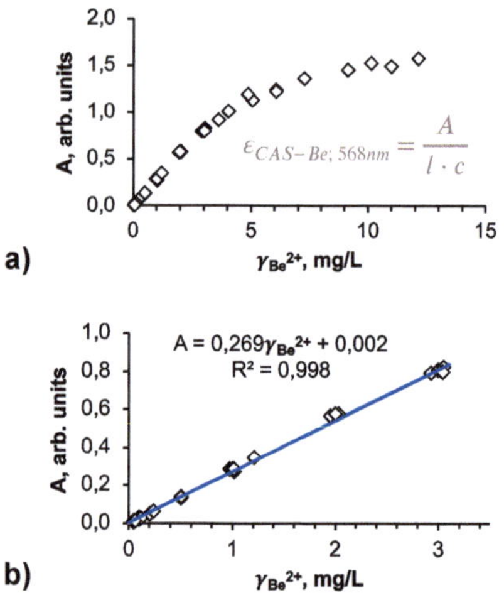

Fig. 2. **Be-CAS complex photometric calibration curves for a) all range, b) linear range**

Control area results

LoQ > 0.13 µg/100 cm²

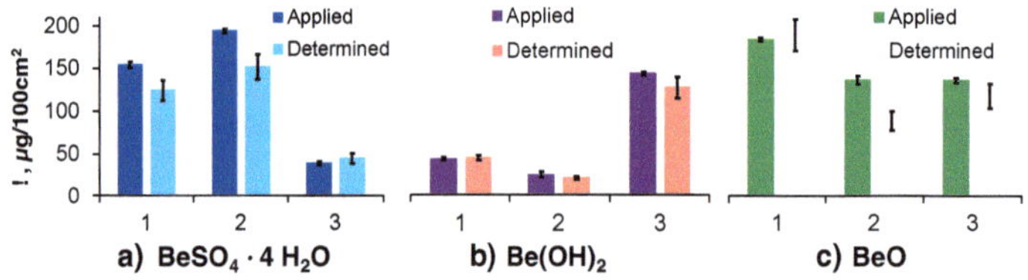

Fig. 1. Applied vs photometrically determined Be^{2+} ion concentration per area for a) beryllium sulfate tetrahydrate, b) beryllium hydroxide and c) beryllium oxide

[1] Fouad et al., Arabian Journal of Chemistry. 9. 2016. 235–239

19

Sample measurement results

$$m_{Be^{2+}_{phot.}} = \frac{A-b}{a} \bullet V_{phot}$$

$$m_{Be^{2+}_{sample}} = m_{Be^{2+}_{phot}} \bullet \frac{V_{analyte}}{V_{aliquote}} \bullet 10^{-3}$$

$$\vartheta_{Be} = \frac{m_{Be^{2+}_{sample}}}{k^2} \bullet 100$$

Table No. 1
Light absorbtion data for UoL IoCP surface tests

Label	Sample acquisition location	A_1	A_2	A_3	$A_{avg.}$
L1	**Glove box for work with Be**	**0.013**	**0.014**	**0.015**	**0.014**
L2	Reagent storage	0.000	0.001	0.000	0.000
L3	Be pebble storage area	0.002	0.003	0.001	0.002
L4	Analytical scale	0.001	0.000	0.001	0.001
L5	Thermogravimeter table	0.000	0.001	0.000	0.000
L6	Fume hood with HEPA filter for distillation	-0.001	0.000	-0.002	-0.001
L7	Sample preparation table	0.000	0.000	-0.001	0.000
L8	Fume hood with Vance devices (TDS)	0.001	0.000	0.000	0.000
L9	Photometric determination workstation	0.002	0.001	0.001	0.001

The only contamination was detected in **L1** of **(2.1 ± 0.8) µg/100 cm²**, which is under the USA DoE (Dept. of Energy) safe work with beryllium **3 µg/100 cm²** limit [1].

No Be or BeX contamination was observed elsewhere in the laboratory.

To consider the area non-contaminated the value should be < **0.2 ∝g/100 cm²** as indicated by the USA DoE.

[1] Damian. Toxicology Mechanisms and Methods. 21. 2011. 97–102

20

Conclusions

1. Neutron-irradiated Be contains Co-60 giving tracing possibilities but requiring safe handling.
2. More than **99.8% of tritium is released** by **heating** neutron-irradiated Be **up to 1300 K** in airflow.
3. **Neutron-irradiated Be still contains tritium**: ~ 0.2 % of the total initial tritium inventory after pre-treatment.
4. A **quantitative Be detection method** is described for all dissolvable Be compounds (including actual sample collection from workstations at UoL) with a linear range of of **0.050 mg/L** to **3.5 mg/L (LoQ > 0.13 µg/100 cm^2).**
5. During work with beryllium the only contaminated area at UoL was found to be a glove box intended for Be-related work (L1) with **(2.1 ± 0.8) µg/100 cm^2 below** the USA DoE **3 µg/100 cm^2** limit. No contamination elsewhere was observed.
6. Intermetallic alloys of Be should be tested with the Be detection method to evaluate Ti or V impact on Be contamination quantification on surfaces.

21

UNIVERSITY OF LATVIA
INSTITUTE OF CHEMICAL PHYSICS

BeWS-16
2023

UNIVERSITY OF LATVIA
FACULTY OF CHEMISTRY

Thank you for your attention!

rudolfs_janis.zabolockis@lu.lv, elina.pajuste@lu.lv

16th International Workshop on Beryllium Technology
19-20 October 2023, Santander, Spain

6.5 Microstructural examination of neutron irradiated beryllium and beryllides

Michael Klimenkov, Michael Dürrschnabel, Ramil Gaisin und Pavel Vladimirov

The current interest in the mechanical properties and microstructure of neutron-irradiated beryllium and beryllides refers to their planned application in the Helium-Cooled Pebble Bed (HCPB) European concept of a breeding blanket of DEMO [1]. Titanium beryllide $TiBe_{12}$ shows higher radiation resistance and less swelling than pure beryllium. The study of temperature and dose dependences of the beryllium and titanium beryllide swelling by neutron irradiation is crucial for their future application. Irradiation experiments in high-neutron-flux nuclear research reactors yield information about the microstructural evolution of beryllium under fusion-relevant conditions (temperature, damage dose, helium and tritium productions) excluding 14 MeV neutrons impact which is not present in the neutron spectra of fission reactors. The HIDOBE-02 irradiation campaign accomplished at the HFR, Petten corresponds to 1246.5 Full Power Days at a reactor power level of 45 MW in the temperature range from 410 °C to 680 °C. Transmission electron microscopy (TEM) was employed to study the evolution of voids during neutron irradiation at different temperatures. The targeted preparation of specimens was performed using a focused ion beam (FIB). TEM study shows the formation of radiation-induced hexagonal flat gas bubbles inside beryllium and beryllides. The results demonstrate that void swelling in beryllides is significantly lower than that in pure beryllium. The study includes the evaluation of bubbles for both phases and the formation of structural defects, such as twins, inside beryllides. The preferred formation of bubbles along grain boundaries and dislocation lines was observed. Analytical investigations using electron energy loss spectroscopy show the presence of He and H^3 inside the bubbles.

REFERENCES

[1] M. Zmitko, Fusion Engineering, 136, (2018), 1376-1385

Presenting Author: Michael Klimenkov
Presenting Author's Affiliation: Karlsruhe Institute of Technology, Germany
Email: michael.klimenkov@kit.edu

Karlsruhe Institute of Technology

Microstructural examination of neutron irradiated beryllium and beryllides

M. Klimenkov, M. Dürrschnabel, P. Vladimirov, N. Zimber, R. Gaisin, V. Chakin

Institute for Applied Materials - Applied Materials Physics, Karlsruhe Institute of Technology (KIT), Karlsruhe, Germany.

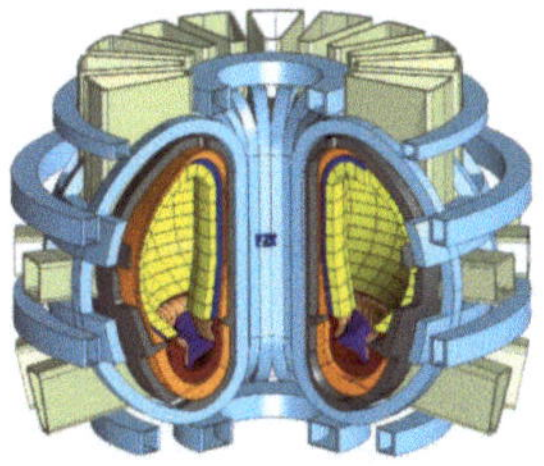

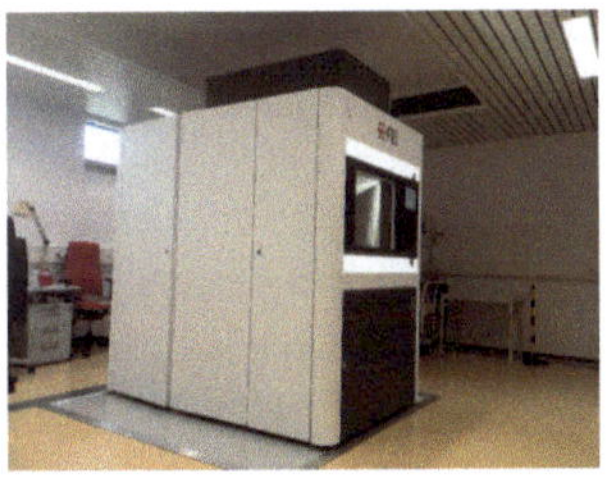

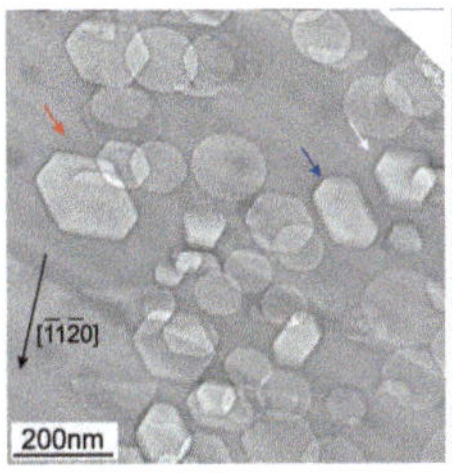

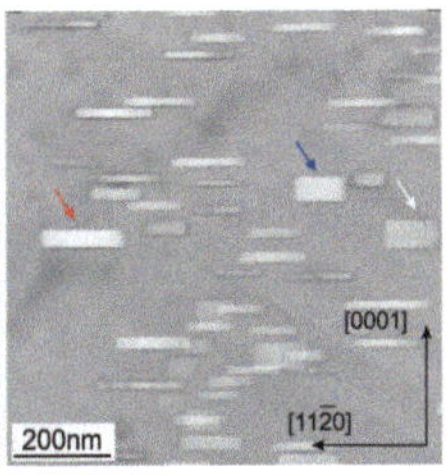

KIT – The Research University in the Helmholtz Association

www.kit.edu

Content

2 Dr. Michael Klimenkov Institute of Applied Materials – Applied Materials Physics (IAM-AWP)

Application of beryllium and beryllides in fusion technology

Karlsruhe Institute of Technology

- As neutron multiplier material in different tritium-breeding blanket concepts for the future demonstration fusion power plant DEMO.

Helium Cooled Pebble Bed Blanket

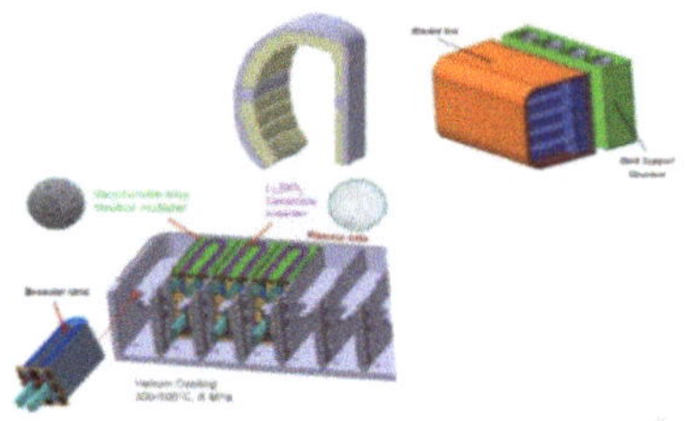

Production of helium and tritium (^{3}H) by transmutation reactions

$^{9}Be + n \rightarrow 2\,^{4}He + 2n$ (E_n>2.7 MeV)

$^{9}Be + n \rightarrow {}^{7}Li + {}^{3}He$ (E_n>10.7 MeV)

$^{7}Li + n \rightarrow {}^{4}He + {}^{3}H + n$

$^{9}Be + n \rightarrow {}^{6}He + {}^{3}H$ (E_n> 0.6 MeV)

$^{6}He \rightarrow {}^{6}Li + b^{-} + \nu$

$^{6}Li + n \rightarrow {}^{4}He + {}^{3}H$

Prediction of irradiation resistance of beryllium pebbles under close-to-fusion conditions.

- operation temperature,
- accumulated damage dose,
- amount of helium and tritium generated by neutron-induced transmutation.

3 Dr. Michael Klimenkov Institute of Applied Materials – Applied Materials Physics (IAM-AWP) IAM

Neutron irradiation programs 2005 - 2016

Karlsruhe Institute of Technology

High Dose Beryllium irradiation program **(HIDOBE-I)** (2005-2007)
High-Flux Reactor (HFR), Petten, Netherlands
Dose 18 dpa, 3000 appm ^{4}He, 300 appm ^{3}H
Irradiation temperatures: 410°C, 480°C, 590°C, 700°C

High Dose Beryllium irradiation program **(HIDOBE-II)** (2005-2011)
High-Flux Reactor (HFR), Petten, Netherlands
38 dpa displacement per atom **5900** appm ^{4}He, **640** appm ^{3}H
Irradiation temperatures: 370°C, 440°C, 540°C, 650°C

Materials: beryllium and beryllides (438°C, 525°C, 664°C, 768°C)

Be pebbles

∅ 1mm
∅ 2mm

1mm pebble

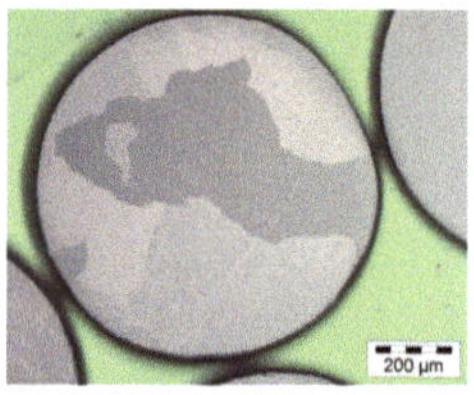

100-300µm grain size

4 Dr. Michael Klimenkov Institute of Applied Materials – Applied Materials Physics (IAM-AWP) IAM

HIDOBE-II
Karlsruhe Institute of Technology
High Dose Beryllium irradiation program (HIDOBE-II) (2005-2011)
High-Flux Reactor (HFR), Petten, Netherlands
38 dpa displacement per atom 5900 appm ^{4}He, 640 appm ^{3}H
Irradiation temperatures: 370°C, 440°C, 540°C, 650°C
TEM lamellae were prepared from irradiated tungsten using focused ion beam (FEI SCIOS) in the FML at KIT.
Transmission Electron Microscope Talos F200X G2 / 200 kV FEG
5
Dr. Michael Klimenkov
Institute of Applied Materials – Applied Materials Physics (IAM-AWP)
IAM

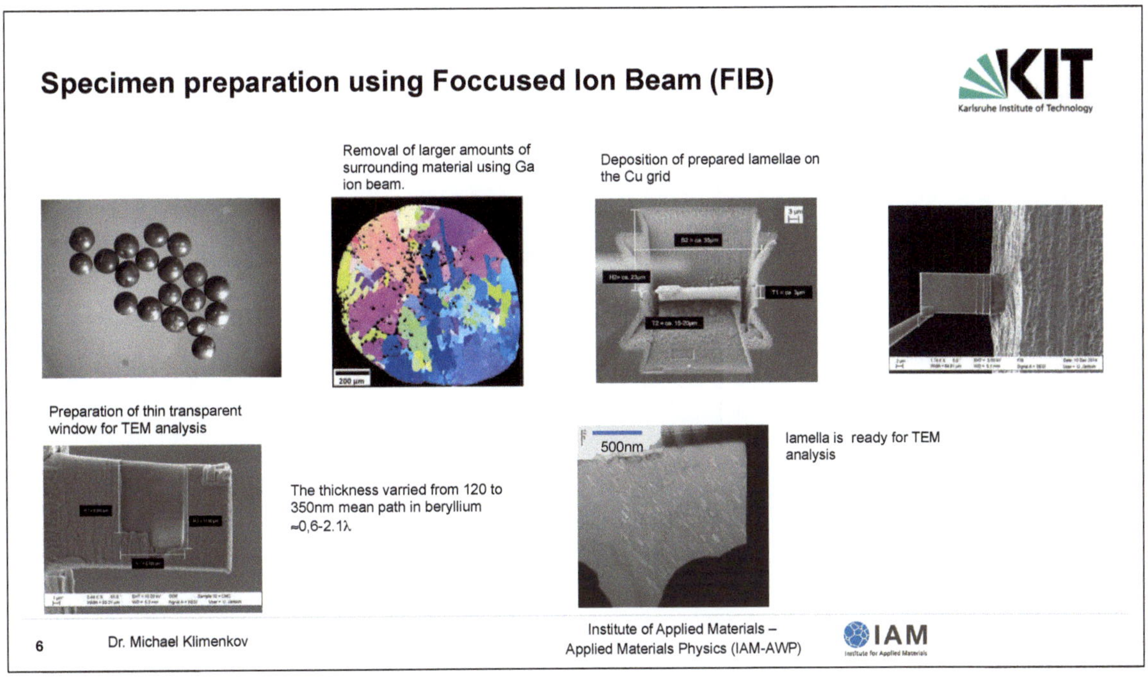
Specimen preparation using Foccused Ion Beam (FIB)
Karlsruhe Institute of Technology
Removal of larger amounts of surrounding material using Ga ion beam.
Deposition of prepared lamellae on the Cu grid
200 µm
Preparation of thin transparent window for TEM analysis
The thickness varried from 120 to 350nm mean path in beryllium ≈0,6-2.1λ
500nm
lamella is ready for TEM analysis
6
Dr. Michael Klimenkov
Institute of Applied Materials – Applied Materials Physics (IAM-AWP)
IAM

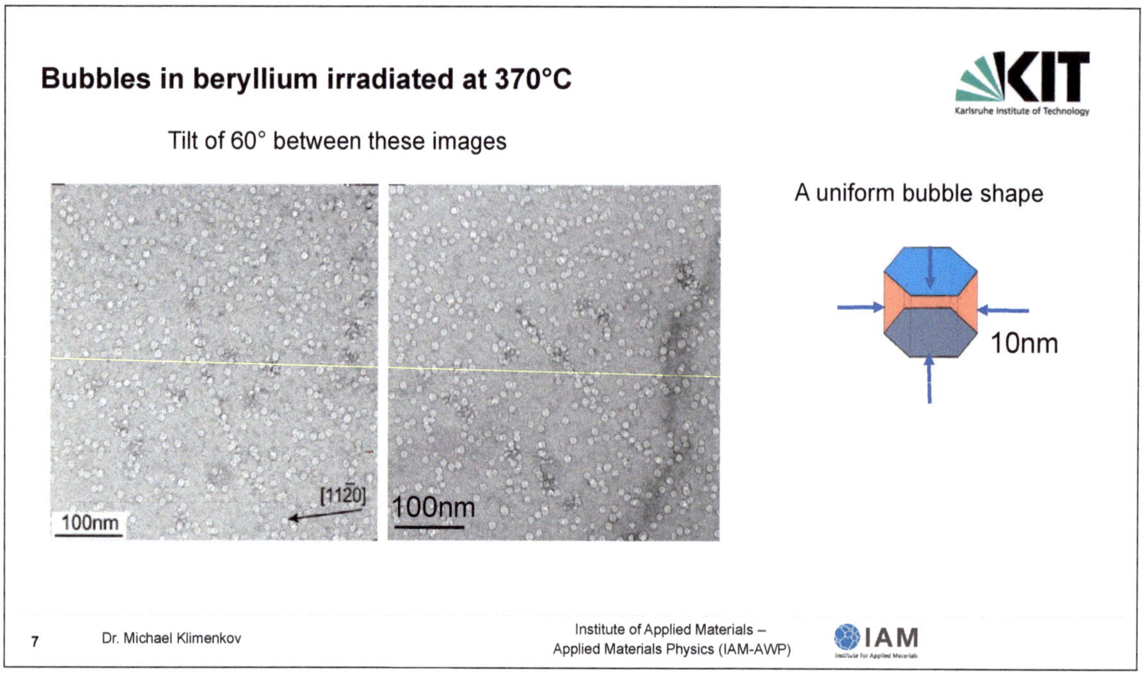
Bubbles in beryllium irradiated at 370°C
Tilt of 60° between these images
A uniform bubble shape
10nm
100nm
100nm
7
Dr. Michael Klimenkov
Institute of Applied Materials –
Applied Materials Physics (IAM-AWP)
IAM
Karlsruhe Institute of Technology

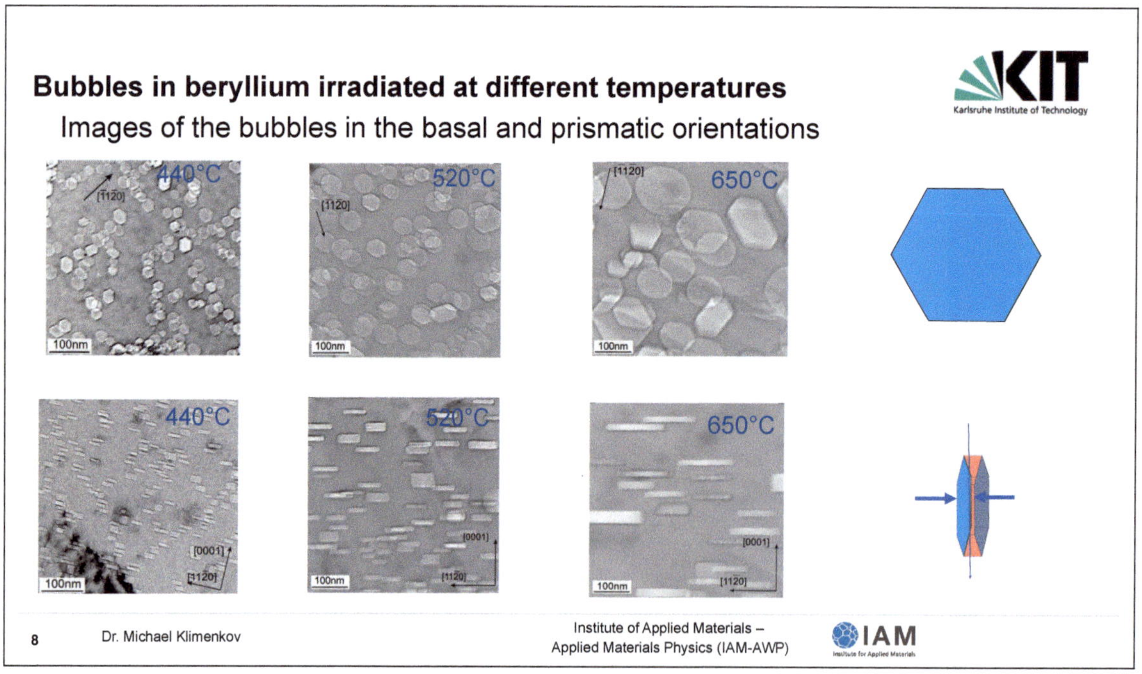
Bubbles in beryllium irradiated at different temperatures
Images of the bubbles in the basal and prismatic orientations
440°C
520°C
650°C
440°C
520°C
650°C
100nm
[0001]
8
Dr. Michael Klimenkov
Institute of Applied Materials –
Applied Materials Physics (IAM-AWP)
IAM
Karlsruhe Institute of Technology

Tomography of bubbles in Be in TEM

66° total tilt (from -32° to +34°)

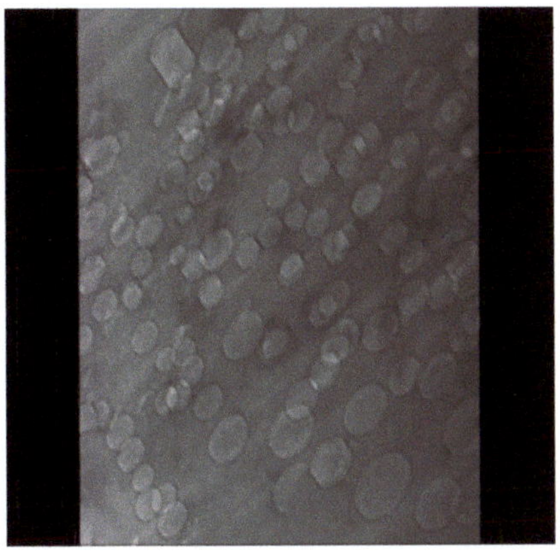

step ~3°
23 images

9 Dr. Michael Klimenkov Institute of Applied Materials – Applied Materials Physics (IAM-AWP)

Quantitative evaluation

Effect of radiation dose on the microstructure of irradiated beryllium

Bubble diameter

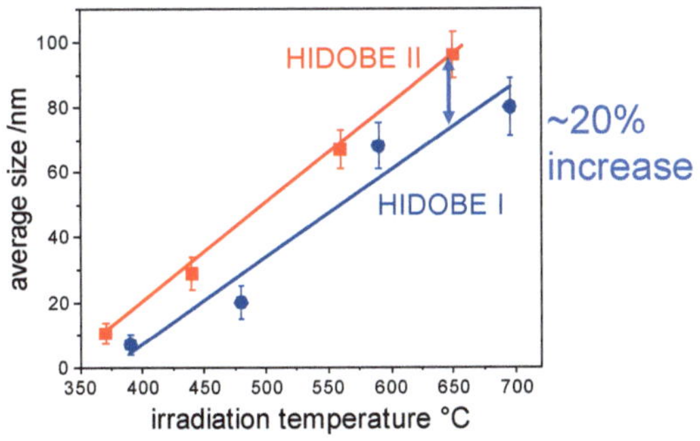

Swelling culculated based on TEM imaging

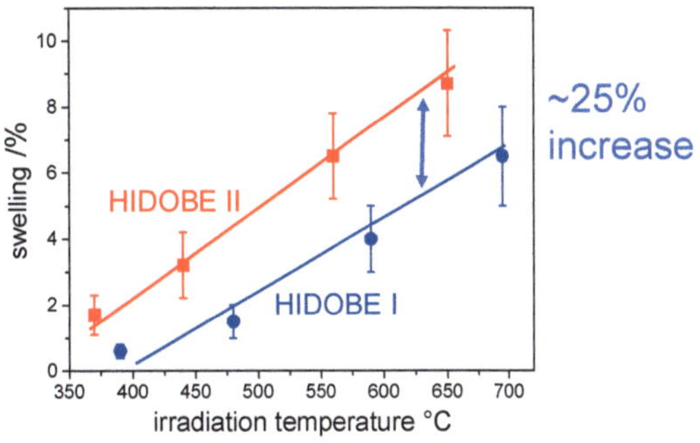

Increasing damage/He production by a factor of 2 results in a ~25% change in microstructure.

10 Dr. Michael Klimenkov Institute of Applied Materials – Applied Materials Physics (IAM-AWP)

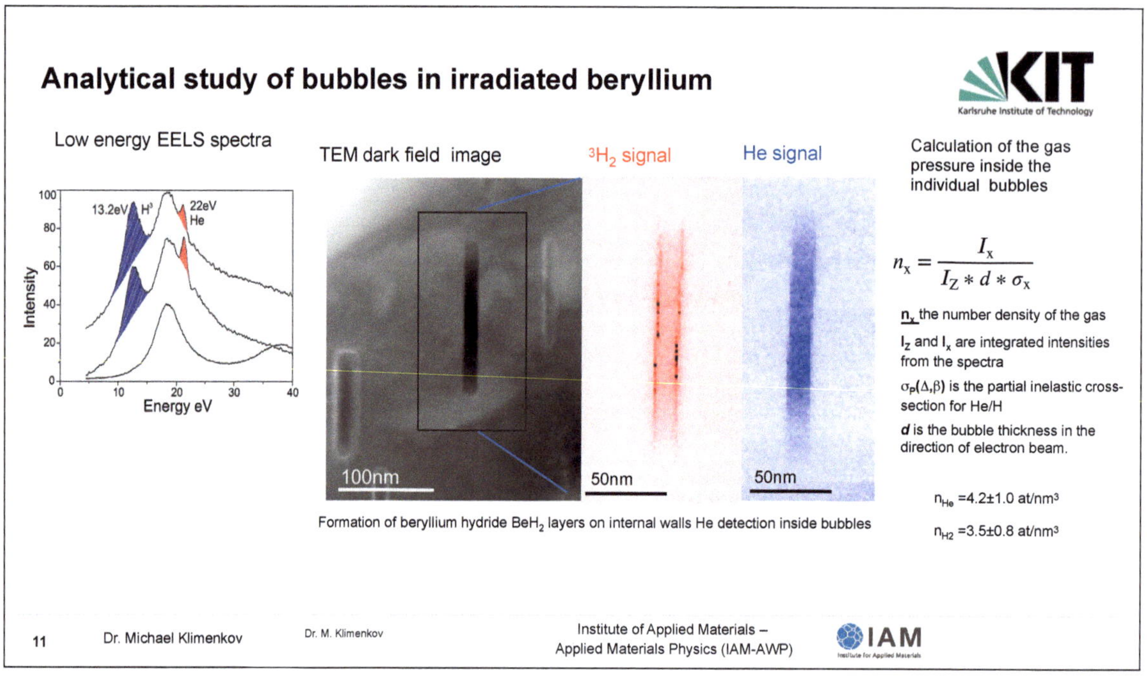
Analytical study of bubbles in irradiated beryllium
KIT
Karlsruhe Institute of Technology
Low energy EELS spectra
13.2eV H
22eV He
Intensity
Energy eV
TEM dark field image
$^3H_2$ signal
He signal
100nm
50nm
50nm
Formation of beryllium hydride BeH_2 layers on internal walls He detection inside bubbles
Calculation of the gas pressure inside the individual bubbles
$n_x = \frac{I_x}{I_Z * d * \sigma_x}$
n_x the number density of the gas
I_z and I_x are integrated intensities from the spectra
$\sigma_P(\Delta,\beta)$ is the partial inelastic cross-section for He/H
d is the bubble thickness in the direction of electron beam.
n_{He} =4.2±1.0 at/nm³
n_{H2} =3.5±0.8 at/nm³
11
Dr. Michael Klimenkov
Dr. M. Klimenkov
Institute of Applied Materials – Applied Materials Physics (IAM-AWP)
IAM

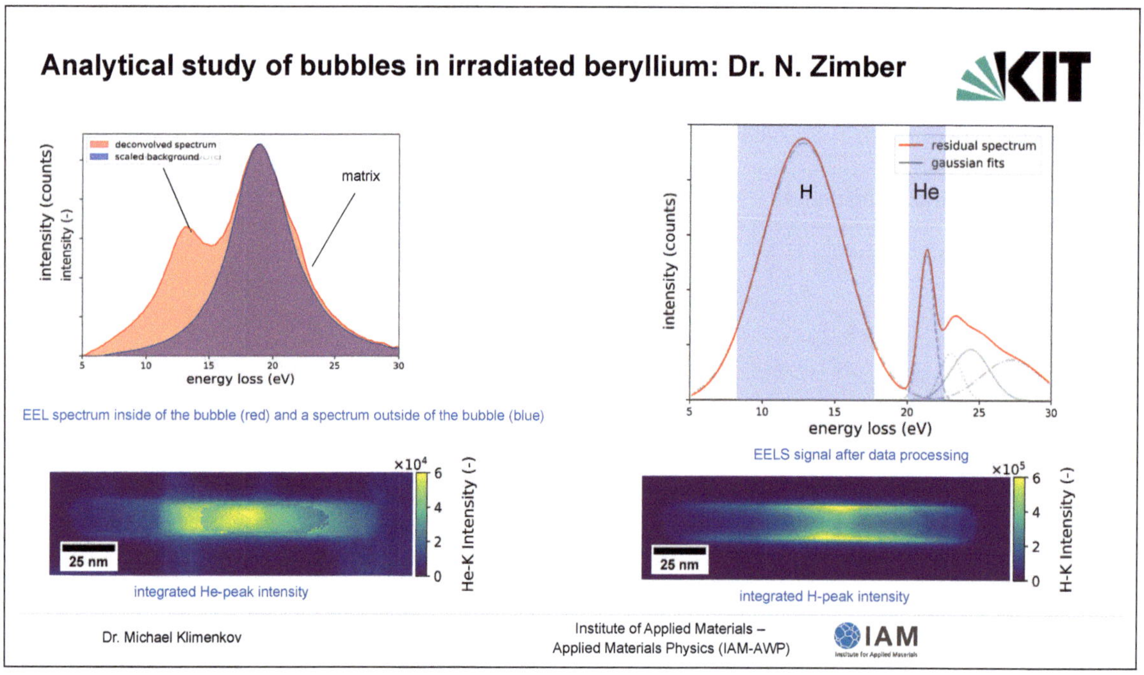
Analytical study of bubbles in irradiated beryllium: Dr. N. Zimber
KIT
deconvolved spectrum
scaled background
matrix
intensity (counts)
intensity (-)
energy loss (eV)
EEL spectrum inside of the bubble (red) and a spectrum outside of the bubble (blue)
residual spectrum
gaussian fits
H
He
intensity (counts)
energy loss (eV)
EELS signal after data processing
25 nm
×10⁴
He-K Intensity (-)
integrated He-peak intensity
25 nm
×10⁵
H-K Intensity (-)
integrated H-peak intensity
Dr. Michael Klimenkov
Institute of Applied Materials – Applied Materials Physics (IAM-AWP)
IAM

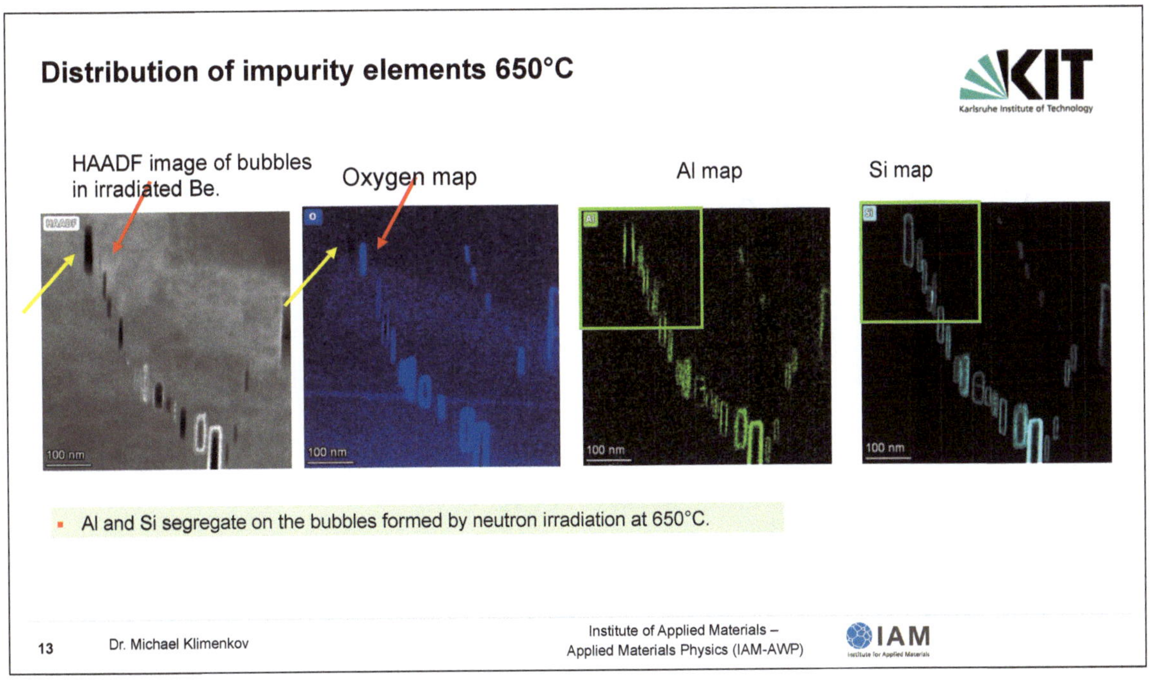
Distribution of impurity elements 650°C
KIT
Karlsruhe Institute of Technology
HAADF image of bubbles in irradiated Be.
Oxygen map
Al map
Si map
HAADF
O
Al
Si
100 nm
100 nm
100 nm
100 nm
Al and Si segregate on the bubbles formed by neutron irradiation at 650°C.
13
Dr. Michael Klimenkov
Institute of Applied Materials –
Applied Materials Physics (IAM-AWP)
IAM

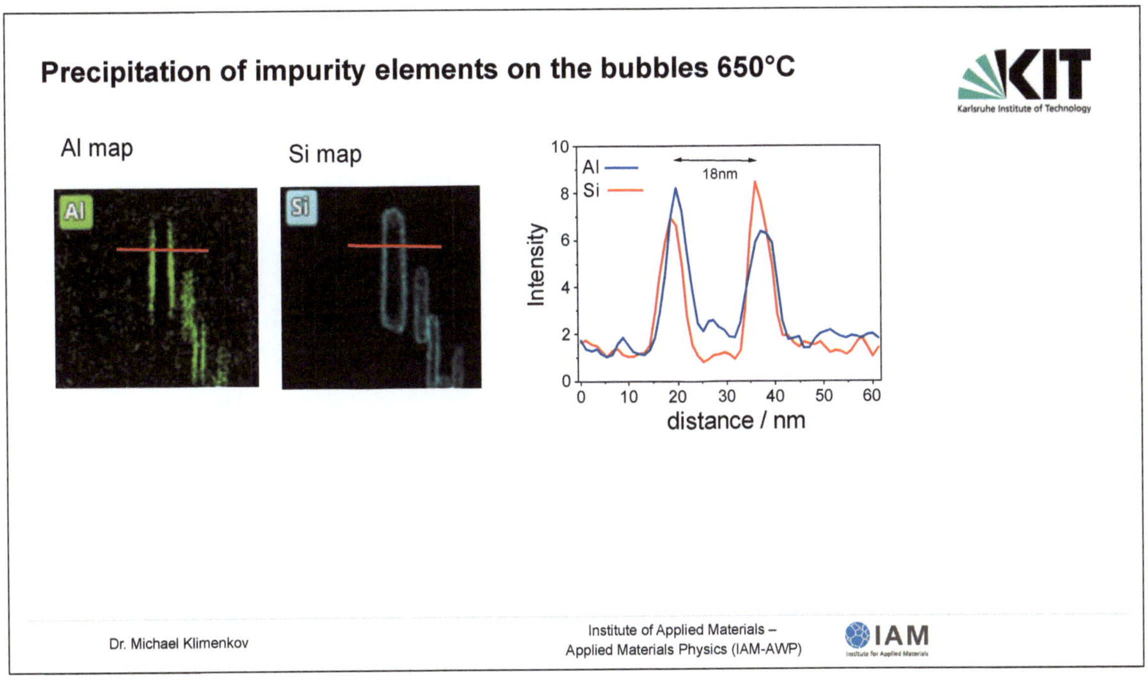
Precipitation of impurity elements on the bubbles 650°C
KIT
Karlsruhe Institute of Technology
Al map
Si map
Al
Si
Al
Si
18nm
Intensity
0
2
4
6
8
10
0
10
20
30
40
50
60
distance / nm
Dr. Michael Klimenkov
Institute of Applied Materials –
Applied Materials Physics (IAM-AWP)
IAM

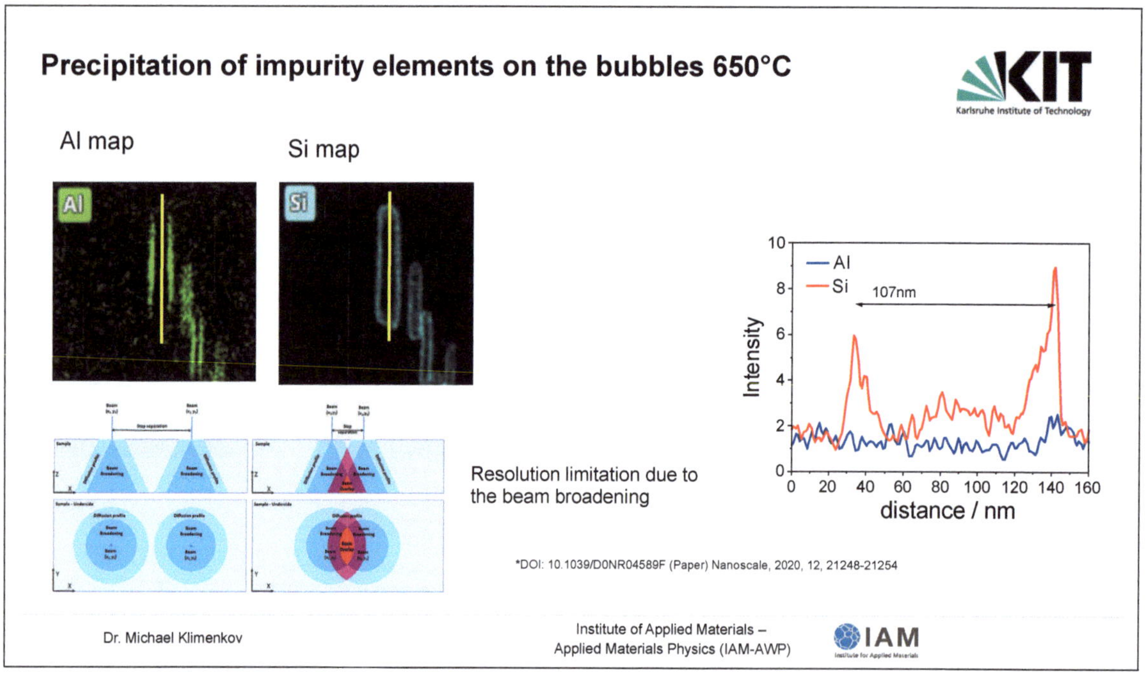
Precipitation of impurity elements on the bubbles 650°C
Al map
Si map
Al
Si
Resolution limitation due to the beam broadening
Intensity
distance / nm
107nm
*DOI: 10.1039/D0NR04589F (Paper) Nanoscale, 2020, 12, 21248-21254
Dr. Michael Klimenkov
Institute of Applied Materials – Applied Materials Physics (IAM-AWP)

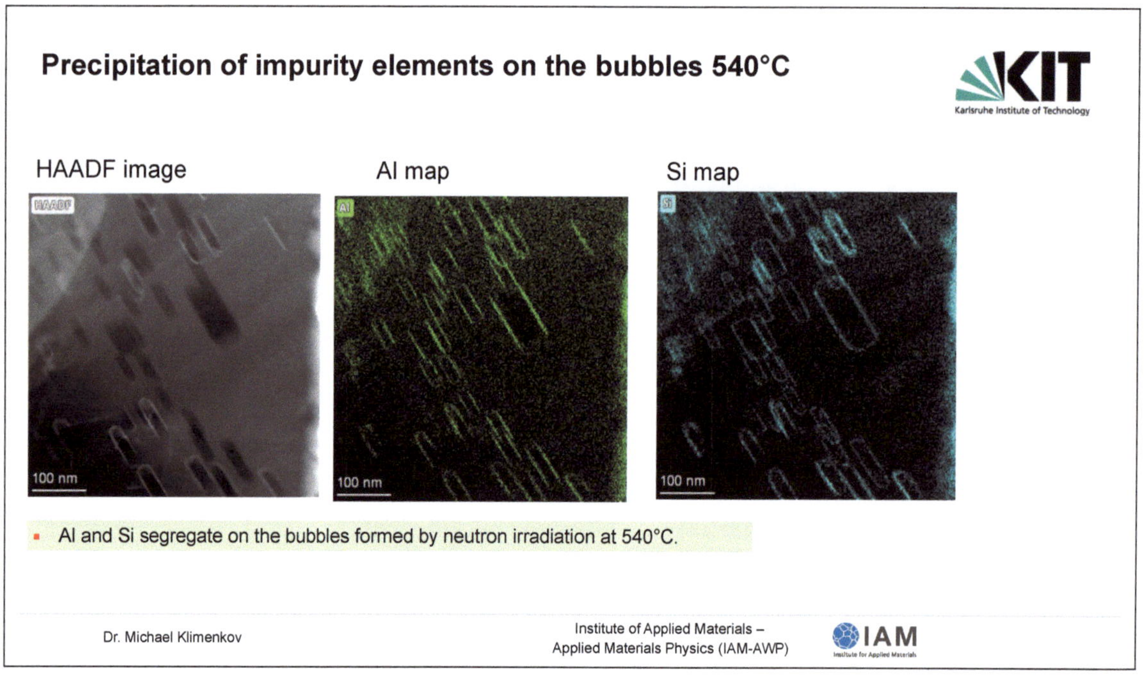
Precipitation of impurity elements on the bubbles 540°C
HAADF image
Al map
Si map
100 nm
Al and Si segregate on the bubbles formed by neutron irradiation at 540°C.
Dr. Michael Klimenkov
Institute of Applied Materials – Applied Materials Physics (IAM-AWP)

Microstructure of neutron irradiated beryllides

HAADF image | Al map | Si map

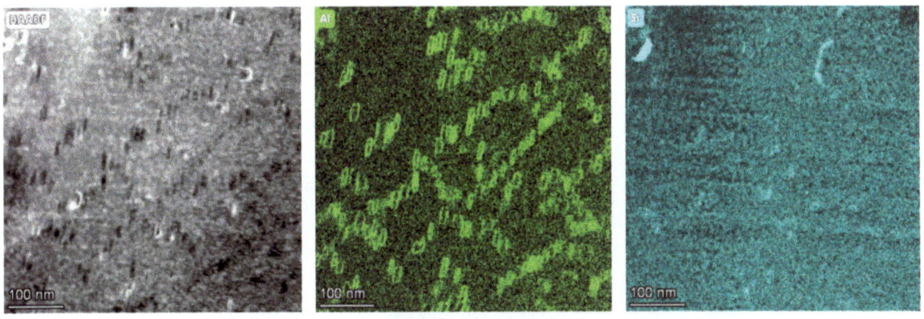

- Both Al and Si tend to segregate on the bubbles formed by neutron irradiation.
- Si is not present on the bubbles formed at irradiation temperatures below 500 °C - the temperature below which Si containing precipitates does not dissolve
- Al is dissolved inside Be in the unirradiated state and .

Dr. Michael Klimenkov

Institute of Applied Materials –
Applied Materials Physics (IAM-AWP)

Content

- Introduction
- Microstructural examination of neutron irradiated beryllium at IAM-AWP (HIDOBE I and HIDOBE II)
- Detection of He and ^{3}H inside bubbles
- Distribution of impurity atoms in irradiated Be
- **TEM examination of neutron irradiated $B_{12}Ti$ phase**
- Conclusions

17 Dr. Michael Klimenkov

Institute of Applied Materials –
Applied Materials Physics (IAM-AWP)

Microstructure of neutron irradiated beryllides

KIT Karlsruhe Institute of Technology

TEM lamella prepared from beryllide disk

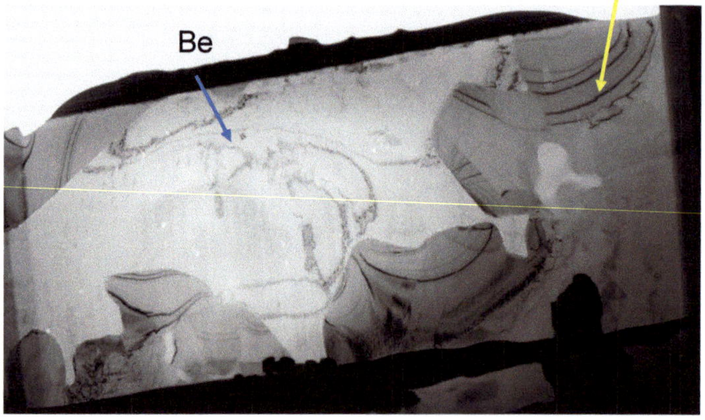

High Dose Beryllium irradiation program (**HIDOBE-II**) (2005-2011)

High-Flux Reactor (HFR), Petten, Netherlands

38 dpa displacement per atom **5900** appm ^{4}He, **640** appm ^{3}H

Irradiation temperatures: 370°C, 440°C, 540°C, 650°C

Beryllide disks contain both pure Be and beryllide grains

The beryllide grains are visible with a darker contrast compare to the pure Be in TEM images

Dr. Michael Klimenkov — Institute of Applied Materials – Applied Materials Physics (IAM-AWP) — IAM

Microstructure of neutron irradiated beryllides

KIT Karlsruhe Institute of Technology

Phase identification using electron diffraction

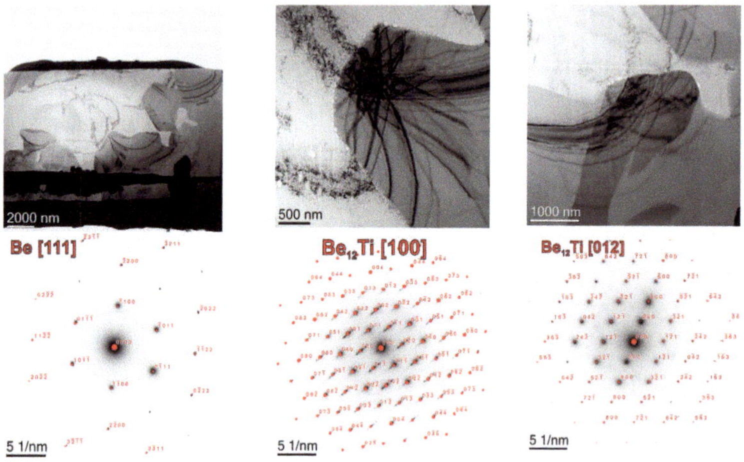

Analysis of electron diffraction images show:

Pure Be has a hexagonal structure.

Beryllide grains contain of $Be_{12}Ti$ phase.

Dr. Michael Klimenkov — Institute of Applied Materials – Applied Materials Physics (IAM-AWP) — IAM

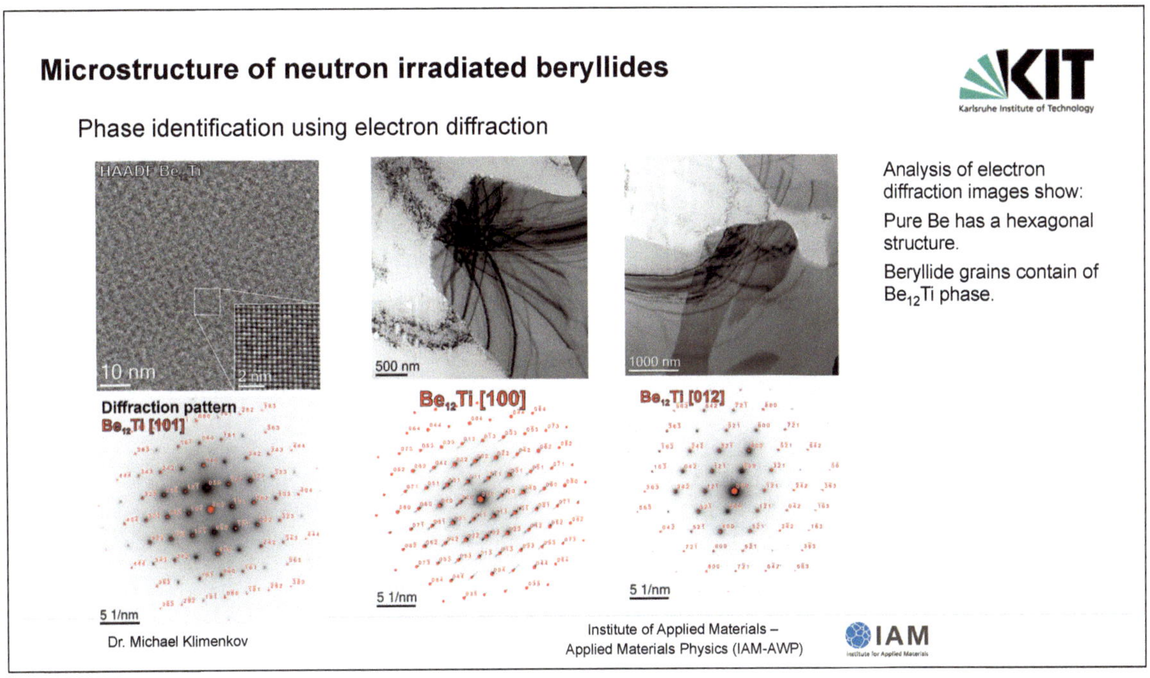
Microstructure of neutron irradiated beryllides
Phase identification using electron diffraction
HAADF Be-Ti
10 nm
2 nm
500 nm
1000 nm
Diffraction pattern
$Be_{12}Ti$ [101]
$Be_{12}Ti$ [100]
$Be_{12}Ti$ [012]
5 1/nm
5 1/nm
5 1/nm
Analysis of electron diffraction images show:
Pure Be has a hexagonal structure.
Beryllide grains contain of $Be_{12}Ti$ phase.
Dr. Michael Klimenkov
Institute of Applied Materials – Applied Materials Physics (IAM-AWP)

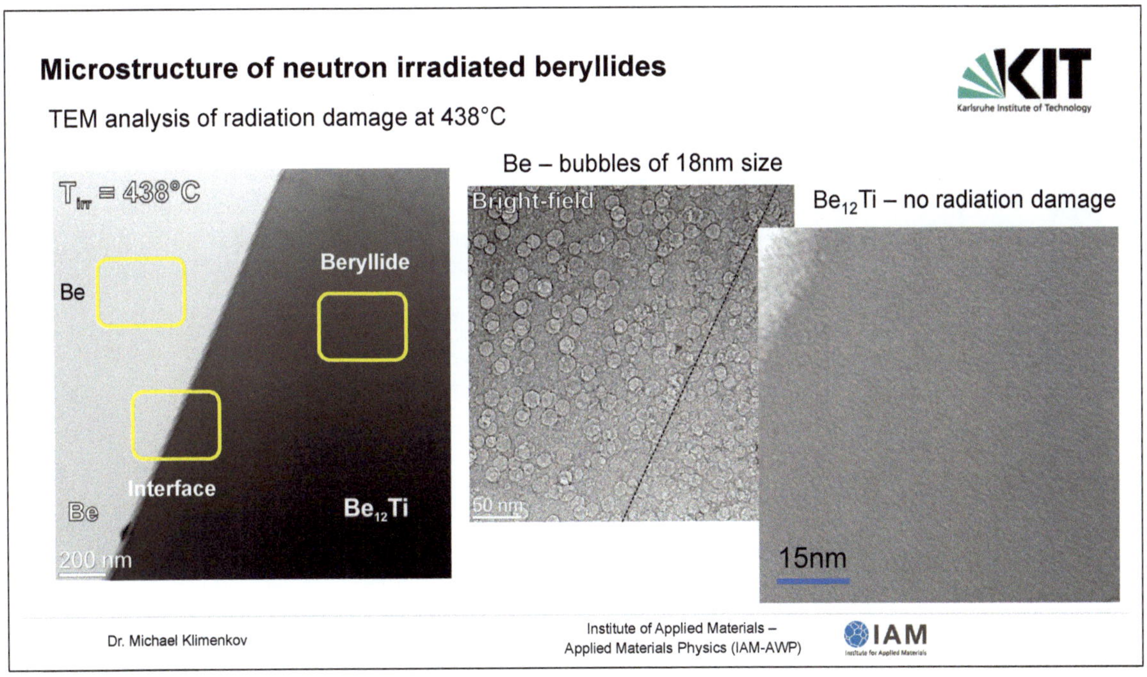
Microstructure of neutron irradiated beryllides
TEM analysis of radiation damage at 438°C
T_{irr} = 438°C
Be
Beryllide
Interface
Be
$Be_{12}Ti$
200 nm
Be – bubbles of 18nm size
Bright-field
50 nm
$Be_{12}Ti$ – no radiation damage
15nm
Dr. Michael Klimenkov
Institute of Applied Materials – Applied Materials Physics (IAM-AWP)

Microstructure of neutron irradiated beryllides

TEM analysis of radiation damage at 438°C

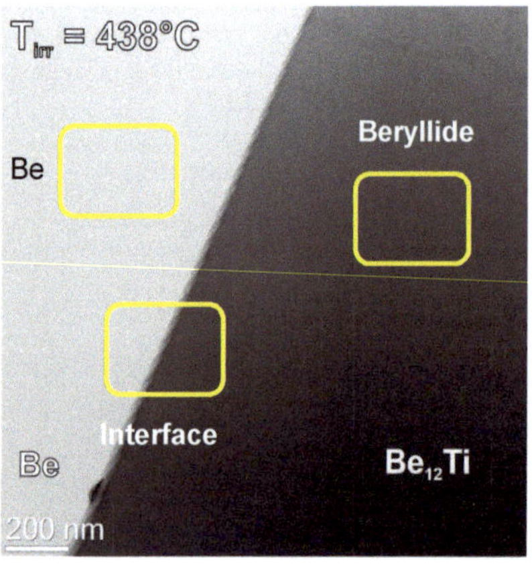

Bubbles of about 7nm size

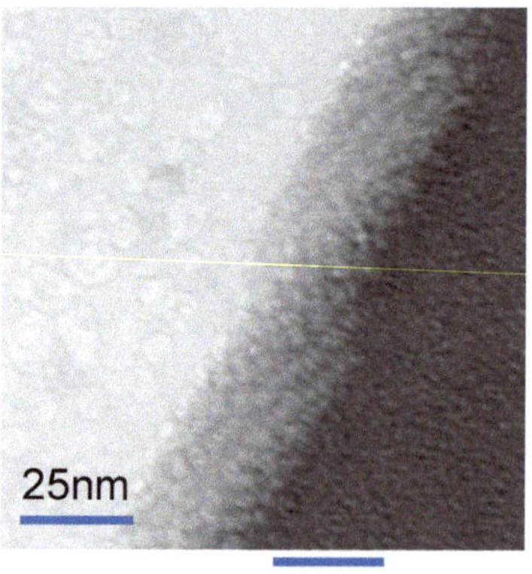

Dr. Michael Klimenkov

Institute of Applied Materials –
Applied Materials Physics (IAM-AWP)

Microstructure of neutron irradiated beryllides

TEM analysis of radiation damage at 525°C

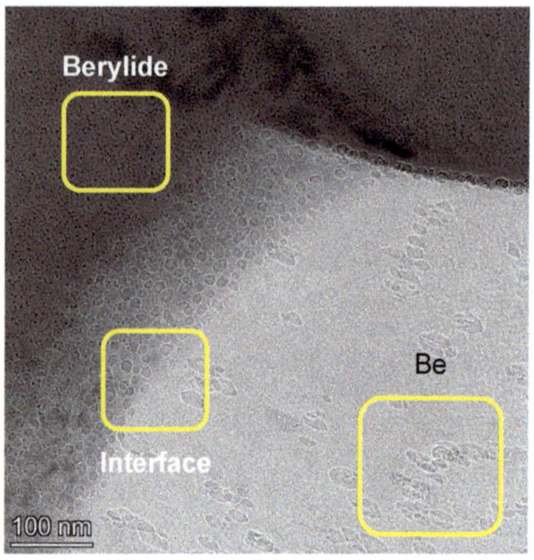

Be – bubbles of 35nm size

$Be_{12}Ti$ ~2-5nm bubbles

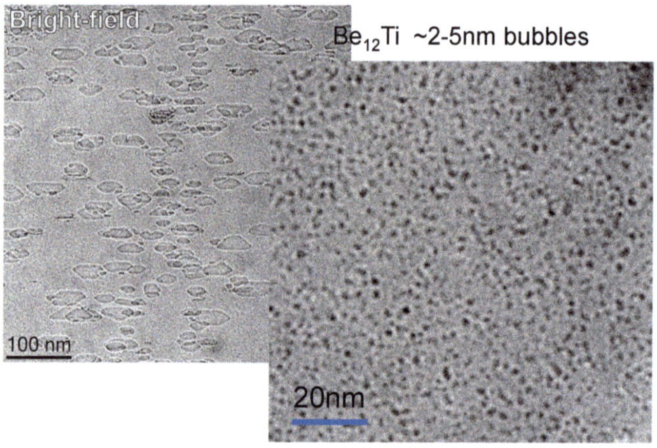

Dr. Michael Klimenkov

Institute of Applied Materials –
Applied Materials Physics (IAM-AWP)

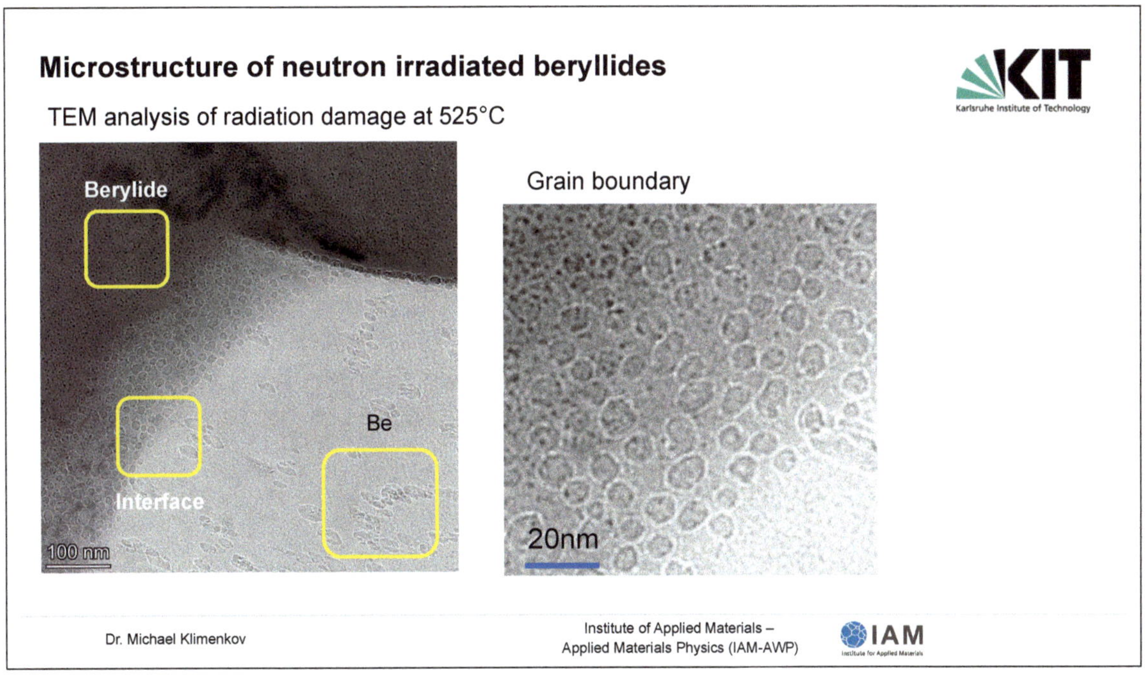

Microstructure of neutron irradiated beryllides
TEM analysis of radiation damage at 525°C
KIT
Karlsruhe Institute of Technology
Berylide
Be
Interface
100 nm
Grain boundary
20nm
Dr. Michael Klimenkov
Institute of Applied Materials –
Applied Materials Physics (IAM-AWP)
IAM

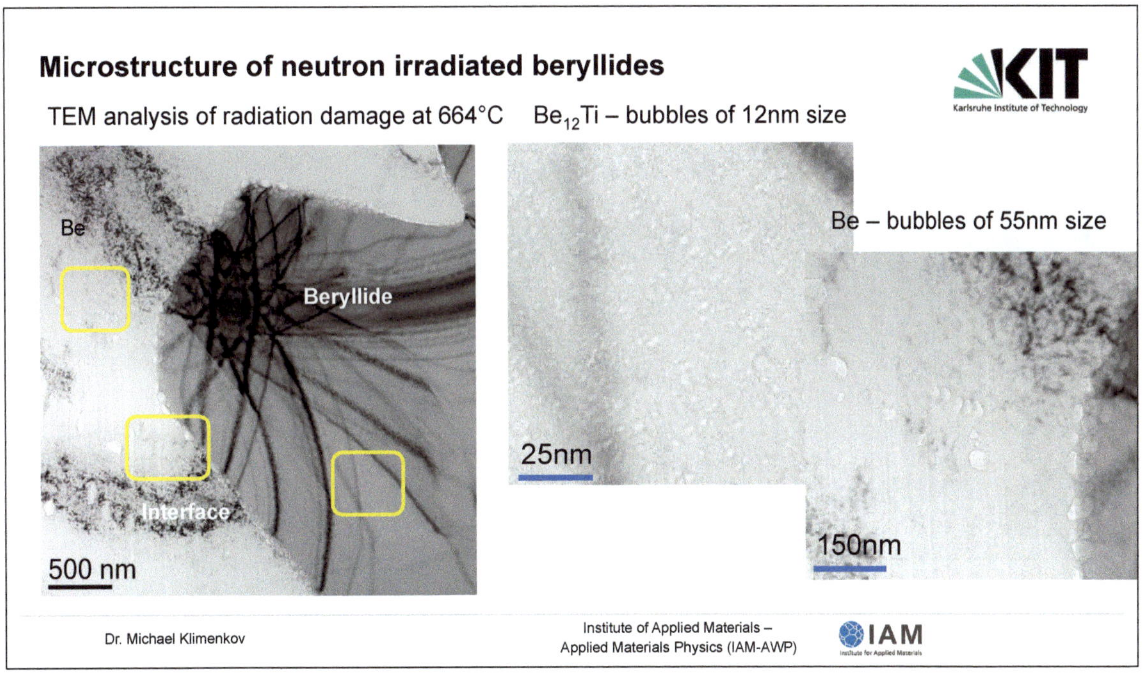

Microstructure of neutron irradiated beryllides
TEM analysis of radiation damage at 664°C
Be12Ti – bubbles of 12nm size
KIT
Karlsruhe Institute of Technology
Be
Beryllide
Interface
500 nm
25nm
Be – bubbles of 55nm size
150nm
Dr. Michael Klimenkov
Institute of Applied Materials –
Applied Materials Physics (IAM-AWP)
IAM

Microstructure of neutron irradiated beryllides 664°C

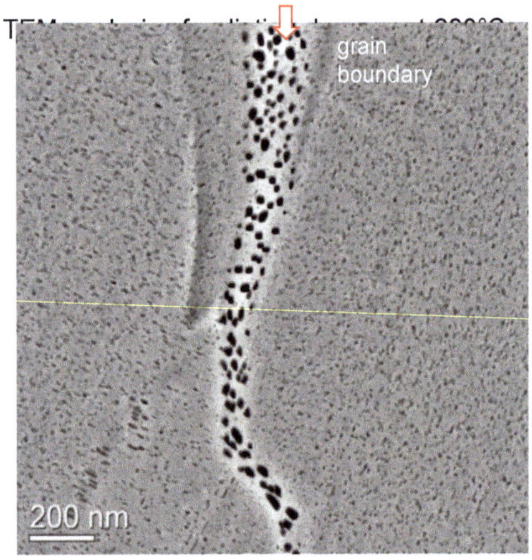

- Bubbles in $Be_{12}Ti$ are log normally distributed and have an average size of about 14 nm.
- The bubble number density can be estimated using STEM-EELS (t ≈ 250 nm) → $6.86\ 10^{21}\ m^{-3}$
- Bubbles at the grain boundary show a ~3 times larger size

Dr. Michael Klimenkov

Institute of Applied Materials –
Applied Materials Physics (IAM-AWP)

Microstructure of neutron irradiated beryllides

TEM analysis of radiation damage at 768°C

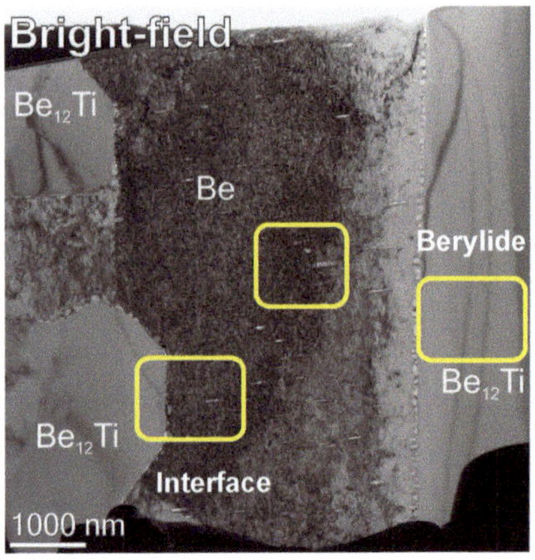

Size of the bubbles up to 300nm

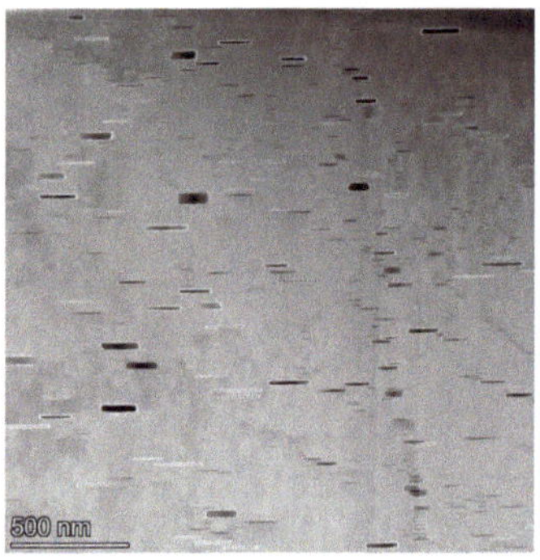

Dr. Michael Klimenkov

Institute of Applied Materials –
Applied Materials Physics (IAM-AWP)

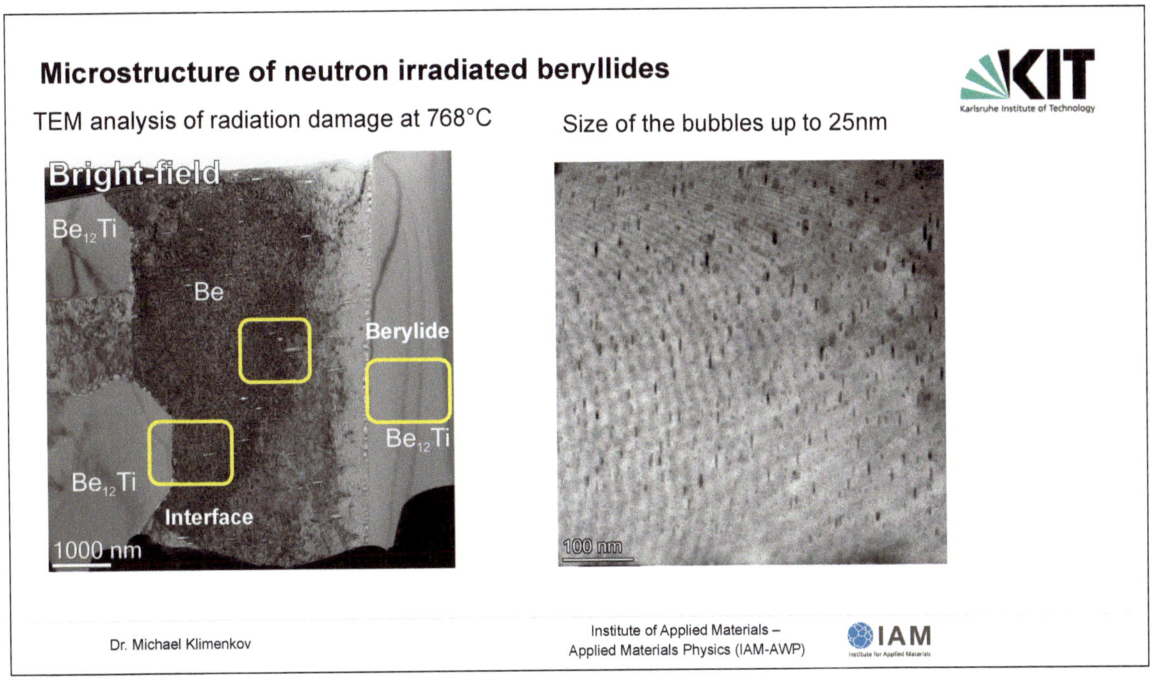
Microstructure of neutron irradiated beryllides
KIT
Karlsruhe Institute of Technology
TEM analysis of radiation damage at 768°C
Size of the bubbles up to 25nm
Bright-field
$Be_{12}Ti$
Be
Berylide
$Be_{12}Ti$
$Be_{12}Ti$
Interface
1000 nm
100 nm
Dr. Michael Klimenkov
Institute of Applied Materials –
Applied Materials Physics (IAM-AWP)
IAM

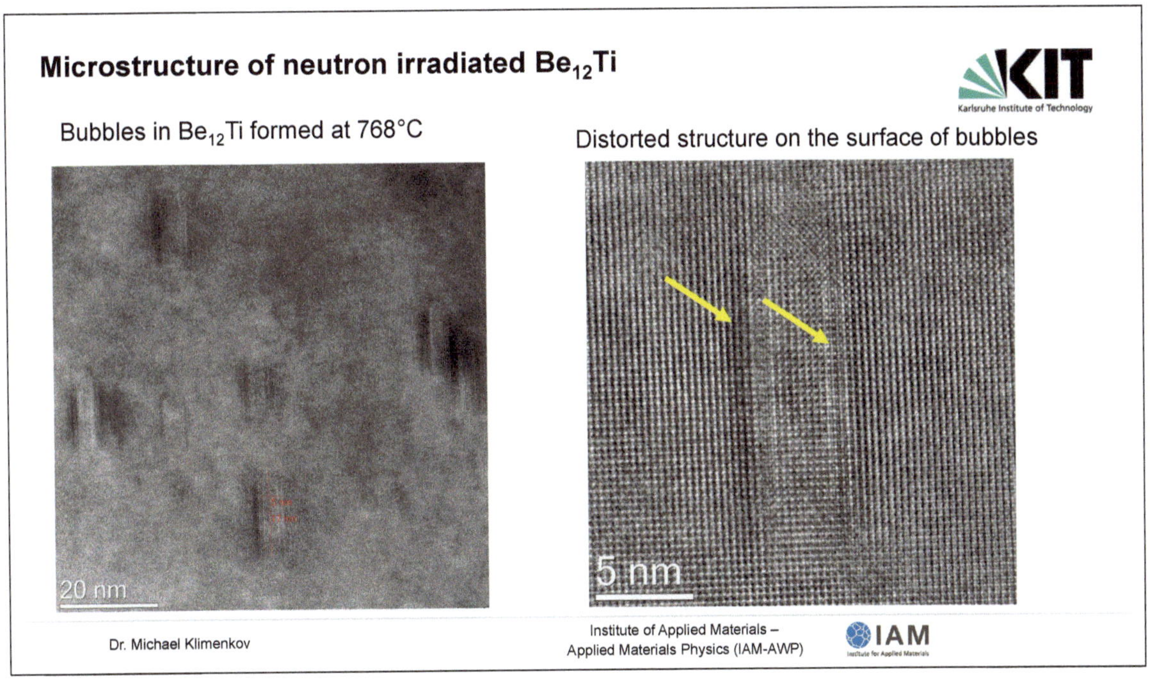
Microstructure of neutron irradiated $Be_{12}Ti$
KIT
Karlsruhe Institute of Technology
Bubbles in $Be_{12}Ti$ formed at 768°C
Distorted structure on the surface of bubbles
20 nm
5 nm
Dr. Michael Klimenkov
Institute of Applied Materials –
Applied Materials Physics (IAM-AWP)
IAM

Microstructure of neutron irradiated beryllides

KIT Karlsruhe Institute of Technology

Effect of radiation damage and temperature on the microstructure of irradiated beryllium and $B_{12}Ti$ phase

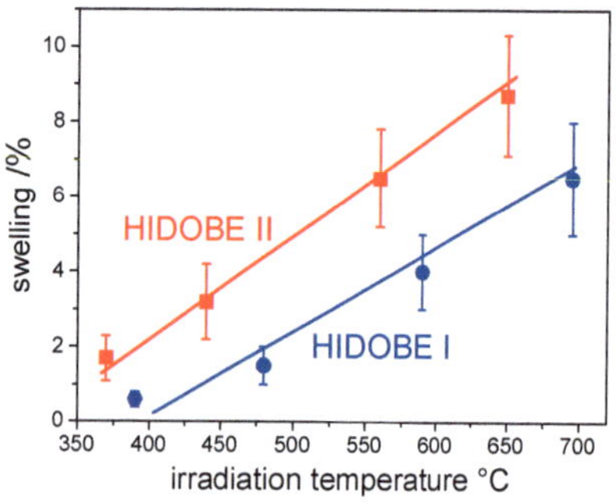

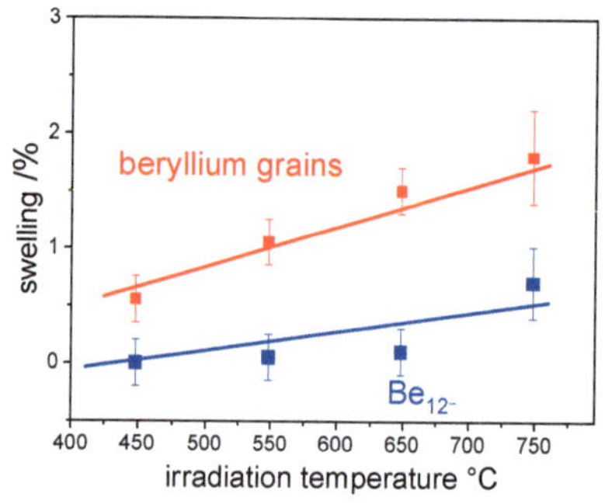

Dr. Michael Klimenkov

Institute of Applied Materials – Applied Materials Physics (IAM-AWP)

IAM

STEM-EELS $Be_{12}Ti$ T_{irr} = 768°C

KIT Karlsruhe Institute of Technology

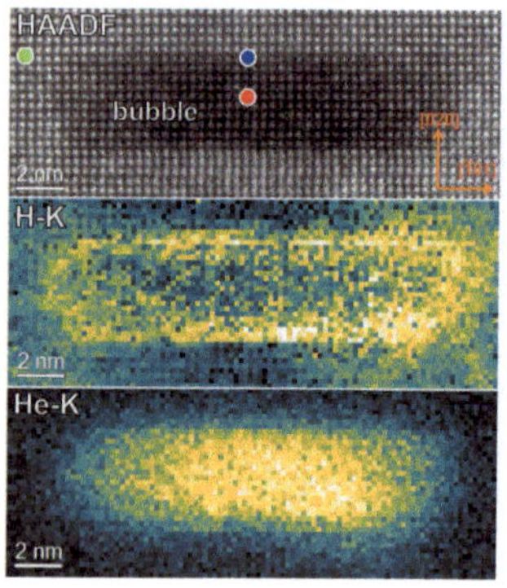

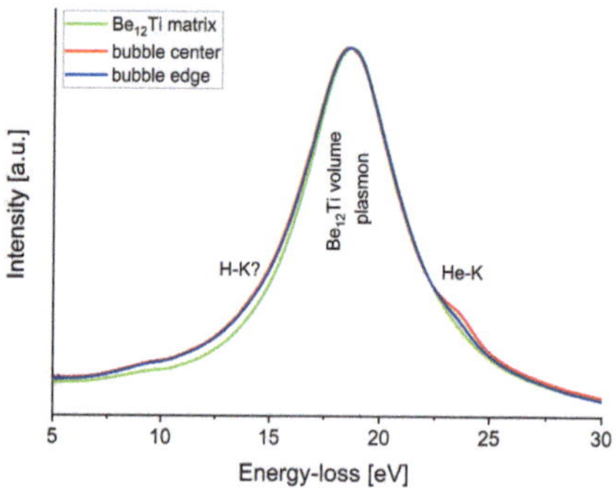

- ^{3}H is preferentially located on bubble surfaces.
- He is in the bubble center.
- Situation in $Be_{12}Ti$ less clear than in pure Be due to experimental limitations.

Dr. Michael Klimenkov

Institute of Applied Materials – Applied Materials Physics (IAM-AWP)

IAM

Summary

- A detailed TEM investigation of beryllium and $Be_{12}Ti$ neutron irradiated at 4 different temperatures was carried out.
- Bubbles produced inside $Be_{12}Ti$ are about 5 times smaller than inside pure Be.
- STEM-EDX shows the presence of 1.5 nm thin Si and Al layers on bubble surfaces.
- STEM-EELS quantification reveals phase composition of $Be_{12}Ti$ close to the nominal values.
- Effects of neutron irradiation on the microstructure of $Be_{12}Ti$ significantly lower than for pure beryllium

29 Dr. Michael Klimenkov Institute of Applied Materials – Applied Materials Physics (IAM-AWP)

SESSION 7

Modeling & experimental validation

7.1 DFT study of helium behavior in titanium beryllides

Dmitry Bachurin, Christopher Stihl, Pavel Vladimirov

Beryllium is planned to be used as a neutron multiplier in future fusion devices such as ITER and DEMO. However, pure beryllium has limitations that can be overcome by using, in particular, titanium beryllides, which have excellent properties, including high strength at elevated temperatures, excellent oxidation resistance, lower tritium retention, and lower swelling. This makes them an attractive candidate material for advanced neutron multiplier breeding blankets for European DEMO concepts. When high-energy neutrons interact with beryllides, transmutation events occur, transforming beryllium into helium and beta-radioactive tritium. In general, swelling of materials is associated with the formation of helium bubbles. At present, there is a lack of theoretical data to fully understand why titanium berylides show lower swelling with respect to pure beryllium. Experimental studies demonstrate that under certain conditions, $Be_{12}Ti$ may contain inclusions of other phases such as Be_2Ti, $Be_{17}Ti_2$ as well as phases of pure metals Be and Ti.

In this regard, these three titanium beryllides were chosen for investigations. All calculations were performed by means of density-functional theory method. In the present work, solution energies of a helium atom located in all non-equivalent interstitial sites of the studied titanium beryllides were computed. The formation energy of all types of vacancies in the titanium beryllides was found to be higher than that in pure beryllium. The binding energy of single helium atom located both inside and outside the vacancies was calculated. The obtained results contribute to a deeper understanding of the origin of the lower swelling of titanium beryllides and for a further study of interstitial helium diffusion at the atomic level.

Presenting Author: Dmitry Bachurin
Presenting Author's Affiliation: Karlsruhe Institute of Technology, Germany
Email: dmitry.bachurin@kit.edu

KIT
Karlsruher Institut für Technologie
BeWS-16
2023
DFT study of helium behavior in titanium beryllides
Dmitry Bachurin, Christopher Stihl, Pavel Vladimirov
KIT – Die Forschungsuniversität in der Helmholtz-Gemeinschaft
www.kit.edu

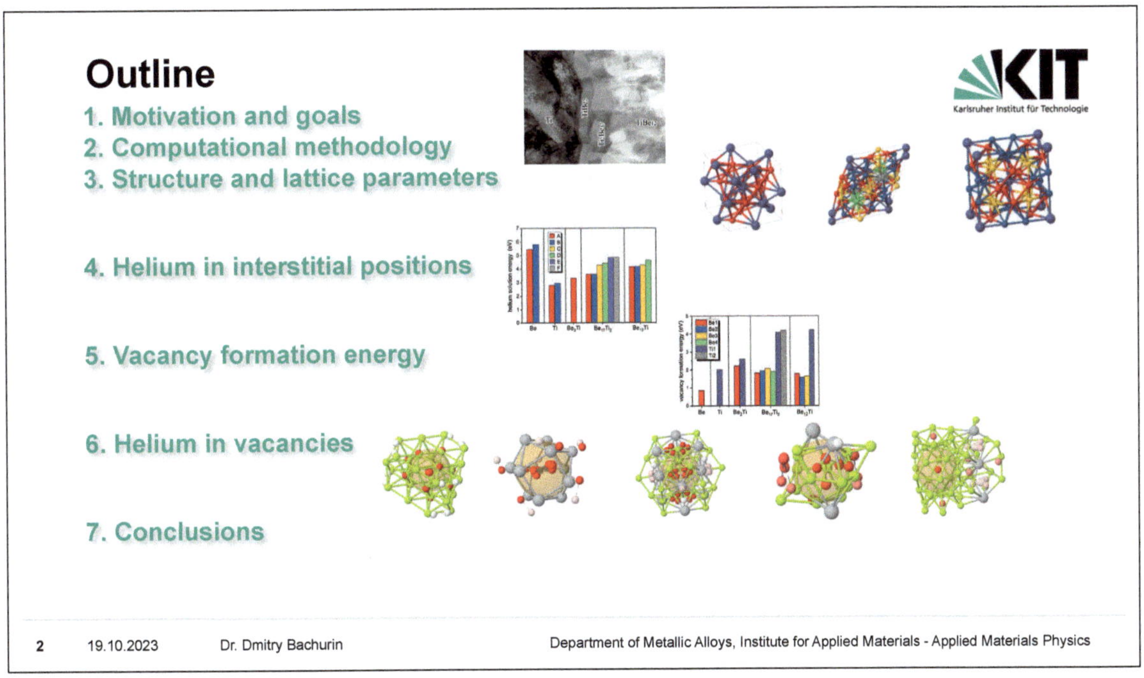
Outline
1. Motivation and goals
2. Computational methodology
3. Structure and lattice parameters
4. Helium in interstitial positions
5. Vacancy formation energy
6. Helium in vacancies
7. Conclusions
KIT
Karlsruher Institut für Technologie
2
19.10.2023
Dr. Dmitry Bachurin
Department of Metallic Alloys, Institute for Applied Materials - Applied Materials Physics

Beryllium in fusion reactor

- Beryllium is considered as effective neutron multiplier for helium cooled pebble bed (HCPB) breeding blanket
- Hydrogen isotopes and impurities are implanted into Be breeding blanket
- He and T are produced in Be under neutron irradiation

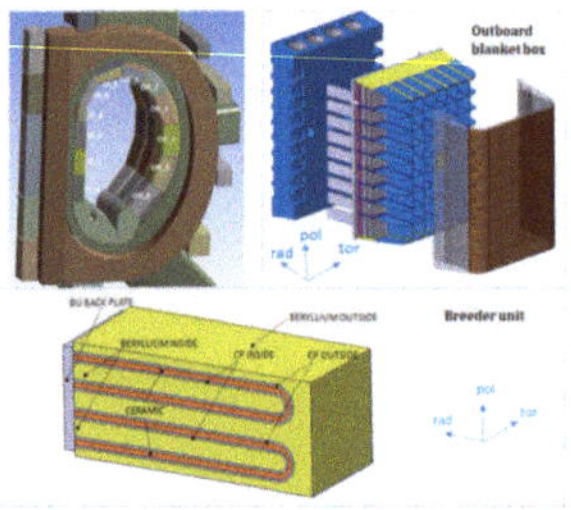

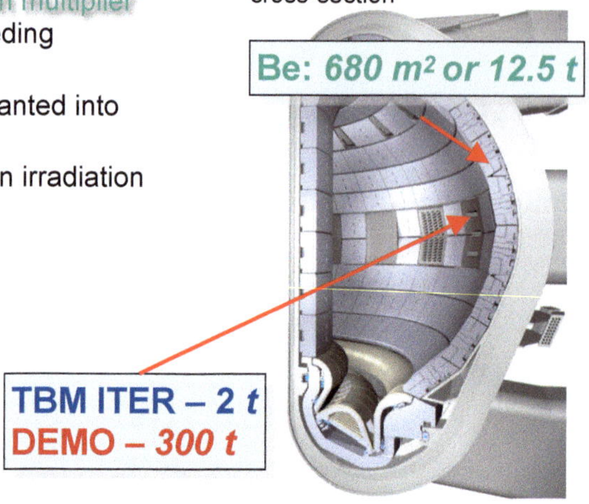

K. Ioki et al. Nucl.Fusion **41**(3) 2001 265-275

Motivation and goals

Motivation: Intermetallic beryllium compounds such as $Be_{12}Ti$, $Be_{12}V$ and $Be_{13}Zr$ are considered as possible candidates for fusion applications, namely as neutron multiplier for the DEMO breeder blanket to be used instead of pure beryllium.

Superior properties of beryllides: higher melting point, higher oxidation resistance, lower tritium retention, lower swelling, good compatibility with the structural materials, **but** lower neutron multiplication efficiency.

Goals: Elucidation of the origin of the superior properties of titanium beryllides (Be_2Ti, $Be_{17}Ti_2$, $Be_{12}Ti$) affecting the tritium release at lower temperatures and investigation of tritium and helium behavior in titanium beryllides.

Approach: DFT methods

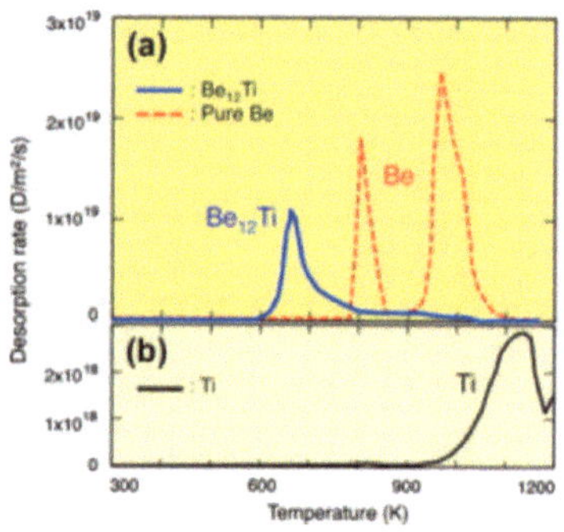

M. Nakamichi et al. JNM 442, 1-3, S465-S471 (2013)

Microstructure of titanium beryllide

Microstructure of Be-Ti composite after extrusion and hot isostatic pressing at 900°C

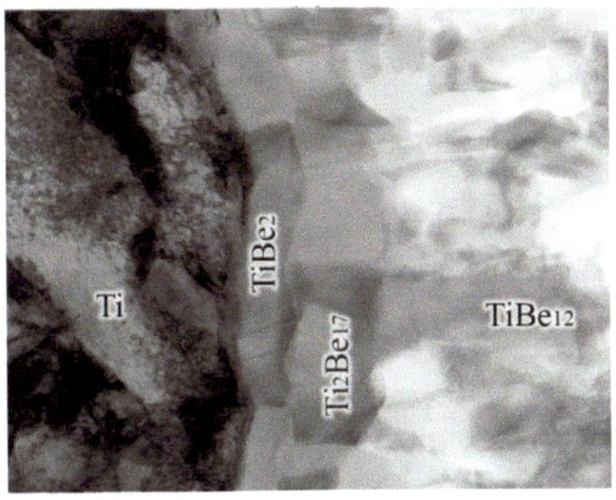

R.G. Gaisin, V. Chakin, M. Duerrschnabel et al. Nucl Mater Energy 24 (2020) 100771.

The structure along with the main phase of $Be_{12}Ti$ contains inclusions of other phases: Be_2Ti, $Be_{17}Ti_2$, pure Ti and Be

- Be_2Ti in the form of a thin layer surrounding the pure Ti phase;
- $Be_{17}Ti_2$ in the form of small particles located at the prior Ti phase locations;
- pure Ti and Be phases, which do not dissolve completely

5 19.10.2023 Dr. Dmitry Bachurin Department of Metallic Alloys, Institute for Applied Materials - Applied Materials Physics

Computational methodology

- Static ab-initio calculations (using VASP)
- Projector augmented wave potential (PAW)
- Generalized gradient approximation (GGA)
- Fermi broadening: 0.2 eV
- Cut-off energy: 487 eV

Helium solution energy

$$E_s = E_{total}^{Be+He} - E_{total}^{Be} - E_{ref}^{He}$$

E_{total}^{Be+He} and E_{total}^{Be} are the total energies of the simulation cells with and without helium atom

$E_{ref}^{He} = 0.001\text{eV}$ is the energy of a single helium atom

Helium binding energy

$$E_b = E(He+V) - E(V) + E(He_I) - E_{bulk}$$

$E(He+V)$ and $E(V)$ are the total energies of the simulation cells with vacancy and one helium atom and only with a vacancy

$E(He_I)$ is the energy of helium atom in interstitial position

E_{bulk} is the energy of the bulk

6 19.10.2023 Dr. Dmitry Bachurin Department of Metallic Alloys, Institute for Applied Materials - Applied Materials Physics

Structure and lattice parameters

KIT Karlsruher Institut für Technologie

	Be_2Ti cubic	$Be_{17}Ti_2$ hexagonal	$Be_{12}Ti$ tetragonal
	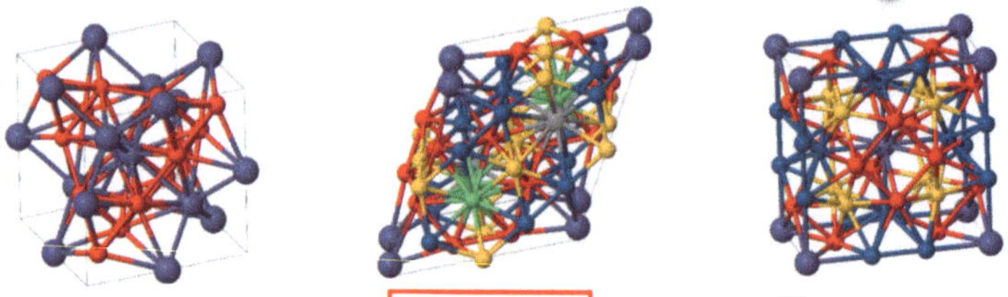		
Lattice parameters	*a* = 4.54Å	*a* = 7.33Å *c* = 7.16Å	*a* = 7.23Å *c* = 4.15Å
Computational cell	32 Be atoms 16 Ti atoms	34 Be atoms 4 Ti atoms	48 Be atoms 4 Ti atoms
Cryst. non-eq. atomic sites	2 (1Be +1Ti)	6 (4Be +2Ti)	4 (3Be +1Ti)

7 19.10.2023 Dr. Dmitry Bachurin Department of Metallic Alloys, Institute for Applied Materials - Applied Materials Physics

Helium in interstitial positions

KIT Karlsruher Institut für Technologie

Helium atom was placed in the center of all possible non-equivalent interstitial sites

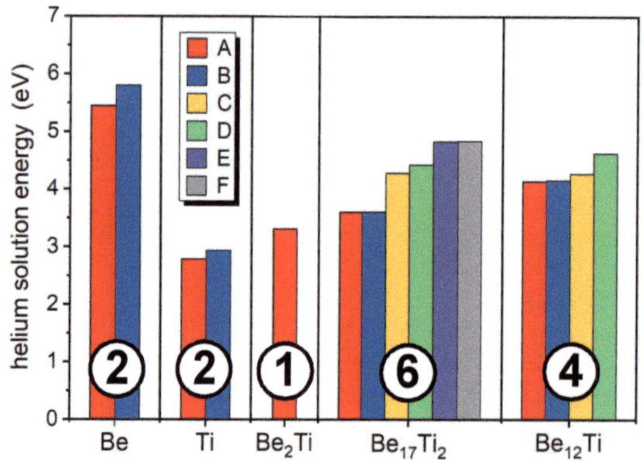

Positive solution energy (endothermic)
Negative solution energy (exothermic)

will be used as input to calculate helium diffusion in titanium beryllides

Hydrogen diffusion paths $Be_{12}Ti$

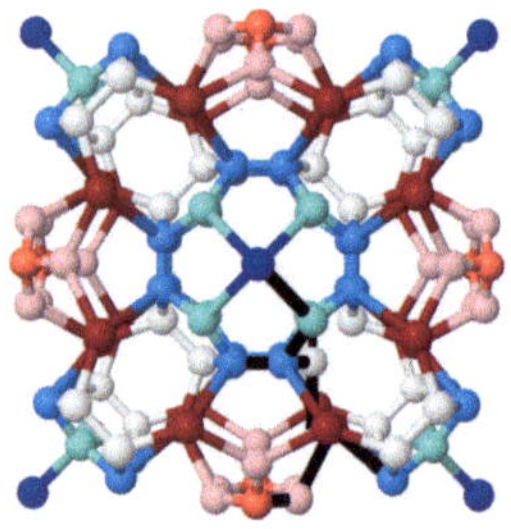

D.V. Bachurin, C. Stihl, P.V. Vladimirov, Comp. Mat. Sci. 187 (2021) 109921.

8 19.10.2023 Dr. Dmitry Bachurin Department of Metallic Alloys, Institute for Applied Materials - Applied Materials Physics

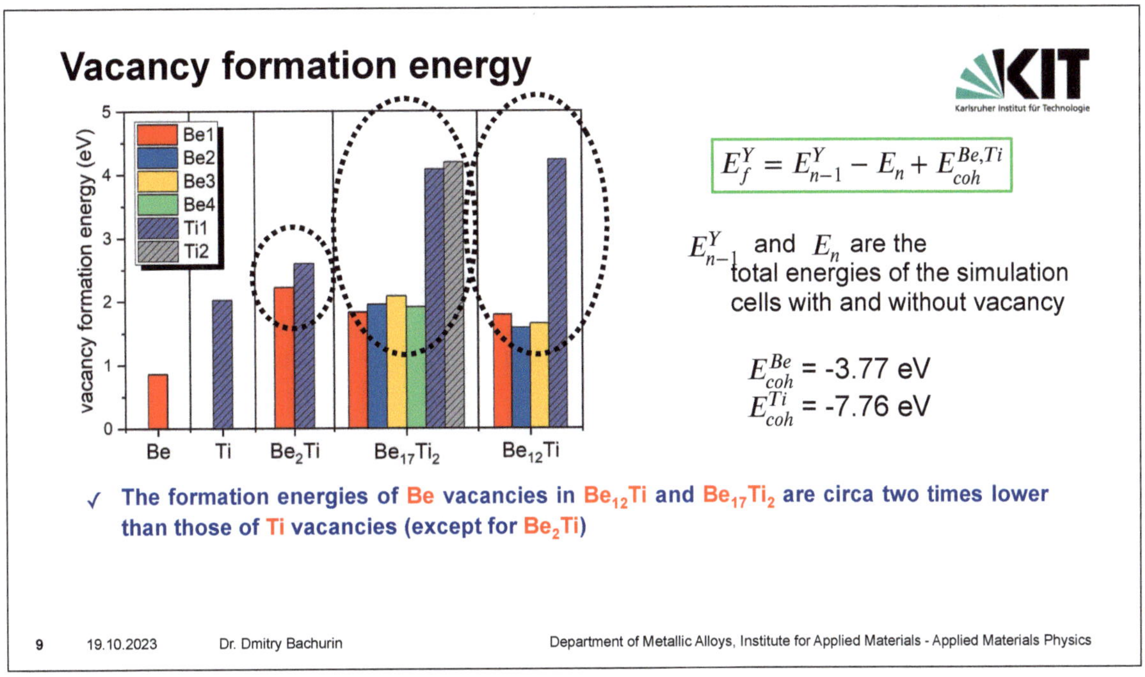
Vacancy formation energy
KIT
Karlsruher Institut für Technologie
vacancy formation energy (eV)
0
1
2
3
4
5
Be1
Be2
Be3
Be4
Ti1
Ti2
Be
Ti
Be2Ti
Be17Ti2
Be12Ti
$E_f^Y = E_{n-1}^Y - E_n + E_{coh}^{Be,Ti}$
E_{n-1}^Y and E_n are the total energies of the simulation cells with and without vacancy
E_{coh}^{Be} = -3.77 eV
E_{coh}^{Ti} = -7.76 eV
The formation energies of Be vacancies in Be12Ti and Be17Ti2 are circa two times lower than those of Ti vacancies (except for Be2Ti)
9
19.10.2023
Dr. Dmitry Bachurin
Department of Metallic Alloys, Institute for Applied Materials - Applied Materials Physics

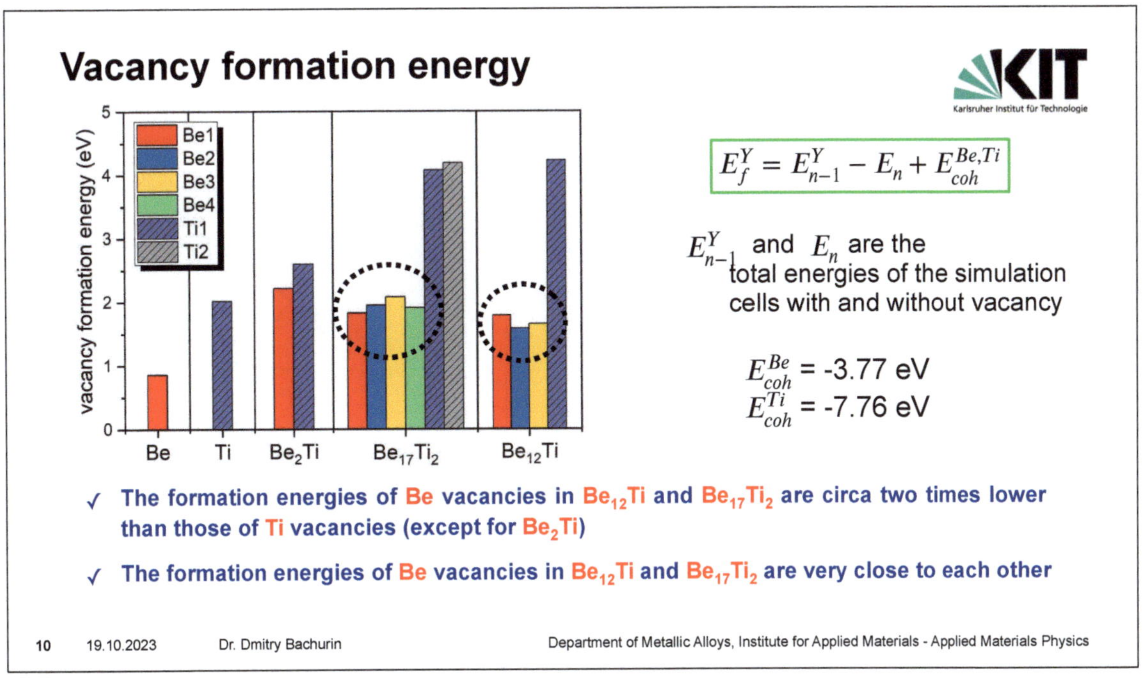
Vacancy formation energy
KIT
Karlsruher Institut für Technologie
vacancy formation energy (eV)
0
1
2
3
4
5
Be1
Be2
Be3
Be4
Ti1
Ti2
Be
Ti
Be2Ti
Be17Ti2
Be12Ti
$E_f^Y = E_{n-1}^Y - E_n + E_{coh}^{Be,Ti}$
E_{n-1}^Y and E_n are the total energies of the simulation cells with and without vacancy
E_{coh}^{Be} = -3.77 eV
E_{coh}^{Ti} = -7.76 eV
The formation energies of Be vacancies in Be12Ti and Be17Ti2 are circa two times lower than those of Ti vacancies (except for Be2Ti)
The formation energies of Be vacancies in Be12Ti and Be17Ti2 are very close to each other
10
19.10.2023
Dr. Dmitry Bachurin
Department of Metallic Alloys, Institute for Applied Materials - Applied Materials Physics

Binding energy of He atom in vacancy

KIT Karlsruher Institut für Technologie

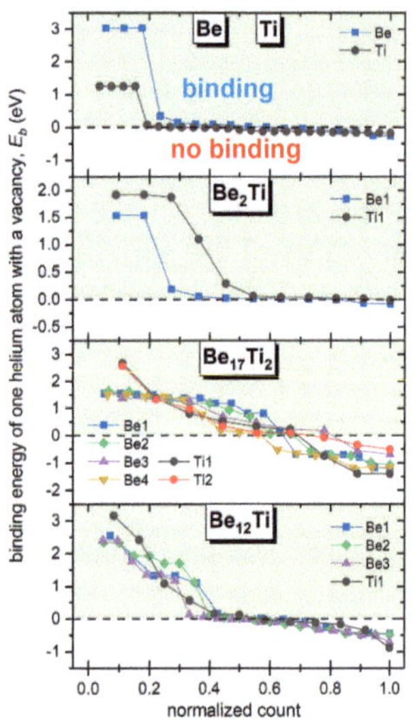

- Binding energy of He atom in Be vacancy is circa 1.8eV higher than that in Ti vacancy
- (Be_2Ti) Binding energies of He atom with Be and Ti vacancies are 1.53 and 1.92eV, respectively
- ($Be_{17}Ti_2$) Binding energies of He atom with Be vacancies are 1.53, 1.62, 1.37 and 1.49eV, with Ti vacancies are 2.66 and 2.57eV
- ($Be_{12}Ti$) Binding energies of He atom with Be vacancies are 2.53, 2.32 and 2.39eV, with Ti vacancy is 3.14eV
- Binding energy of He atom with Ti vacancy in titanium beryllides is always higher than with Be vacancy

11 19.10.2023 Dr. Dmitry Bachurin Department of Metallic Alloys, Institute for Applied Materials - Applied Materials Physics

Binding energy of helium atom in vacancy

KIT Karlsruher Institut für Technologie

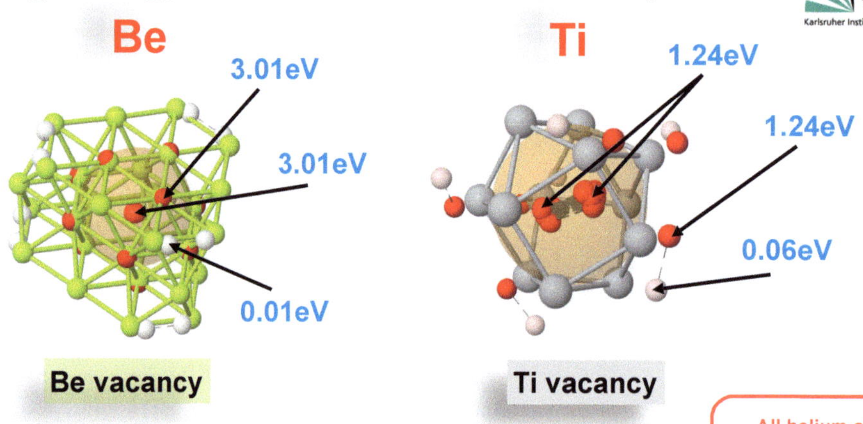

- ✓ Binding energy of He atom with Be vacancy is significantly higher than that in Ti
- ✓ Some helium atoms located outside a vacancy are trapped by it

All helium atoms with negative binding energy were made invisible

12 19.10.2023 Dr. Dmitry Bachurin Department of Metallic Alloys, Institute for Applied Materials - Applied Materials Physics

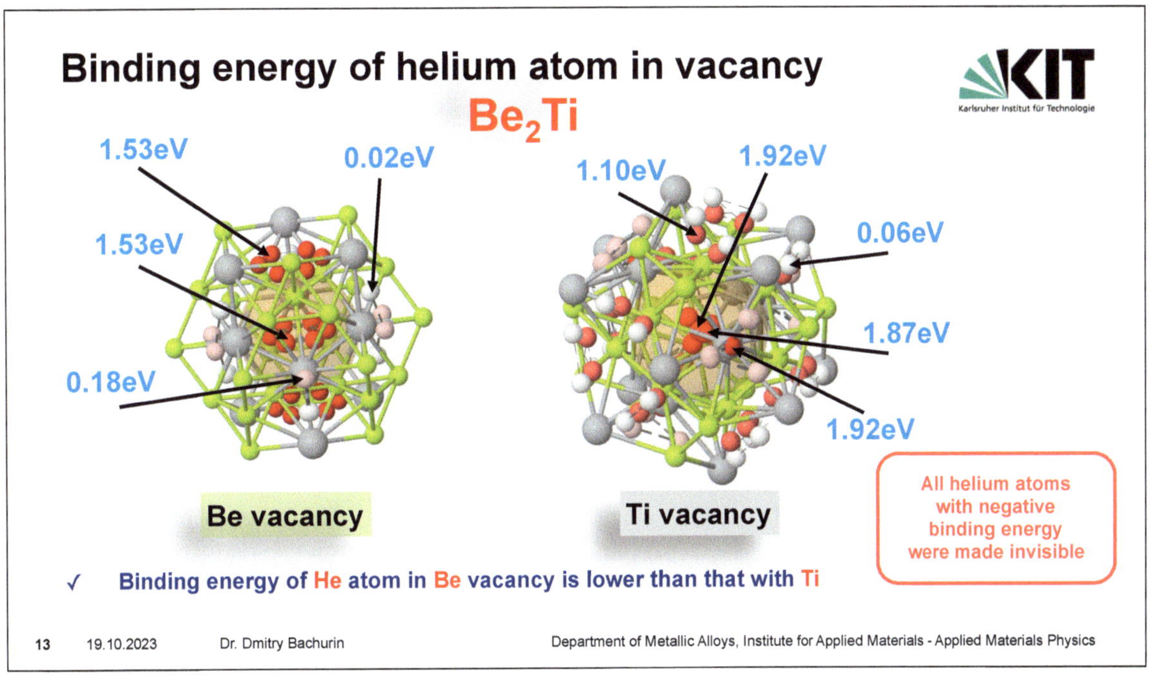
Binding energy of helium atom in vacancy
Be_2Ti
KIT
Karlsruher Institut für Technologie
1.53eV
0.02eV
1.53eV
0.18eV
1.10eV
1.92eV
0.06eV
1.87eV
1.92eV
Be vacancy
Ti vacancy
All helium atoms with negative binding energy were made invisible
✓ Binding energy of He atom in Be vacancy is lower than that with Ti
13 19.10.2023 Dr. Dmitry Bachurin Department of Metallic Alloys, Institute for Applied Materials - Applied Materials Physics

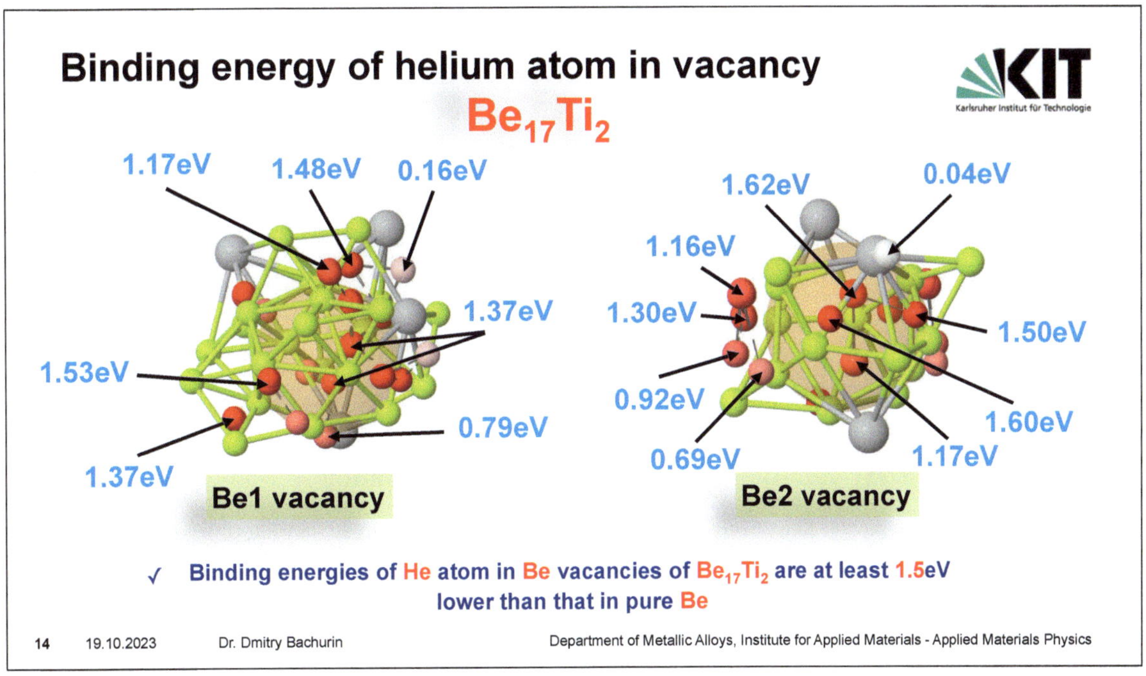
Binding energy of helium atom in vacancy
$Be_{17}Ti_2$
KIT
Karlsruher Institut für Technologie
1.17eV
1.48eV
0.16eV
1.37eV
1.53eV
0.79eV
1.37eV
Be1 vacancy
1.62eV
0.04eV
1.16eV
1.30eV
1.50eV
0.92eV
1.60eV
0.69eV
1.17eV
Be2 vacancy
✓ Binding energies of He atom in Be vacancies of $Be_{17}Ti_2$ are at least 1.5eV lower than that in pure Be
14 19.10.2023 Dr. Dmitry Bachurin Department of Metallic Alloys, Institute for Applied Materials - Applied Materials Physics

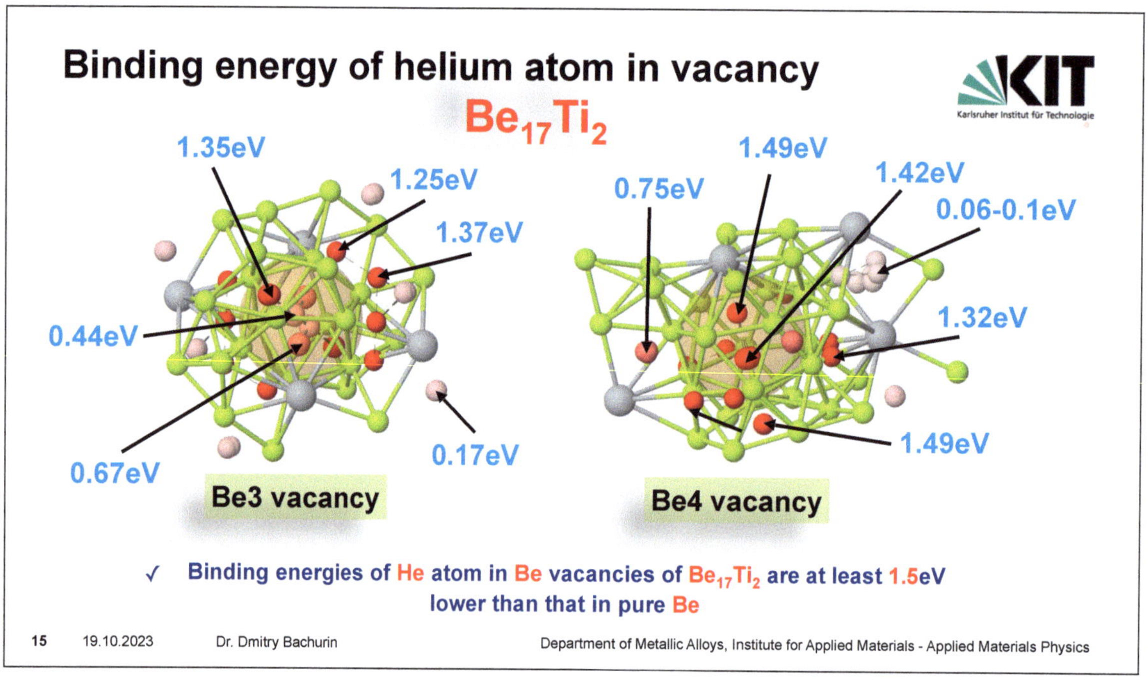
Binding energy of helium atom in vacancy
Be17Ti2
KIT
Karlsruher Institut für Technologie
1.35eV
1.25eV
1.37eV
0.44eV
0.67eV
0.17eV
Be3 vacancy
1.49eV
0.75eV
1.42eV
0.06-0.1eV
1.32eV
1.49eV
Be4 vacancy
✓ Binding energies of He atom in Be vacancies of Be17Ti2 are at least 1.5eV lower than that in pure Be
15 19.10.2023 Dr. Dmitry Bachurin Department of Metallic Alloys, Institute for Applied Materials - Applied Materials Physics

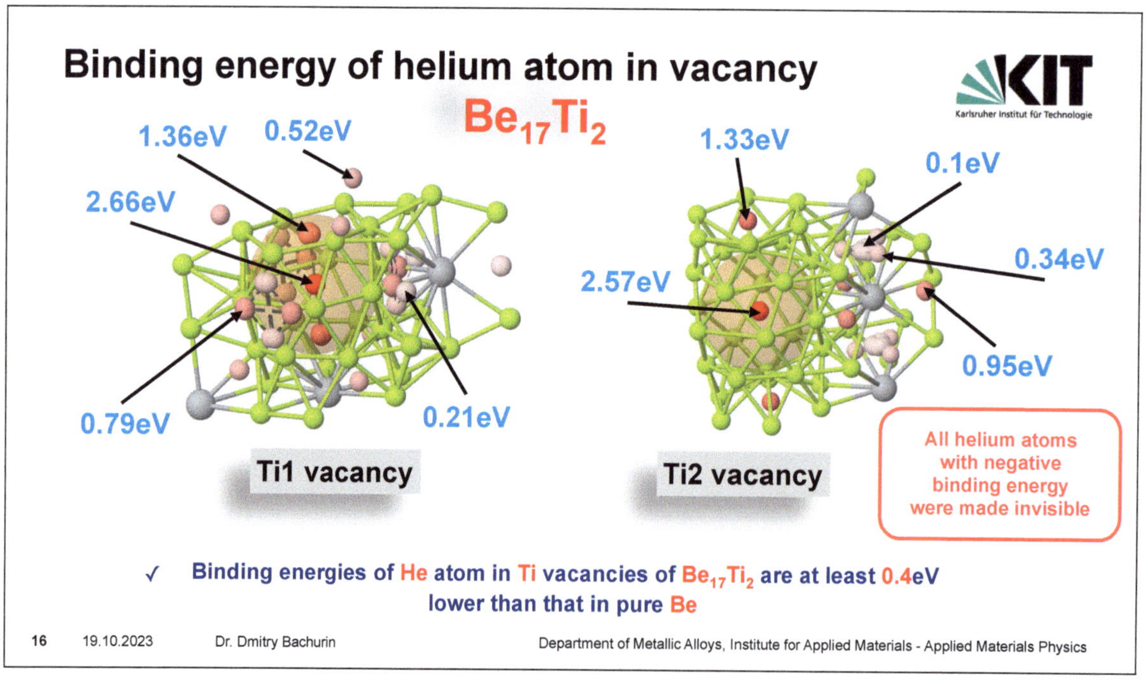
Binding energy of helium atom in vacancy
Be17Ti2
KIT
Karlsruher Institut für Technologie
1.36eV
0.52eV
2.66eV
0.79eV
0.21eV
Ti1 vacancy
1.33eV
0.1eV
0.34eV
2.57eV
0.95eV
Ti2 vacancy
All helium atoms with negative binding energy were made invisible
✓ Binding energies of He atom in Ti vacancies of Be17Ti2 are at least 0.4eV lower than that in pure Be
16 19.10.2023 Dr. Dmitry Bachurin Department of Metallic Alloys, Institute for Applied Materials - Applied Materials Physics

Binding energy of helium atom in vacancy

$Be_{12}Ti$

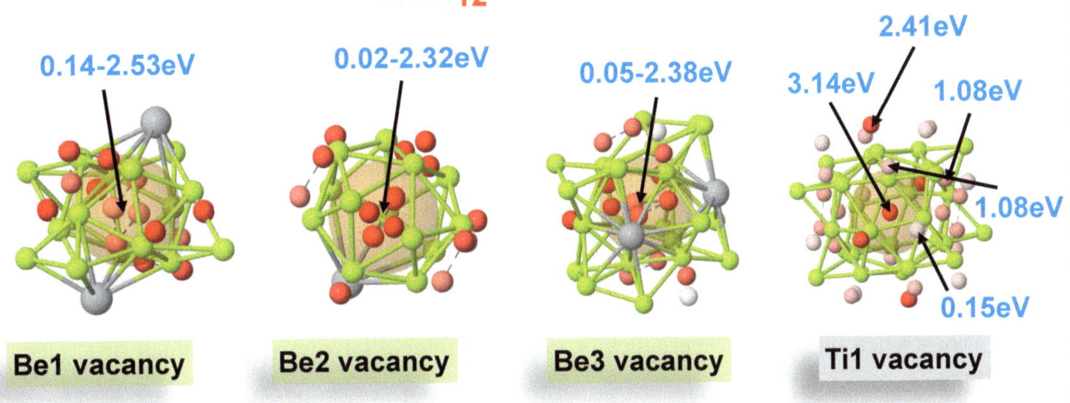

- ✓ Binding energies of He atom in the vacancies of $Be_{12}Ti$ are lower than in pure Be
- ✓ Binding energies of He atom inside the vacancies of $Be_{12}Ti$ are always higher than outside them

17 19.10.2023 Dr. Dmitry Bachurin Department of Metallic Alloys, Institute for Applied Materials - Applied Materials Physics

Conclusions

- Helium solution energy in interstitial positions of titanium beryllides are at least 0.6-2.1eV lower than that for Be indicating easier dissolution of helium atoms in the beryllides.
- Formation energy of Be vacancies is circa two times lower than the corresponding energy for Ti vacancies in $Be_{12}Ti$ and $Be_{17}Ti_2$. In Be_2Ti the formation energies of Be and Ti vacancies are similar to each other.
- Helium is more strongly bound to the titanium vacancy than to the beryllium vacancy in the titanium beryllides. Helium binding energy to a monovacancy in the titanium beryllides $Be_{12}Ti$ and $Be_{17}Ti_2$ is of the same order as in pure beryllium (with the exception of Be_2Ti).
- Some helium atoms may be trapped by a vacancy without being directly inside it.
- A helium atom in the vicinity of a vacancy causes displacement of neighboring Be or Ti atoms resulting in vacancy migration.
- The data obtained will be further used for the calculation of helium diffusion coefficients in titanium beryllium.

18 19.10.2023 Dr. Dmitry Bachurin Department of Metallic Alloys, Institute for Applied Materials - Applied Materials Physics

7.2 Hydrogen Diffusion in Titanium Beryllides

Pavel Vladimirov, Dmitry Bachurin, Christopher Stihl

Beryllium-based intermetallic alloys are alternative candidates for solid neutron multiplier material replacing pure beryllium in a tritium-breeding blanket of the demonstration fusion reactor DEMO. Preselection of the most suitable crystal structure from a variety of phases acquired by beryllides is a primary task for material scientists and designers. Although retaining significantly less tritium after neutron irradiation than beryllium, tritium inventory in beryllides presents a substantial concern for the safety and postprocessing of radioactive wastes.

In this work, hydrogen diffusion in titanium beryllides ($Be_{12}Ti$, Be_2Ti) is investigated using analytical methods for the calculation of hydrogen diffusion coefficient in intermetallic compounds using migration and solution energies previously obtained by static ab initio calculations. Cubic symmetry of Be_2Ti implies isotropic diffusion, while $Be_{12}Ti$ phase, as expected from its tetragonal crystal structure, shows anisotropic hydrogen diffusion along the c- and a-axis. Particular channels of superfast diffusion were revealed in $Be_{12}Ti$. However, such channels are activated either at high hydrogen concentrations when other more energetically favorable interstitial sites are occupied or at elevated temperatures.

The implications of the obtained results on tritium diffusion in beryllides are discussed with respect to the reduction of the overall tritium inventory in neutron multiplier materials.

Presenting Author: Pavel Vladimirov
Presenting Author's Affiliation: Karlsruhe Institute of Technology, Germany
Email: pavel.vladimirov@kit.edu

KIT
Karlsruhe Institute of Technology

Hydrogen diffusion in beryllides

P. Vladimirov, D. Bachurin and C. Stihl

KIT – The Research University in the Helmholtz Association

www.kit.edu

Outline

KIT
Karlsruhe Institute of Technology

- Introduction
  - Motivation
  - Master equation of diffusion on Bravias and non-Bravias Lattices
- Method
- Results
  - Diffusion in cubic $TiBe_2$
  - Diffusion in tetragonal $TiBe_{12}$
- Conclusions

2 20.10.2023 Pavel Vladimirov - Hydrogen diffusion in beryllides Department of Metallic Alloys, Institute for Applied Materials - Applied Materials Physics IAM

From pebble beds of Be to solid blocks of $TiBe_{12}$

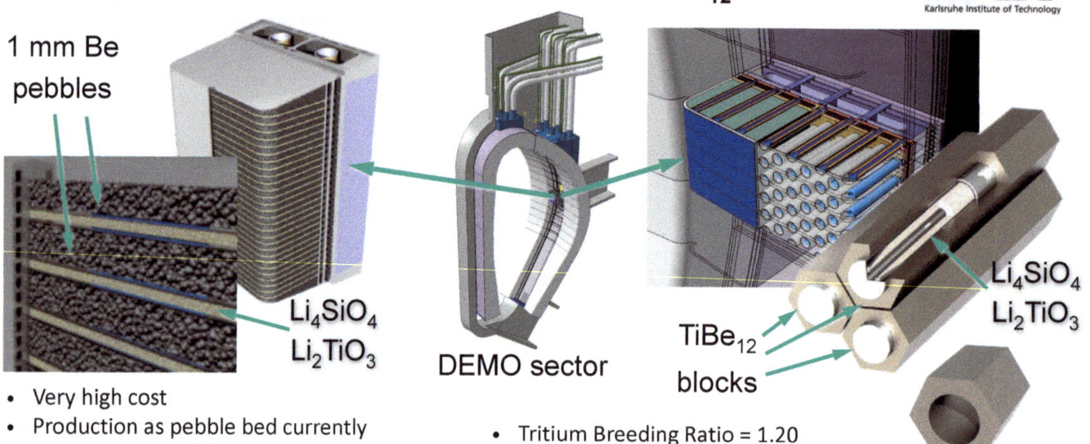

- Very high cost
- Production as pebble bed currently unfeasible for DEMO volumes

- Tritium Breeding Ratio = 1.20
- Lack of production technology for $TiBe_{12}$ blocks

Pavel Vladimirov - Hydrogen diffusion in beryllides

Department of Metallic Alloys, Institute for Applied Materials - Applied Materials Physics

Tritium accumulation in beryllides

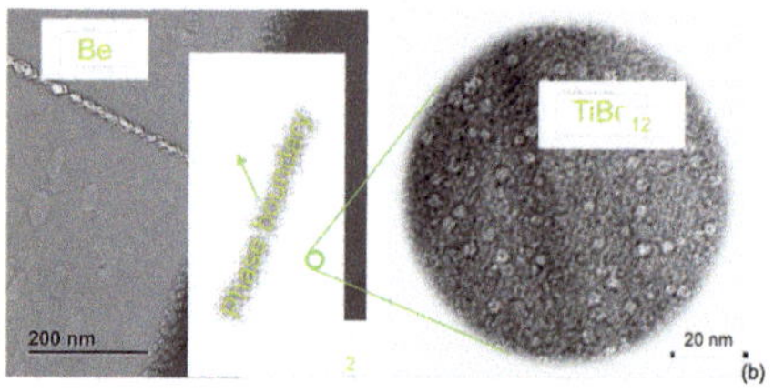

Beryllium and titanium beryllide phases irradiated in HIDOBE-02 at 530°C

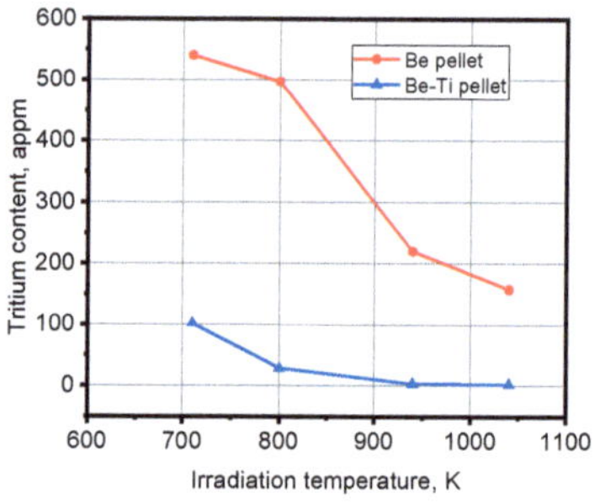

T-retention in Be or Be-Ti phase mixture irradiated at various temperatures

Pavel Vladimirov - Hydrogen diffusion in beryllides

Department of Metallic Alloys, Institute for Applied Materials - Applied Materials Physics

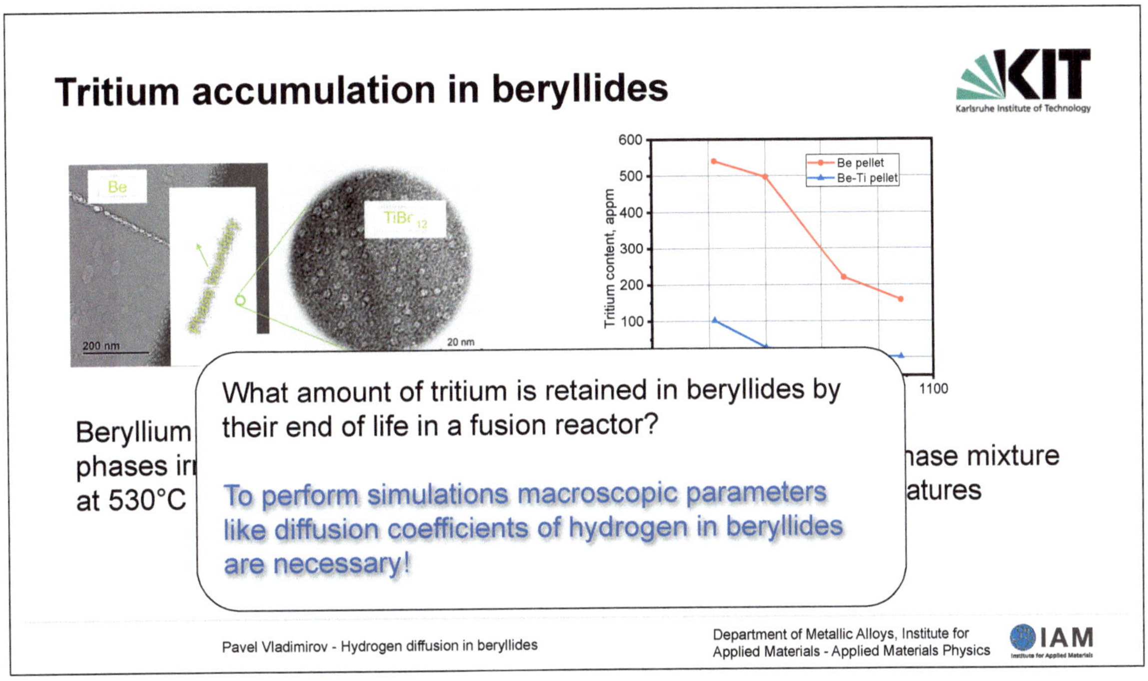
Tritium accumulation in beryllides
KIT
Karlsruhe Institute of Technology
Be
TiBe12
200 nm
20 nm
600
500
400
300
200
100
Tritium content, appm
Be pellet
Be-Ti pellet
1100
Beryllium
phases ir
at 530°C
hase mixture
atures
What amount of tritium is retained in beryllides by their end of life in a fusion reactor?
To perform simulations macroscopic parameters like diffusion coefficients of hydrogen in beryllides are necessary!
Pavel Vladimirov - Hydrogen diffusion in beryllides
Department of Metallic Alloys, Institute for Applied Materials - Applied Materials Physics
IAM

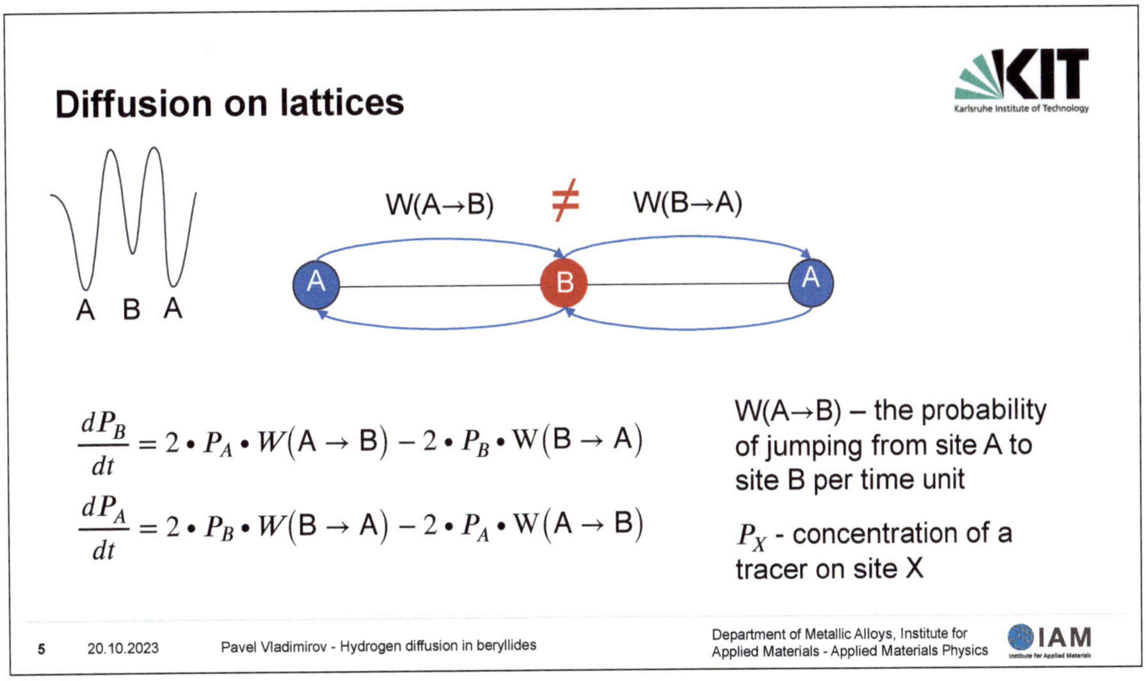
Diffusion on lattices
KIT
Karlsruhe Institute of Technology
A B A
W(A→B) ≠ W(B→A)
A
B
A
$\frac{dP_B}{dt} = 2 \bullet P_A \bullet W(A \to B) - 2 \bullet P_B \bullet W(B \to A)$
$\frac{dP_A}{dt} = 2 \bullet P_B \bullet W(B \to A) - 2 \bullet P_A \bullet W(A \to B)$
W(A→B) – the probability of jumping from site A to site B per time unit
P_X - concentration of a tracer on site X
5
20.10.2023
Pavel Vladimirov - Hydrogen diffusion in beryllides
Department of Metallic Alloys, Institute for Applied Materials - Applied Materials Physics
IAM

Master equation of diffusion

$$\frac{dP_i}{dt} = \underbrace{\sum_{j=1}^{N} P_j \bullet W(j \to i)}_{\text{Incoming flux}} - \underbrace{P_i \sum_{k=1}^{N} W(i \to k)}_{\text{Outcoming flux}}$$

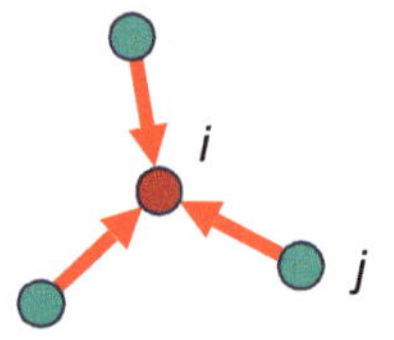

Incoming fluxes

Formally, the summation is over all sites in the jumping network, but in reality, it is only over interconnected sites. For not connected sites the transition (jump) probability is zero.

6 20.10.2023 Pavel Vladimirov - Hydrogen diffusion in beryllides Department of Metallic Alloys, Institute for Applied Materials - Applied Materials Physics

Master equation of diffusion

$$\frac{dP_i}{dt} = \underbrace{\sum_{j=1}^{N} P_j \bullet W(j \to i)}_{\text{Incoming flux}} - \underbrace{P_i \sum_{k=1}^{N} W(i \to k)}_{\text{Outcoming flux}}$$

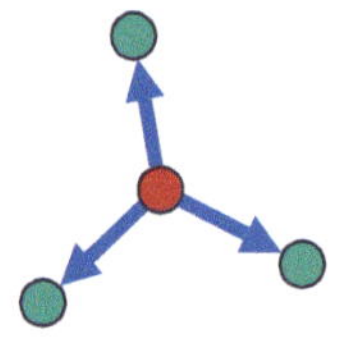

Outcoming fluxes

Formally, the summation is over all sites in the jumping network, but in reality, it is only over interconnected sites. For not connected sites the transition (jump) probability is zero.

W(i→j) denotes the conditional probability for jumping from the site *i* to the site *j*

7 20.10.2023 Pavel Vladimirov - Hydrogen diffusion in beryllides Department of Metallic Alloys, Institute for Applied Materials - Applied Materials Physics

Diffusion on non-Bravias lattices

KIT Karlsruhe Institute of Technology

Bravias lattice = one atom per unit cell

non-Bravias lattice = more than one atom per unit cell

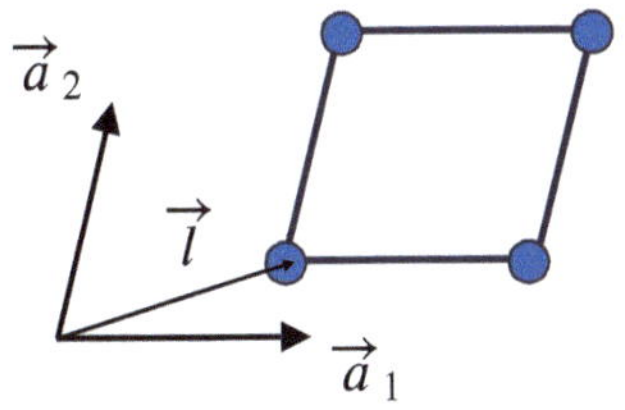

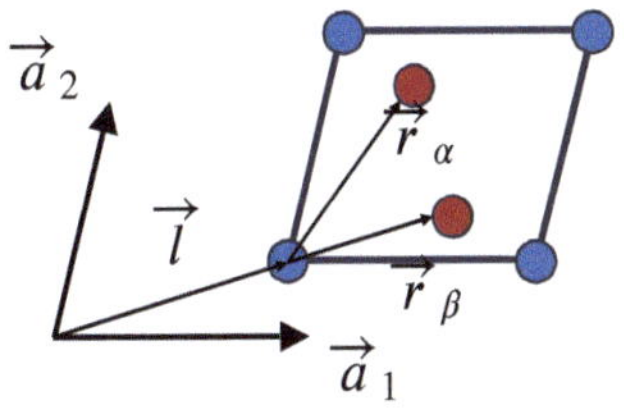

$$\vec{R}\left(\vec{l}\right) = l_1\vec{a}_1 + l_2\vec{a}_2 + l_3\vec{a}_3$$

$$\vec{R}_\alpha\left(\vec{l}\right) = l_1\vec{a}_1 + l_2\vec{a}_2 + l_3\vec{a}_3 + \vec{r}_\alpha$$

8 20.10.2023 Pavel Vladimirov - Hydrogen diffusion in beryllides

Department of Metallic Alloys, Institute for Applied Materials - Applied Materials Physics IAM

Diffusion on non-Bravias lattices

KIT Karlsruhe Institute of Technology

Master equation:

$$\frac{dP_\alpha(\boldsymbol{l},t)}{dt} = \sum_{\beta=1}^{N}\sum_{\boldsymbol{l}'}[W_{\beta\alpha}^{\boldsymbol{l}'\boldsymbol{l}}P_\beta(\boldsymbol{l}',t) - W_{\alpha\beta}^{\boldsymbol{l}\boldsymbol{l}'}P_\alpha(\boldsymbol{l},t)]$$

Summation is going over all sites β which are connected with the given site α

Here $P_\alpha(\boldsymbol{l},t)$ is a probability to find a particle at time t at sublattice site α within unit cell which position is determined by vector $\boldsymbol{l}$.

$W_{\beta\alpha}^{\boldsymbol{l}'\boldsymbol{l}}$ is a jump rate which is a probability to jump from lattice site $^{\boldsymbol{l}'}_{\beta}$ to site $^{\boldsymbol{l}}_{\alpha}$.

$P_\alpha(\boldsymbol{l},t) = p_\alpha \cdot e^{i\boldsymbol{k}\boldsymbol{R}_\alpha(\boldsymbol{l})-\lambda t}$

Fourier & Laplace transformations for analysis of relaxation times of the system: $1/\lambda$

9 20.10.2023 Pavel Vladimirov - Hydrogen diffusion in beryllides

Department of Metallic Alloys, Institute for Applied Materials - Applied Materials Physics IAM

Master equation of diffusion

KIT Karlsruhe Institute of Technology

The master equation in vector/matrix form:

$$\frac{d\vec{P}}{dt} = \hat{A} \bullet \vec{P}$$

$$-\lambda \vec{P} = \hat{A} \bullet \vec{P}$$

$$\left(\hat{A} + \lambda \hat{I}\right) \bullet \vec{P} = 0$$

$$det\left(\hat{A} + \lambda \hat{I}\right) = 0$$

This is a linear system of ODEs and its solution depends on the properties of the matrix $\hat{A}$.

If $\hat{A}$ is a symmetric matrix $N \times N$ it has N real eigenvalues and N corresponding real orthogonal eigenvectors.

10 20.10.2023 Pavel Vladimirov - Hydrogen diffusion in beryllides

Department of Metallic Alloys, Institute for Applied Materials - Applied Materials Physics IAM

Master equation of diffusion

KIT Karlsruhe Institute of Technology

The master equation in vector/matrix form:

$$\frac{d\vec{P}}{dt} = \hat{A} \bullet \vec{P}$$

$$-\lambda \vec{P} = \hat{A} \bullet \vec{P}$$

$$\left(\hat{A} + \lambda \hat{I}\right) \bullet \vec{P} = 0$$

Some tricks of "trade":

- Utilize crystal periodicity and reduce matrix size from N (number of sites to n (number of sites inside unit cell)
- Utilize Laplace transformation to convert system of ODEs to a system of linear equations
- Use symmetry so much as you can to collect symmetric sites together

10 20.10.2023 Pavel Vladimirov - Hydrogen diffusion in beryllides

Department of Metallic Alloys, Institute for Applied Materials - Applied Materials Physics IAM

Calculation method (I)

$$A_{\alpha\beta}(\theta) = -\sum_{h}\left[W_{\beta\alpha}(\boldsymbol{h})\, e^{ikh} - W_{\alpha\beta}(\boldsymbol{h})\delta_{\alpha\beta}\right]$$

RHS

Unit vector $\vec{\xi} \perp$ crystal plane

$\vec{a}_1$ is outside of a plane

Diffusion coefficient along direction $\vec{\xi}$

$$D_{\xi\xi} = -\frac{\Delta^2}{2}\left[\frac{d^2}{d\theta^2}det\left(\hat{A}(\theta)\right)\right]_{\theta=0} / det\left(\hat{B}\right)$$

where $\theta = \left(\vec{a}_1 \vec{\xi}\right) \bullet t = \Delta \bullet t$

$\hat{B}$ is $\hat{A}(\theta = 0)$ in which one row is replaced with ones

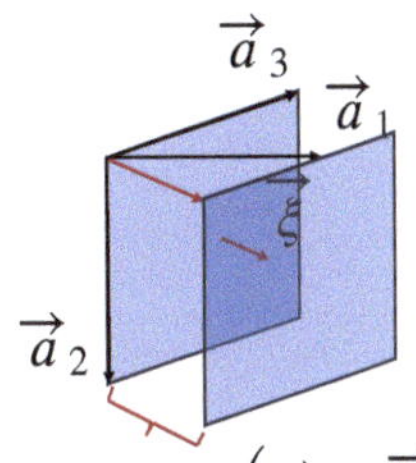

$\Delta = \left(\vec{a}_1 \bullet \vec{\xi}\right)$

interplanar distance

After Shunya Ishioka & Masahiro Koiwa (1985) *Diffusion coefficient in crystals with multiple jump frequencies*, Phil. Mag. A, 52:2, 267-277, DOI: 10.1080/01418618508237623

11 20.10.2023 Pavel Vladimirov - Hydrogen diffusion in beryllides Department of Metallic Alloys, Institute for Applied Materials - Applied Materials Physics

Calculation method (III)

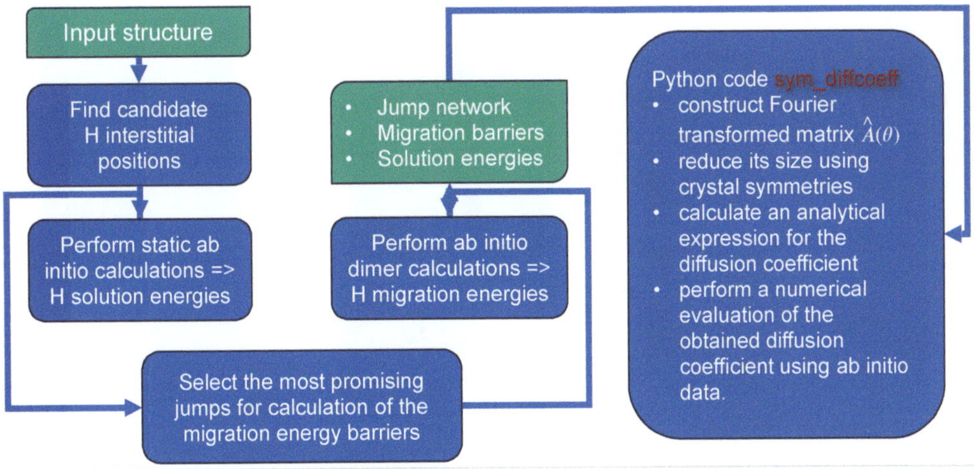

12 20.10.2023 Pavel Vladimirov - Hydrogen diffusion in beryllides Department of Metallic Alloys, Institute for Applied Materials - Applied Materials Physics

Karlsruhe Institute of Technology

Hydrogen diffusion in $TiBe_2$

13 20.10.2023 Pavel Vladimirov - Hydrogen diffusion in beryllides

Department of Metallic Alloys, Institute for Applied Materials - Applied Materials Physics

---

Karlsruhe Institute of Technology

Analytical description of diffusion in $TiBe_2$

Hydrogen interstitial diffusion in $TiBe_2$

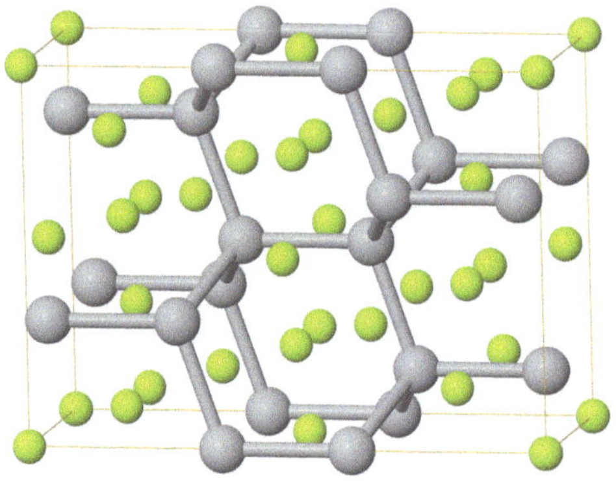

14 25.05.2022 Pavel Vladimirov - Hydrogen diffusion in beryllides

Department of Metallic Alloys, Institute for Applied Materials - Applied Materials Physics

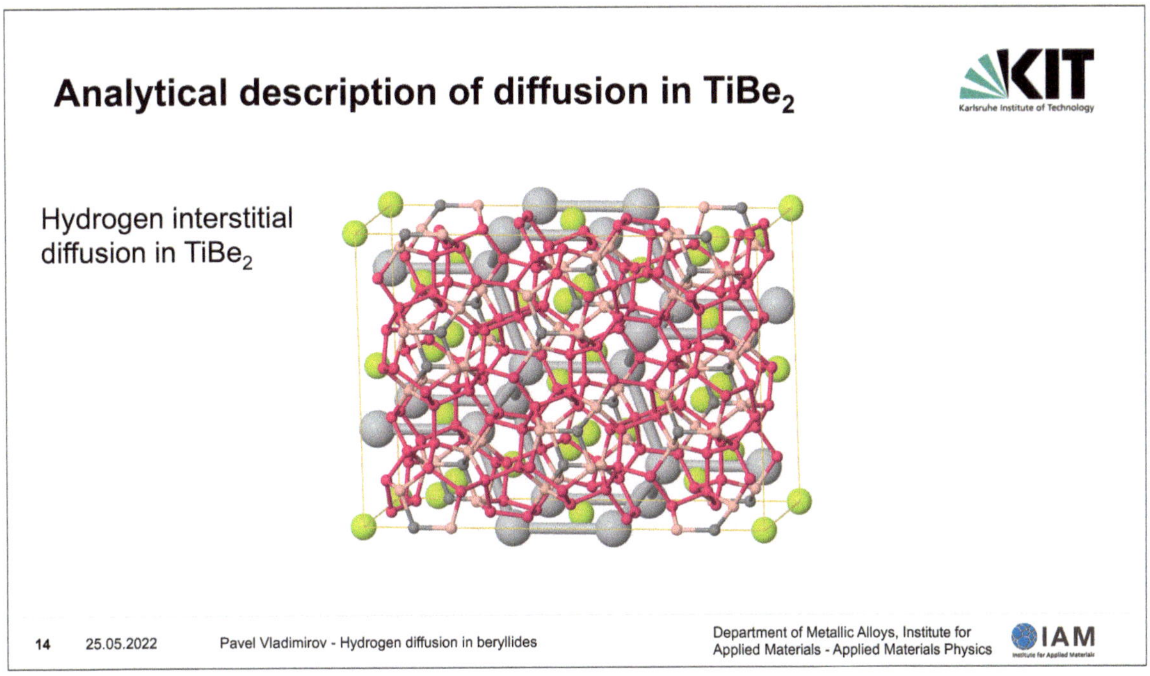
Analytical description of diffusion in $TiBe_2$
Karlsruhe Institute of Technology
Hydrogen interstitial diffusion in $TiBe_2$
14
25.05.2022
Pavel Vladimirov - Hydrogen diffusion in beryllides
Department of Metallic Alloys, Institute for Applied Materials - Applied Materials Physics
IAM

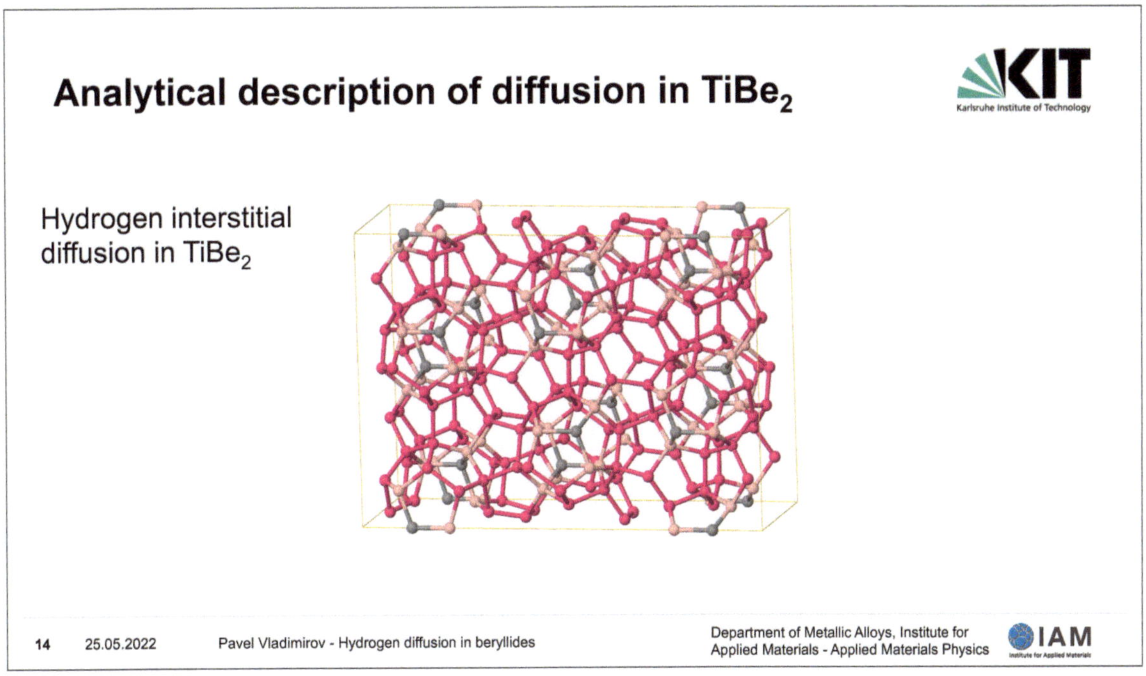
Analytical description of diffusion in $TiBe_2$
Karlsruhe Institute of Technology
Hydrogen interstitial diffusion in $TiBe_2$
14
25.05.2022
Pavel Vladimirov - Hydrogen diffusion in beryllides
Department of Metallic Alloys, Institute for Applied Materials - Applied Materials Physics
IAM

Analytical expression for diffusion coefficient in $TiBe_2$

Karlsruhe Institute of Technology

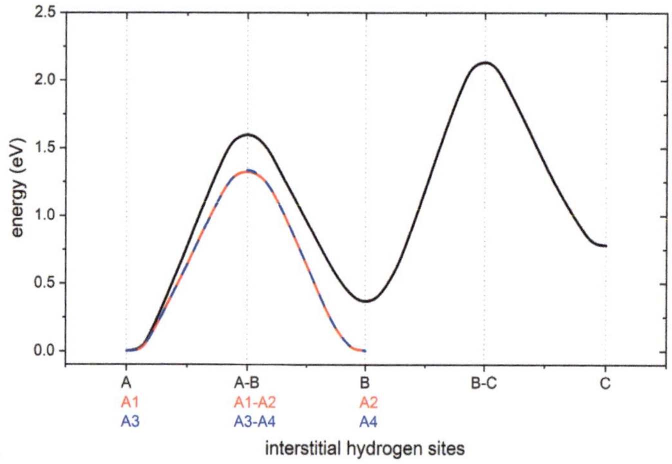

#	Label	Jump length, Å	Activation energy, eV
0	A→A	0.98	1.32
1	A→B	1.04	1.59
2	B→A	1.04	1.23
3	B→C	1.04	1.35
4	C→B	1.04	1.76
5	A→A	1.13	1.33

$$W_{AA}(1.13) \approx W_{AA}(0.98)$$

5-frequencies model

15 20.10.2023 Pavel Vladimirov - Hydrogen diffusion in beryllides

Department of Metallic Alloys, Institute for Applied Materials - Applied Materials Physics IAM

Diffusion coefficient along <111> in $TiBe_2$

Karlsruhe Institute of Technology

5-frequencies model

$$D_{\xi\xi} = D_{\langle 111\rangle\langle 111\rangle}$$

$$= \left(\frac{a}{16}\right)^2 \cdot \frac{W_{CB}(40W_{AA}W_{BA} + 27W_{AB}W_{BA} + 9W_{AB}W_{BC})}{(W_{AB}W_{BC} + 4W_{AB}W_{CB} + 12W_{BA}W_{CB})}$$

Diffusion in cubic crystals is isotropic:

$$D_{11} = D_{22} = D_{33}$$

16 20.10.2023 Pavel Vladimirov - Hydrogen diffusion in beryllides

Department of Metallic Alloys, Institute for Applied Materials - Applied Materials Physics IAM

Linear fit to Arrhenius plot for $TiBe_2$

Karlsruhe Institute of Technology

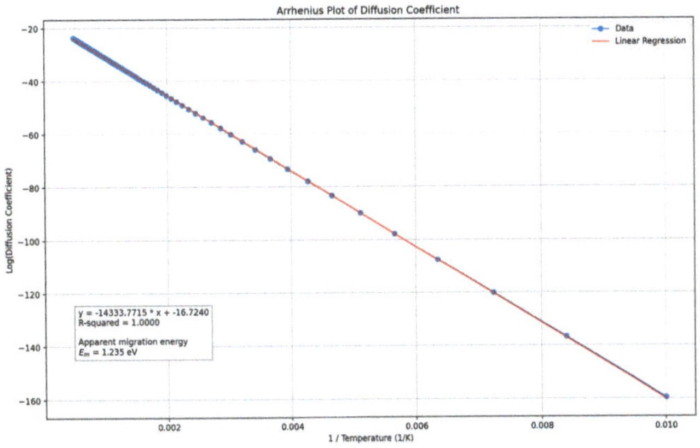

- **Points:** calculated results using an analytical solution
- **Line:** Linear regression fit to the points
- Apparent migration energy $E_m = 1.23\ eV$

17 20.10.2023 Pavel Vladimirov - Hydrogen diffusion in beryllides Department of Metallic Alloys, Institute for Applied Materials - Applied Materials Physics IAM

Karlsruhe Institute of Technology

Diffusion along c-axis in $TiBe_{12}$

18 20.10.2023 Pavel Vladimirov - Hydrogen diffusion in beryllides Department of Metallic Alloys, Institute for Applied Materials - Applied Materials Physics IAM

Interstitial positions for H in Be: $TiBe_{12}$

KIT Karlsruhe Institute of Technology

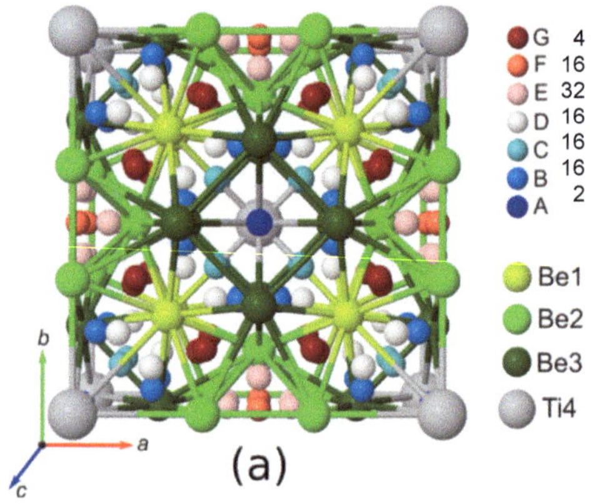

19 20.10.2023 Pavel Vladimirov - Hydrogen diffusion in beryllides

Department of Metallic Alloys, Institute for Applied Materials - Applied Materials Physics IAM

Interstitial positions for H in Be: $TiBe_{12}$

KIT Karlsruhe Institute of Technology

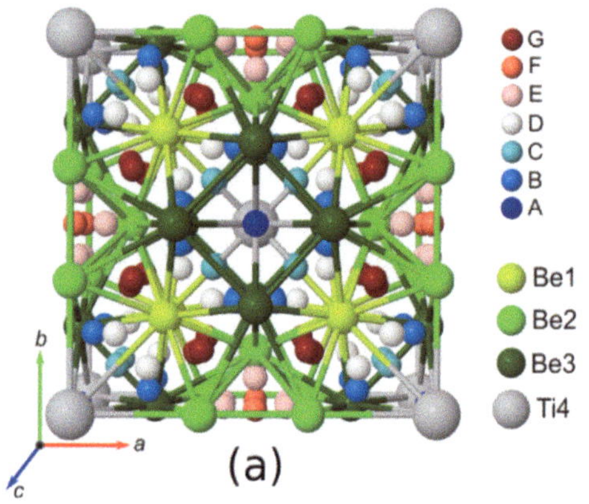

Structure of $TiBe_{12}$ and H interstitial sites

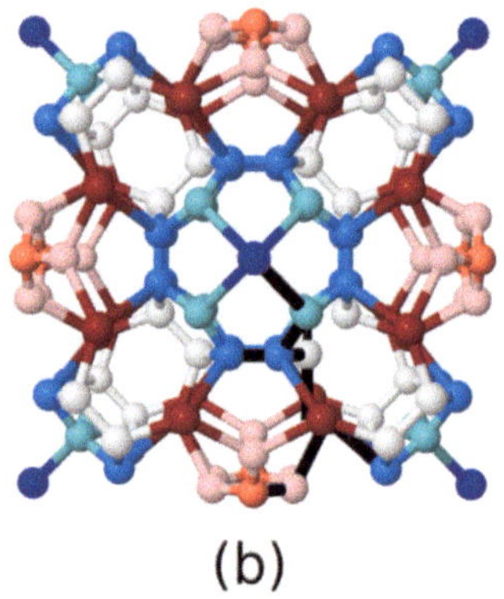

Jump network of H interstitial sites

20 20.10.2023 Pavel Vladimirov - Hydrogen diffusion in beryllides

Department of Metallic Alloys, Institute for Applied Materials - Applied Materials Physics IAM

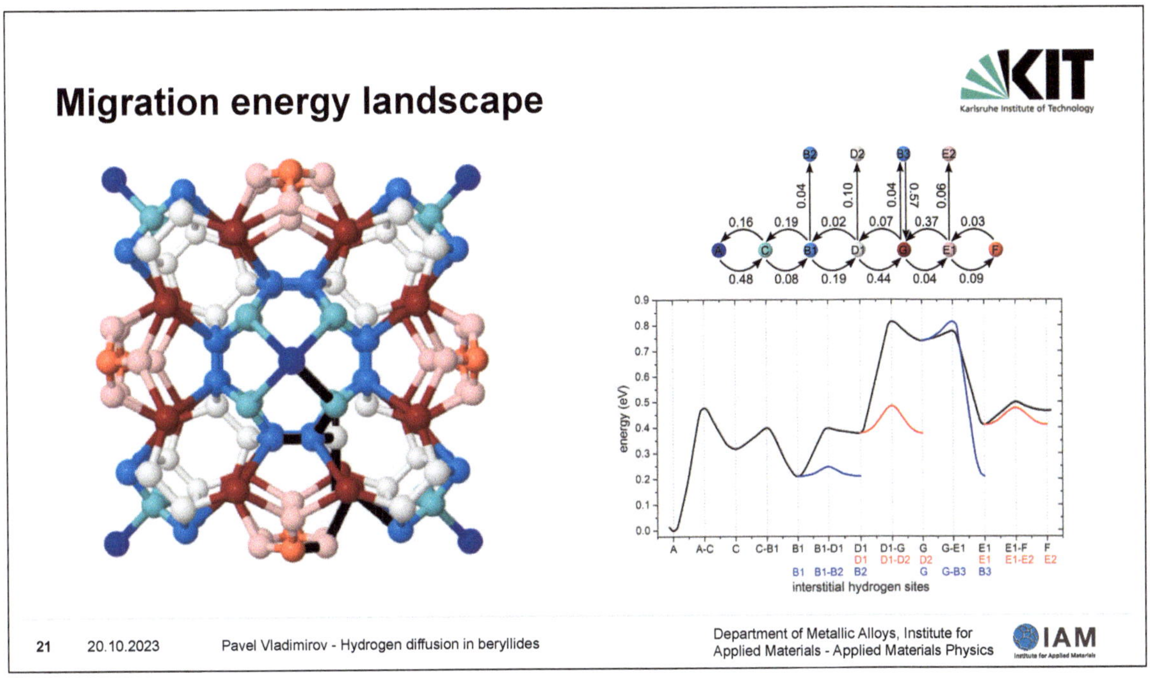
Migration energy landscape
Karlsruhe Institute of Technology
B2
D2
B3
E2
0.04
0.10
0.04
0.57
0.06
0.16
0.19
0.02
0.07
0.37
0.03
A
C
B1
D1
G
E1
F
0.48
0.08
0.19
0.44
0.04
0.09
energy (eV)
A A-C C C-B1 B1 B1-D1 D1 D1-G G G-E1 E1 E1-F F
D1 D1-D2 D2 E1 E1-E2 E2
B1 B1-B2 B2 G G-B3 B3
interstitial hydrogen sites
21 20.10.2023 Pavel Vladimirov - Hydrogen diffusion in beryllides
Department of Metallic Alloys, Institute for Applied Materials - Applied Materials Physics
IAM
Institute for Applied Materials

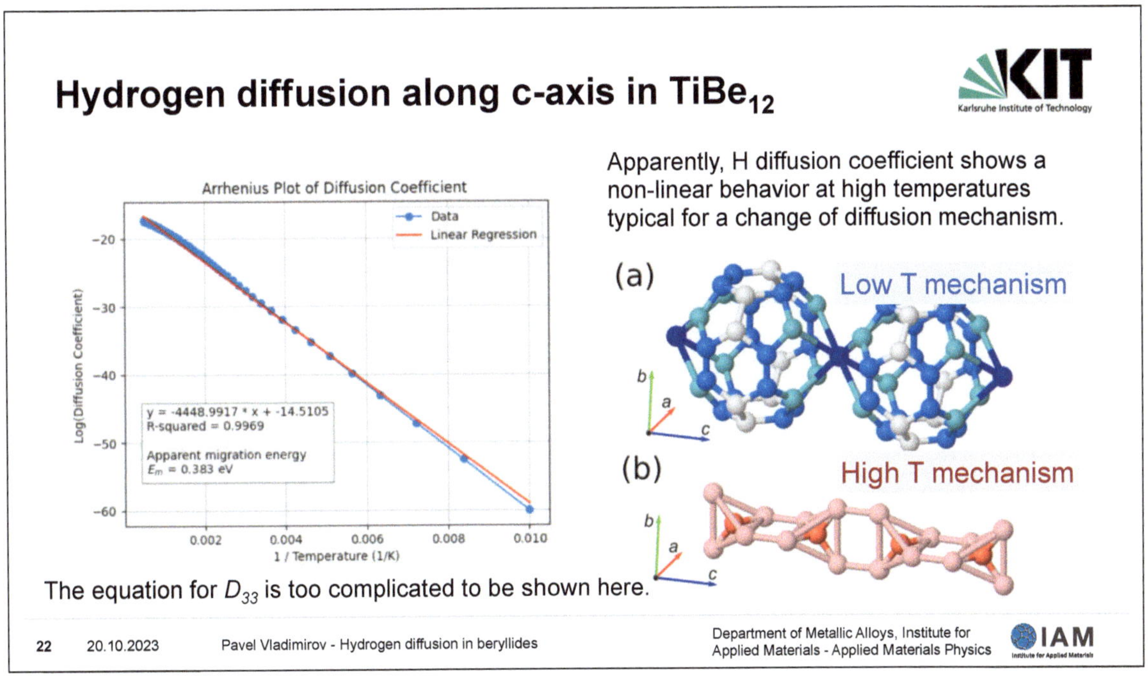
Hydrogen diffusion along c-axis in TiBe12
Karlsruhe Institute of Technology
Apparently, H diffusion coefficient shows a non-linear behavior at high temperatures typical for a change of diffusion mechanism.
Arrhenius Plot of Diffusion Coefficient
Data
Linear Regression
y = -4448.9917 * x + -14.5105
R-squared = 0.9969
Apparent migration energy
Em = 0.383 eV
Log(Diffusion Coefficient)
1 / Temperature (1/K)
(a)
Low T mechanism
(b)
High T mechanism
The equation for D33 is too complicated to be shown here.
22 20.10.2023 Pavel Vladimirov - Hydrogen diffusion in beryllides
Department of Metallic Alloys, Institute for Applied Materials - Applied Materials Physics
IAM
Institute for Applied Materials

Transition rate matrix for $TiBe_{12}$

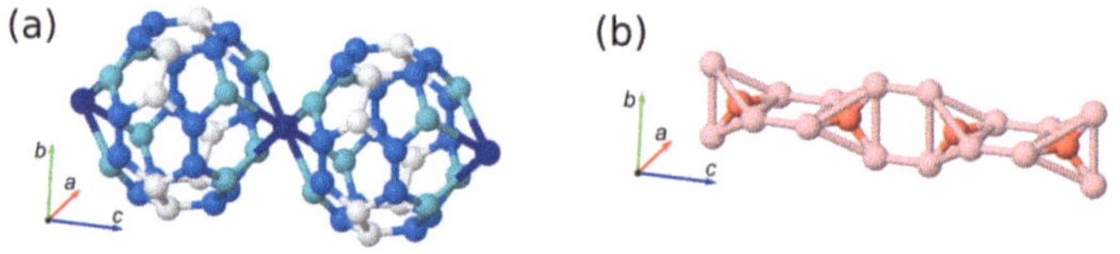

	A	B	C	D	E	F	G
A							
B							
C							
D							
E							
F							
G							

Department of Metallic Alloys, Institute for Applied Materials - Applied Materials Physics

Low temperature diffusion path

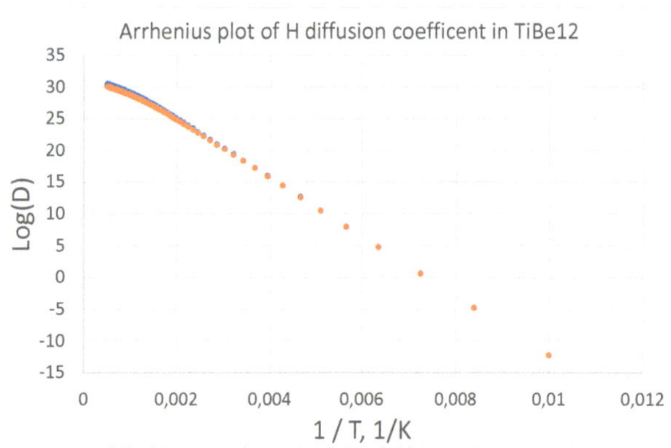

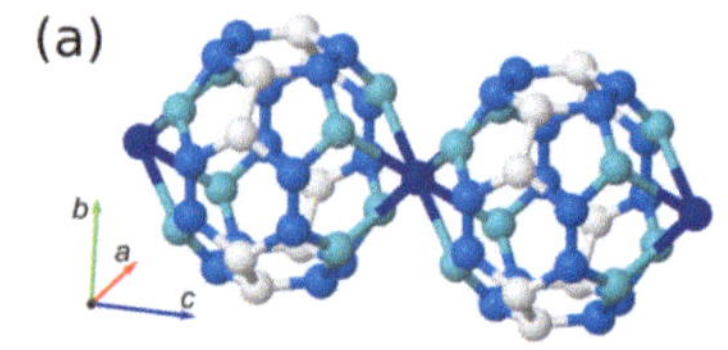

Low T mechanism

Department of Metallic Alloys, Institute for Applied Materials - Applied Materials Physics

Low temperature diffusion path

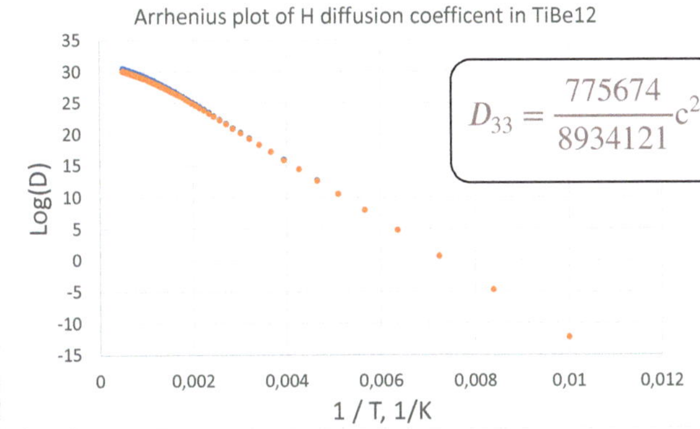

4-frequencies model

$$D_{33} = \frac{775674}{8934121} c^2 \frac{W_{BC} W_{CB} W_{DB}}{\left(W_{BC} W_{CB} + W_{BC} W_{DB} + 2 W_{CB} W_{DB}\right)}$$

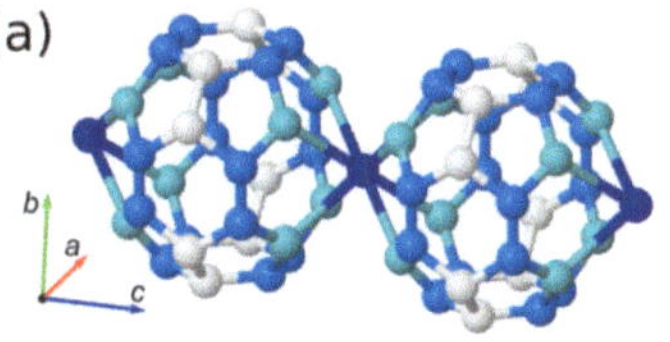

Low T mechanism

24 20.10.2023 Pavel Vladimirov - Hydrogen diffusion in beryllides Department of Metallic Alloys, Institute for Applied Materials - Applied Materials Physics IAM

Conclusions

- Python code for analytical calculation of diffusion coefficients in lattices with multiple jump frequencies was developed and successfully applied to the case of hydrogen diffusion $TiBe_2$ and $TiBe_{12}$.
- Hydrogen diffusion in $TiBe_{12}$ along c-axis is faster than that in $TiBe_2$ which will facilitate hydrogen removal from the $TiBe_{12}$ phase.
- This method is applicable to other systems as well, when the number of distinct sites for the tracer is limited (<10). Otherwise, numerical approximation can be used.
- Generation of analytical expressions for diffusion and other kinetic coefficients opens a way for modelling of hydrogen retention and release in materials as well as other atomic scale processes on the macroscopic scale.

25 20.10.2023 Pavel Vladimirov - Hydrogen diffusion in beryllides Department of Metallic Alloys, Institute for Applied Materials - Applied Materials Physics IAM

SESSION 8

Health & Safety

8.1 Beryllium facility design

Beth Walker

An update on the challenges and successes of the Novanta Beryllium facility in Taunton, UK.
A bespoke building that was designed with Beryllium safety at the core, safety of the staff, the wider environment and customer.

Novanta is shortlisted for a Safety and Health Excellence award in the manufacturing category, due to safety being so intrinsic to the design of the building.

Presenting Author: Beth Walker
Presenting Author's Affiliation: Novanta Inc., Taunton, England, UK
Email: saferenvironmentltd@gmail.com

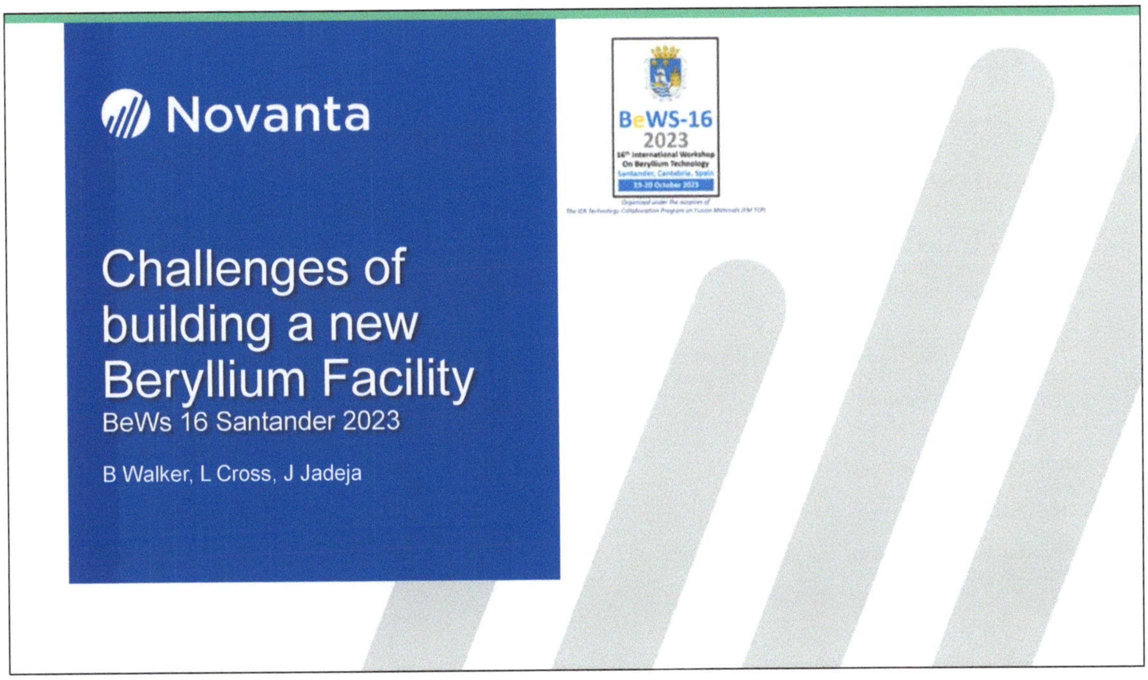
Novanta
BeWS-16
2023
16th International Workshop
On Beryllium Technology
Santander, Cantabria, Spain
Challenges of building a new Beryllium Facility
BeWs 16 Santander 2023
B Walker, L Cross, J Jadeja

In 2020 we took receipt of 4 warehousing units, ready to be converted to a Beryllium manufacturing facility.
This was a challenging but rewarding project from the outset.
2
Novanta

Background

Choosing the right land

- Novanta was based in a 1950's Asbestos 'Northern lights' building that had not undergone major refurbishment for nearly 30 years and the decision was made to build a new, state of the art facility, designed specifically around staff, stakeholder and local environment safety.
- The building chosen was in a sensitive area that required extensive habitation, flooding and ground contamination assessment.
- A permit was required from the Environment Agency.
- The local Highways Agency and Water Company also had to issue permits.

Old Building

Novanta

Airflow

Airflow

- The air flow moves from a positive pressure in the clean room, where the finished product is handled. (+10 pascals)
- To full negative pressure in the beryllium facility (-20 pascals)
- Moderate risk, controlled, but non-respirator areas are between -5 to -15 dependent on risk

Challenges

- The air flow was designed to work using pascals, a measure of pressure.
- This meant establishing the Air changes per hour(ACH) was difficult.
- Testing shows that it is between 8-10 ACH
- The amount of air ingress was adjusted to maintain both Pascals and ACH

Novanta

Air flow and layout

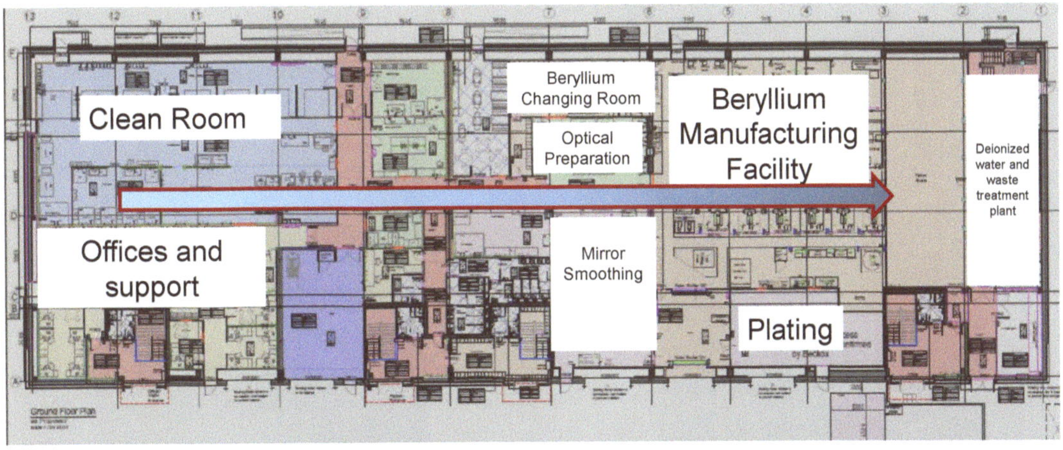

Novanta

Humidity, temperature and pressure control

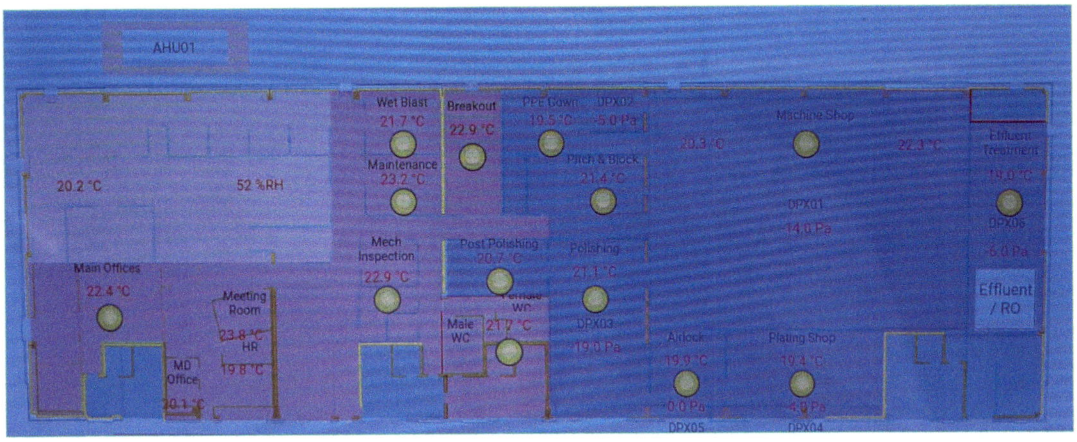

Novanta

Safe air

HEPA Bank for extracted air

Air conditioning for incoming air, including Humidity, purity and flow

Novanta

Water treatment

- The water treatment was especially important, as the effluent discharge is heavily controlled by the Environment Agency and Water Company Permits.
- There are strict tolerances that we must adhere to.
- The water treatment facility was state of the art, never been used before for such a process.
- It uses a mixture of Chemistry and a centrifuge to eliminate coagulated solids.
- This came with challenges!

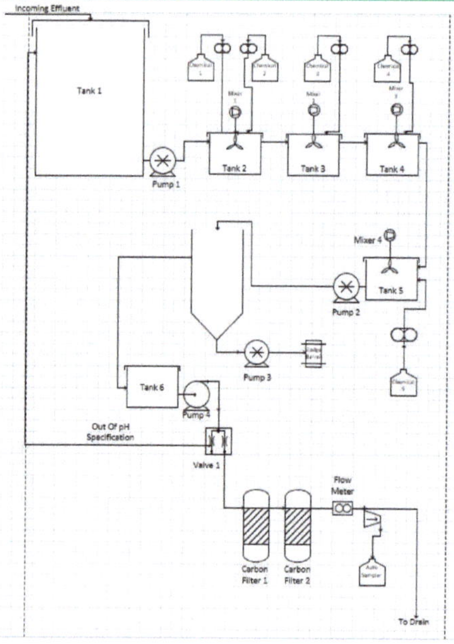

Novanta

Challenges of Building on ‘The Somerset Levels’

- We were determined that the new facility remain in the local area, to maintain the wealth of knowledge already employed.
- This meant we had to ensure the facility would not be compromised in the event of a flood.

Novanta

Flood protection

- The carpark was given a camber that allowed a 15,000-liter capacity.
- Penstocks prevent water leaving the facility, or water from inside, mixing with flood water.
- Floodgates are used around the building to prevent water ingress to the building.
- There is a 350,000 liter, slow release tank under the car park.

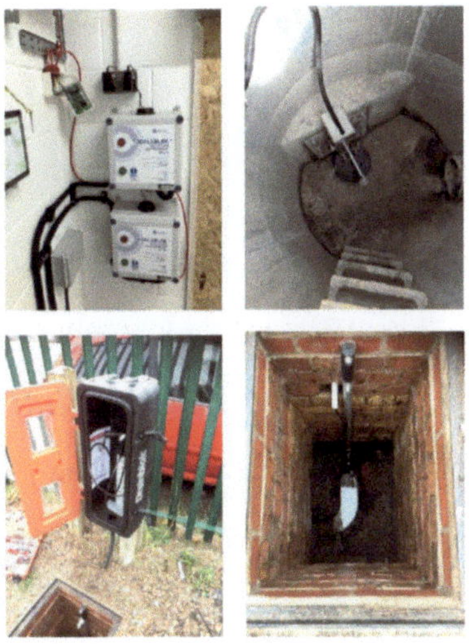

Novanta

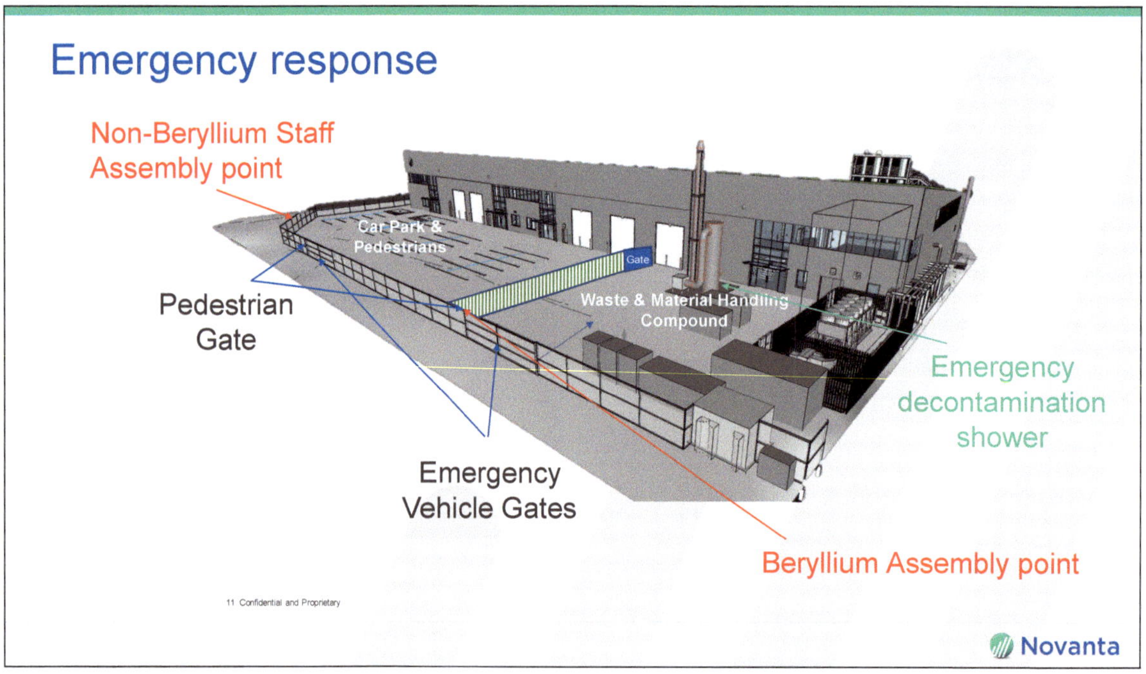
Emergency response
Non-Beryllium Staff
Assembly point
Car Park &
Pedestrians
Gate
Pedestrian
Gate
Waste & Material Handling
Compound
Emergency
decontamination
shower
Emergency
Vehicle Gates
Beryllium Assembly point
11 Confidential and Proprietary
Novanta

Clean room
The decision was taken to build an ISO 8 clean room, despite the requirement being much lower, as this meant the staff had an extra degree of protection wearing full suits.
12 Confidential and Proprietary
Novanta

Beryllium to non-Beryllium communication

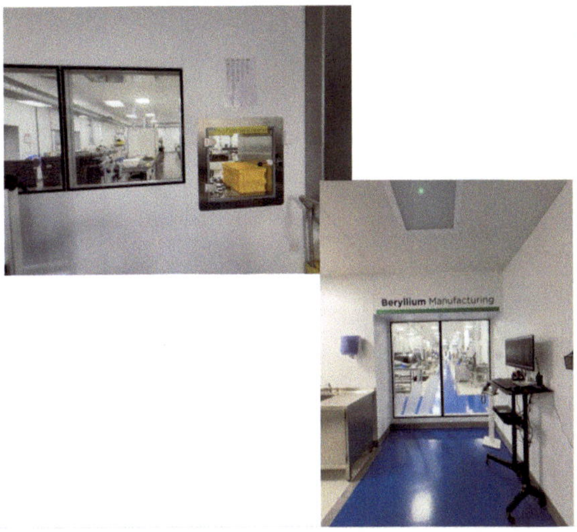

- Direct coms into the room with a large viewing area, so it is easy to communicate, This was always an issue previously.
- Large, extracted, transfer hatches, for safe handover from Be room after clean, and ability to pass paperwork safely to staff if needed (mostly paper free)
- Full intercom and tannoy system for communication and emergency

Novanta

Decontamination

- The first stage is a pre wash.
- This is head to toe.
- There is a mix of soap and water mixing at the brush.
- This stage is completed as many times as required.

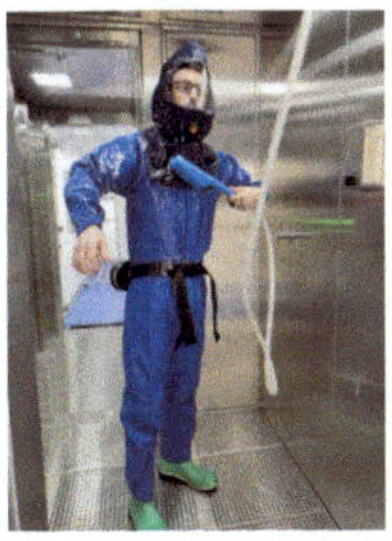

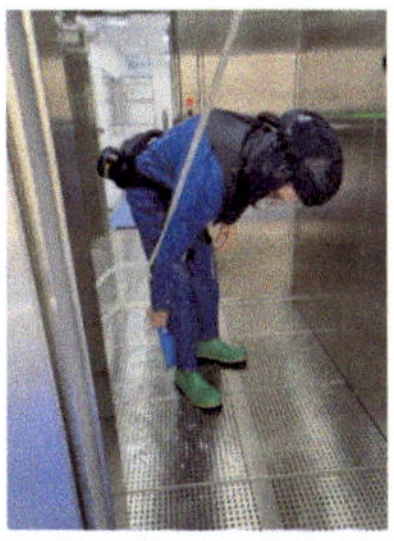

Novanta

Decontamination

- There is then a high-pressure rinse stage.
- This is followed by a drying stage.
- This uses a new technique called an 'air knife'

Novanta

Decontamination

- Once the 3 stage decontamination is completed, staff use the extracted and temperature controlled drying room to remove PPE.

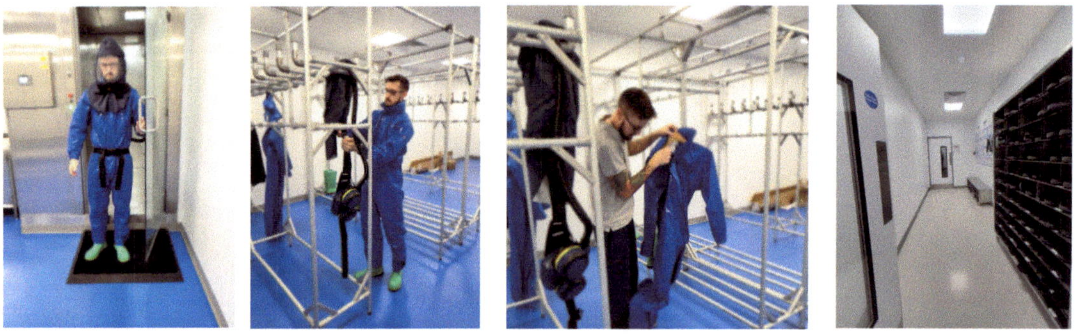

Novanta

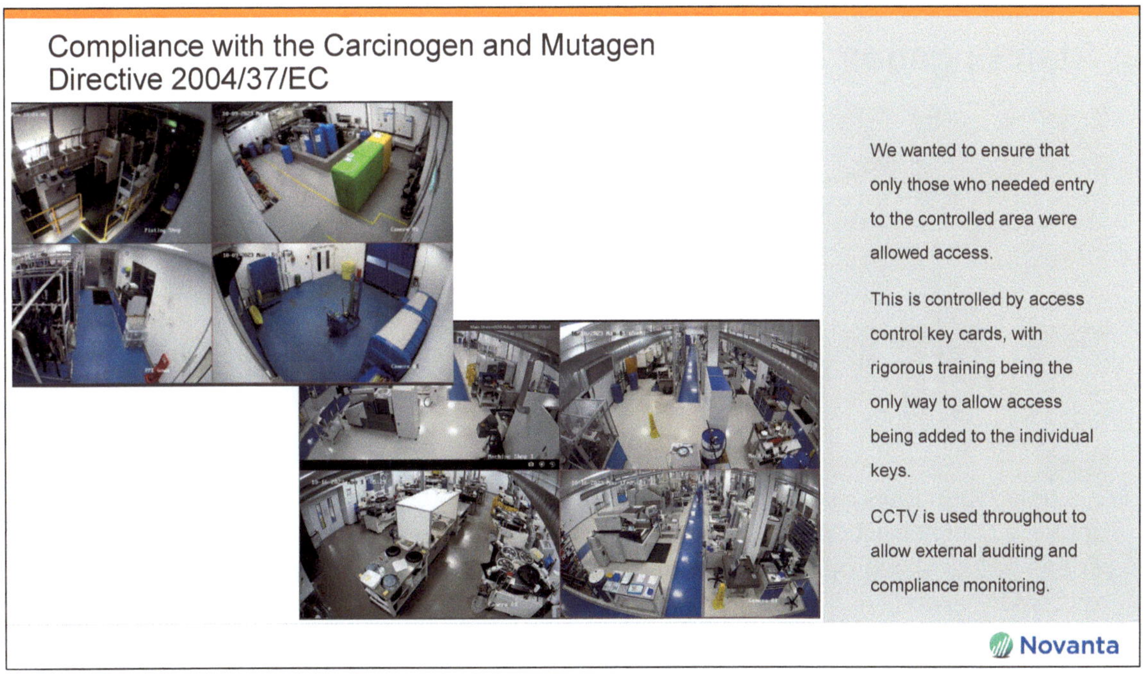
Compliance with the Carcinogen and Mutagen Directive 2004/37/EC
We wanted to ensure that only those who needed entry to the controlled area were allowed access.
This is controlled by access control key cards, with rigorous training being the only way to allow access being added to the individual keys.
CCTV is used throughout to allow external auditing and compliance monitoring.
Novanta

Health and Safety Training facility
Novanta

Staff Engagement

One of the most unexpected challenges was staff engagement

We held Open days, sent updates of construction and 'sneak peaks' using our elf and safety expert!

Novanta

Being recognised as a contribution to UK Health and Safety

- Now the facility is up and running, it can demonstrate the commitment to worker protection and Health & Safety.
- This was recognized at the UK Excellence in Health and Safety Awards earlier this year.

Novanta

8.2 Beryllium Stimulated ELISpot as an Alternative Screening Test for Beryllium Sensitization and Chronic Beryllium Disease

Lisa A Maier, Elizabeth A. Barker, Margaret M. Mroz

The BeELISpot is a unique biomarker of beryllium health effects. The test measures interferon gamma production by lymphocytes in response to beryllium antigen stimulation. This measure of beryllium stimulated cytokine production has been shown to identify beryllium sensitization (BeS) and chronic beryllium disease (CBD) in a clinical setting [1] and as an alternative to the beryllium lymphocyte proliferation test (BeLPT) for workforce medical surveillance [2]. As these studies were promising we went on to optimize the beryllium stimulated ELISpot (BeELISpot) as a clinical test in identifying BeS and CBD. The BeELISpot has advantages over the BeLPT in that it requires less blood (10 mL), has a faster turn-around time (24 hours) and does not require the use of radioactive materials.

METHODS: Forty subjects undergoing clinical evaluation for BeS/CBD had BeELISpot testing. The diagnosis of CBD (n=26) was based on previously defined criteria, including a history of beryllium exposure, confirmed abnormal blood BeLPT results, or abnormal BAL BeLPT, and the presence of granulomatous inflammation and/or mononuclear cell infiltration on lung biopsy. The diagnosis of BeS (n=14) required a history of beryllium exposure, confirmed abnormal blood BeLPT results and no evidence of granulomatous inflammation on transbronchial lung biopsy. BeELISpot plates were set up in duplicate and stimulated with two different concentrations of beryllium. The plates were analyzed by using a CTL Immunospot Analyzer (CellularTechnology Ltd, Cleveland, Ohio), and results were reported as the mean number of spot-forming units (SFU) per well minus background SFU. Normal (NOR) was designated as <5 SFU, borderline (BOR) as $>$ 5 to 15 SFU and abnormal (ABN) $>$ 15 SFU. We calculated the positive predictive value (PPV) of an ABN/BOR BeELISpot.

RESULTS: The BeELISpot median SFU result was 6.25 (range 0-18) for BeS patients compared to 19 (range 0.5 – 636.5) for CBD patients and proved to be significantly different ($p = 0.005$). For BeS/CBD patients combined, BeELISpot identified 70% with ABN/BOR results. For BeS, the PPV was 57% for and for CBD patients, the PPV was 81% but results were not significantly different.

CONCLUSIONS: The optimized BeELISpot was able to identify ABN or BOR result in 81% of CBD patients and 57% of BeS patients. Further work will compare mean SFU to clinical results and enroll beryllium exposed controls to determine sensitivity and specificity of the test. BeELISpot holds promise as a useful biomarker of BeS and CBD in beryllium exposed workforces. The need for less blood, short turn-around-time and no need for radioactive materials combine to compel further study of BeELISpot as a tool for beryllium medical surveillance in occupationally exposed workforces.

REFERENCES:

[1] Pott GB, Palmer BE, Sullivan AK, Silviera L, Maier LA, Newman LS, Kotzin BL, Fontenot AP. Frequency of beryllium-specific, TH1-type cytokine-expressing CD4+ T cells in patients with beryllium-induced disease. J. Allergy Clin. Immunol. (2005) 115:5, 1036-1042.

[2] Martin AK, Mack DG, Falta MT, Mroz MM, Newman LS, Maier LA, Fontenot AP. Beryllium-specific CD4+ T cells in blood as a biomarker of disease progression. J. Allergy Clin. Immunol. (2011) 125:5, 1100-1106.

[3] Mroz MM, Kreiss K, Lezotte DC, Campbell PA, Newman LS. Reexamination of the blood lymphocyte transformation test in the diagnosis of chronic beryllium disease. J Allergy Clin Immunol (1991) 88, 54–60.

Presenting Author: Lisa A Maier, MD, MSPH or Elizabeth A. Barker, MPH
Presenting Author's Affiliation: National Jewish Health, USA

Beryllium Stimulated ELISpot as an Alternative Screening Test for Beryllium Sensitization and Chronic Beryllium Disease

Lisa A. Maier MD MSPH
Elizabeth Barker PMP MPH
Margaret Mroz MSPH
National Jewish Health
Denver, CO USA

BeWS-16
2023

At Risk Industries from Beryllium Exposure

- Defense industries
- Aerospace
- Automotive
- Electronics, Telecommunications, Computer
- Mining and manufacture of beryllium
- Dental
- Foundries/metal reclamation
- Alloy machine shops

Workers at Risk

- Estimates: 10 million U.S. workers exposed
- Estimates: >2000 companies in Quebec, Canada
- International: cases in EU, Australia, Asia
- High risk groups,work processes
  - Machinists
  - Chemical operators
  - Beryllium Oxide, ceramics
  - Research and Development / Engineering
  - Dust disturbers
- Also at risk
  - Bystanders, security guards, front office workers

Health Effects of Beryllium

Routes of Exposure

Breathing

→

← Skin

Illustration by National Jewish Health

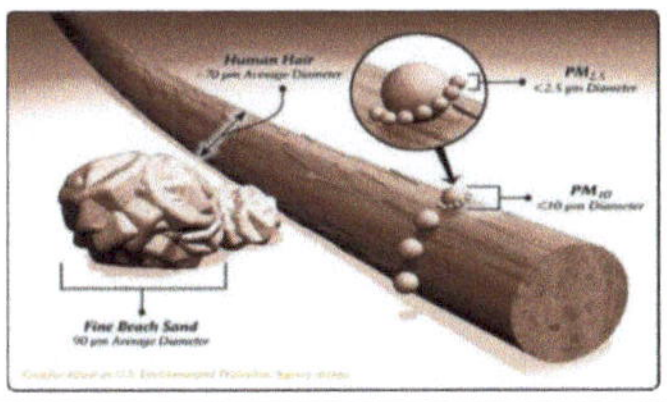

Health Effects

<u>Lungs</u>

1. **Beryllium sensitization (BeS)**
2. **Chronic beryllium disease (CBD)**

- Lung cancer
- Acute beryllium disease

<u>Skin</u>

- Nodules, rash, non-healing lesions
- **Beryllium sensitization (BeS)**

4

Beryllium Sensitization (BeS)

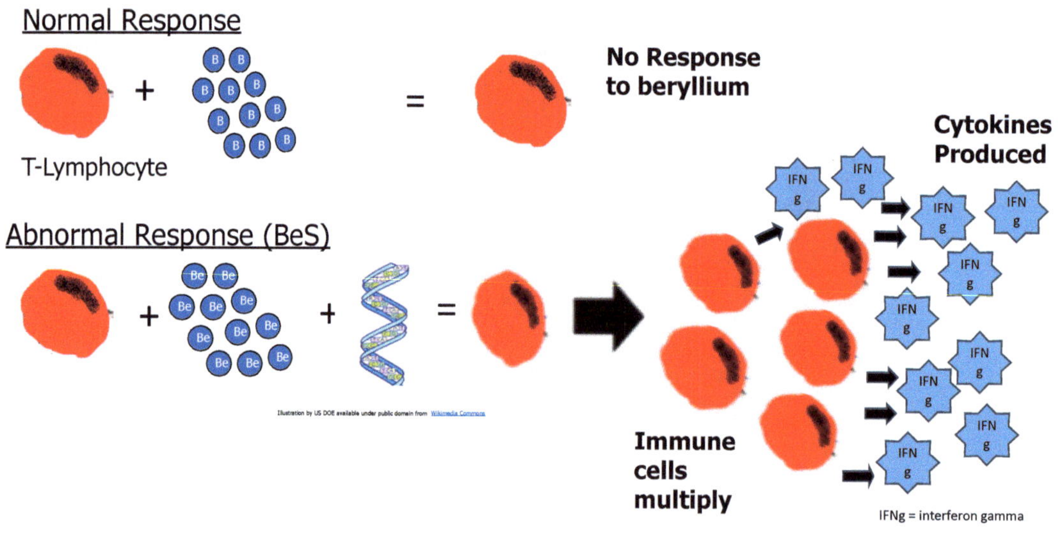

Chronic Beryllium Disease and Exposure

- Process related disease risks (up to 16% prevalence)
- Disease among those with seemingly minimal exposure
- Disease is more common among high exposure jobs
- Exposures below 0.01∝g/m^3: no sensitization seen
- Factors other than *total* beryllium exposure may be important in defining risk:
  - Genetics
  - Features of beryllium particles

Beryllium Sensitization and Chronic Beryllium Disease (CBD) Defined

- BeS
  - 2 Abnormal BeLPTs
  - 1 Abnormal and 1 Borderline
  - 3 Borderline BeLPTs
  - Abnormal BAL BeLPT
- BeLPT is not a perfect test
  - A single abnormal needs to be confirmed (repeated)
  - Only commercially available test for BeS (currently)
- **Asymptomatic**
- Develops within **9 weeks** of first exposure or **years after exposure ceases**
- Medical evaluation for CBD after confirmed BeS is standard medical practice

- CBD
  - Beryllium sensitization AND
  - Granulomas or granulomatous inflammation on TBLB (transbronchial lung biopsy) or
  - Lavage lymphocytosis (>15%) AND Abnormal BAL LPT

  - Possible gas exchange abnormalities
  - Possible spirometric abnormalities
  - Possible radiographic abnormalities

7

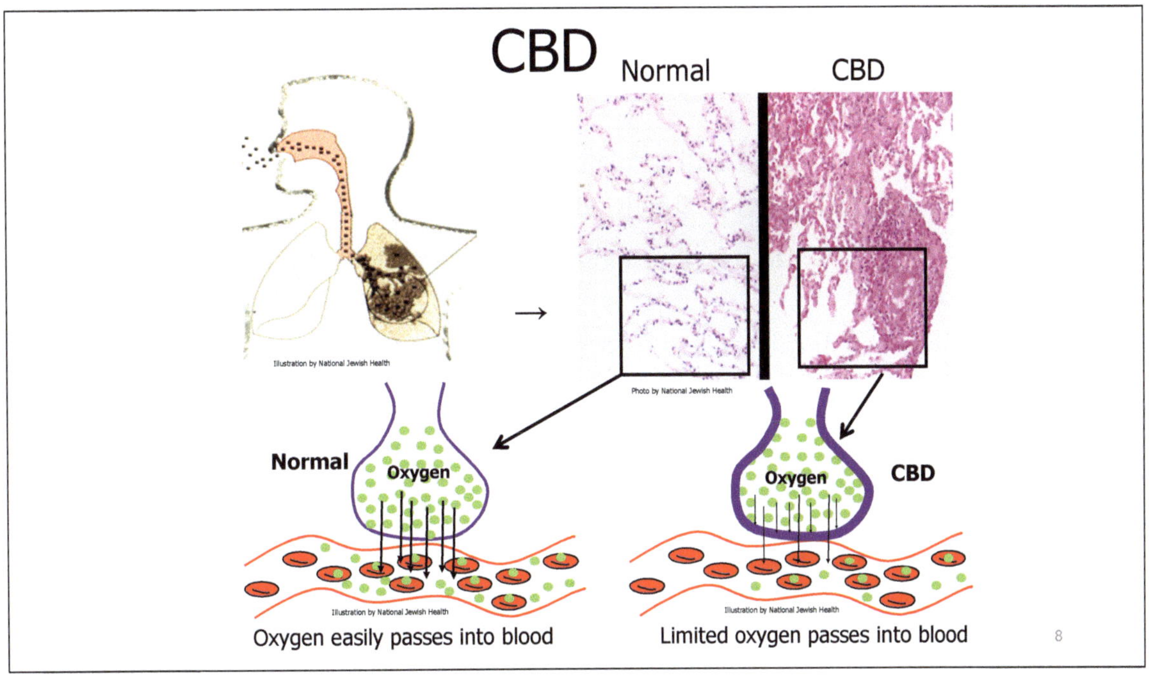

Why Develop the BeElispot ?

Complexities of Diagnosing of Beryllium Sensitization (BeS) or Chronic Beryllium Disease (CBD)

- Proliferation assay does not differentiate between BeS and CBD stages
- CBD diagnosis requires BAL and biopsy
  - Invasive procedures
  - Imprecise since positive result requires presence of granulomatous inflammation <u>and</u> evidence of lymphocytosis ($\geq$ 15%)

Background

- CBD is mediated by CD4+ T cells
- These cells recognize beryllium in association with certain HLA-DP molecules
- The T cells proliferate and secrete TH1-type cytokines (IFN-γ, IL-2)

Hypothesis: Patients with CBD have markedly increased numbers of beryllium-specific cytokine-secreting T-cells in blood compared with BeS patients

BeELISpot

- Unique biomarker of beryllium health effects.
- Measures interferon gamma production by lymphocytes in response to beryllium antigen stimulation.
- This measure of beryllium stimulated cytokine production has been shown to identify beryllium sensitization (BeS) and chronic beryllium disease (CBD) in a clinical setting and as an alternative to the beryllium lymphocyte proliferation test (BeLPT) for workforce medical surveillance.
- BeELISpot has advantages over the BeLPT in that it requires less blood (10mL), has a faster turn-around time (24 hours) and does not require the use of radioactive materials.

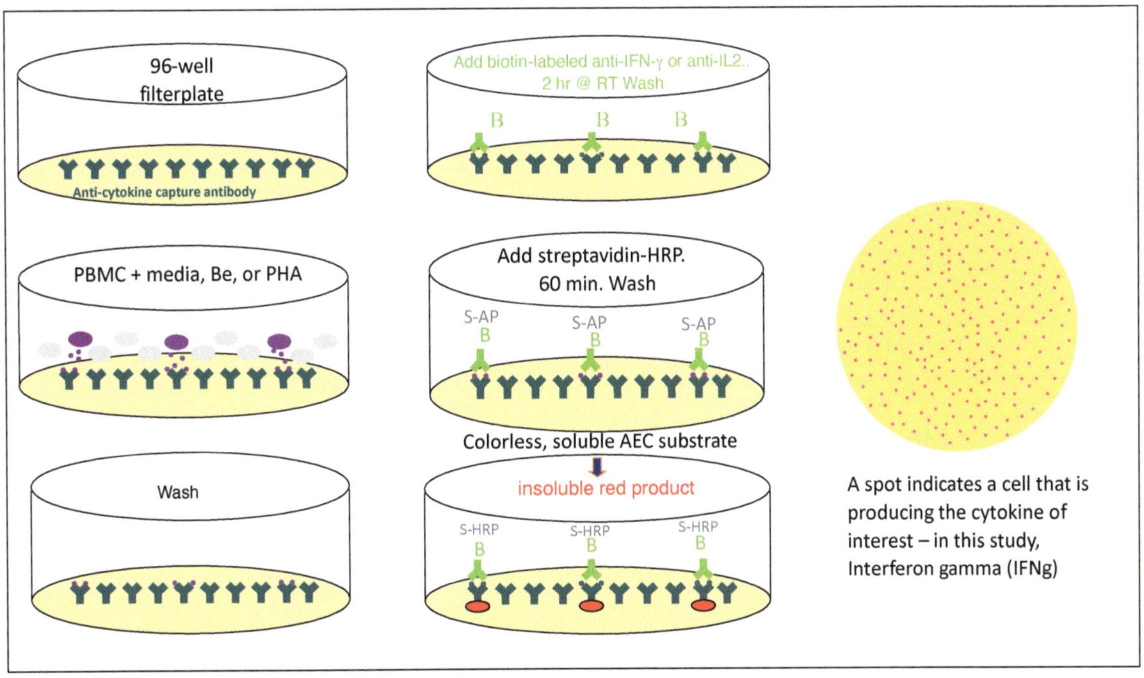

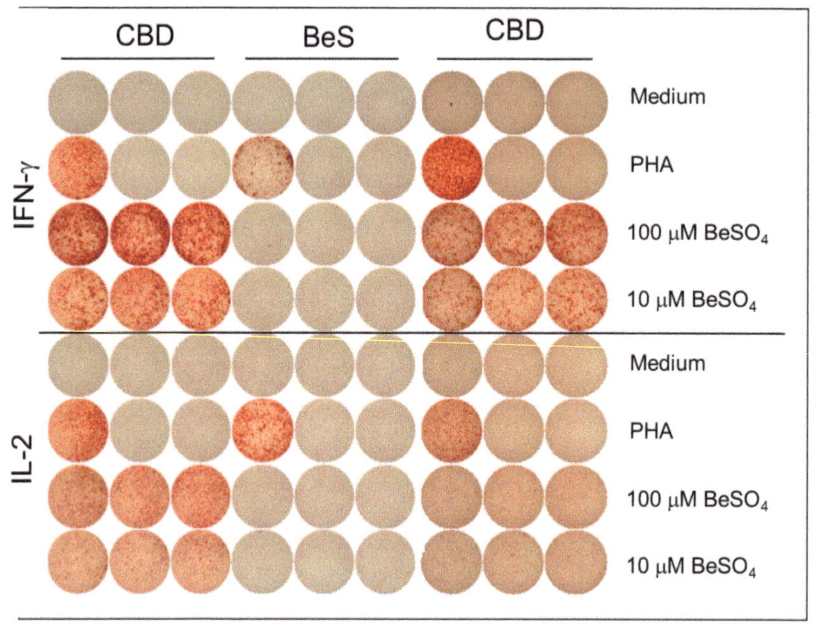

BeELISpot identifies beryllium specific cytokine production in patients with CBD

- This study determined whether the quantity of Be-specific T cells in blood of CBD differs from that found in BeS
- Be-induced T-cell proliferation and TH1-type cytokine secretion were determined in blood cells from 33 patients with CBD and 18 BeS subjects.
- CBD had a significantly elevated number of IFN-γ–producing and IL-2–producing Be-specific CD4+ T cells in blood compared with both BeS and normal control subjects.
- No difference in Be-induced proliferation of blood T cells was seen between BeS patients and patients with CBD.
- Higher numbers of beryllium-specific cells in blood were also associated with the extent of alveolar inflammation, as measured by both bronchoalveolar lavage white blood cell and lymphocyte counts.
- The frequency of beryllium-specific T cells in the blood of beryllium-exposed subjects may be a useful biomarker that helps discriminate between BeS and CBD.

Pott GB, Palmer BE, Sullivan AK, Silviera L, Maier LA, Newman LS, Kotzin BL, Fontenot AP. Frequency of beryllium-specific, TH1-type cytokine-expressing CD4+ T cells in patients with beryllium-induced disease. Journal of Allergy and Clinical Immunology. (2005) 115:5; 1036-1042.

BeELISpot as a medical surveillance tool or a non-invasive marker of CBD

- We determined whether the number of Be-specific T cells in blood predicted the development of CBD in a cohort of Be-exposed subjects.
- Using IFN-γ ELISpot and proliferation-based assays, we determined the frequency and proliferative capacity of Be-responsive T cells in blood.
- Compared with the BeLPT, which detected an abnormal Be-induced proliferative response in 11 (4.2%) of 260 Be-exposed workers, the IFN-γ ELISpot detected a sensitization rate of 10%.
- The transition from BeS to CBD was associated with an increased number of antigen-specific T cells in blood.
- These findings suggest that the frequency of antigen-specific T cells in blood can serve as a noninvasive biomarker to predict disease development and severity of the Be-specific CD4+ T-cell alveolitis.

Martin AK, Mack DG, Falta MT, Mroz MM, Newman LS, Maier LA, Fontenot AP. Beryllium-specific CD4+ T cells in blood as a biomarker of disease progression. Journal of Allergy and Clinical Immunology, (2011)125:5' 1100-1106.

Optimizing the BeELISpot as a clinical test in identifying BeS and CBD.

- 40 subjects undergoing clinical evaluation for BeS/CBD had BeELISpot testing.
- CBD (n=26) diagnosis was based on history of Be exposure, confirmed abnormal blood BeLPT results and/or abnormal BAL BeLPT, and the presence of granulomatous inflammation and/ or mononuclear cell infiltration on lung biopsy.
- BeS (n=14) diagnosis required a history of beryllium exposure, confirmed abnormal blood BeLPT results and no evidence of granulomatous inflammation on transbronchial lung biopsy.
- BeELISpot plates were set up in duplicate and stimulated with two different concentrations of beryllium. The plates were analyzed by using a CTL Immunospot Analyzer (CellularTechnology Ltd, Cleveland, Ohio), results were reported as the mean number of spot-forming units (SFU) per well minus background SFU.
- Normal (NOR) was designated as < 5 SFU, borderline (BOR) as > 5 to 15 SFU and abnormal (ABN) > 15 SFU.
- We calculated the positive predictive value (PPV) of an ABN/BOR BeELISpot.

Results

The BeELISpot median SFU result was 6.25 (range 0-18) for BeS patients compared to 19 (range 0.5 - 636.5) for CBD patients and proved to be significantly different ($p = 0.005$).

For BeS/CBD patients combined, BeELISpot identified 70% with ABN/BOR LPT results. For BeS, the PPV was 57% for and for CBD patients, the PPV was 81% but results were not significantly different.

Conclusions

- The optimized BeELISpot was able to identify ABN or BOR result in 81% of CBD patients and 57% of BeS patients.
- Further work will compare BeELISpot SFU to clinical results and enroll beryllium exposed controls to determine sensitivity and specificity of the test.
- BeELISpot holds promise as a useful biomarker of BeS and CBD in beryllium exposed workforces.
- The need for less blood, short turn-around-time and no need for radioactive materials combine to compel further study of BeELISpot as a tool for beryllium medical surveillance in occupationally exposed workforces.

8.3 Analysis of Surface Sampling Techniques: Discussion of Results from UKAEA Pilot Study and Follow-on Progress

Eilish McKeon, Amy Bleasdale, Lia Au Yeung

At UKAEA, surface sampling for beryllium is performed by Health Physics technicians trained in standard processes. Given the manual nature of the task, natural variations occur from person to person. Current research aims to aid in the development of robotic sampling methods by determining the typical performance of operators.

A pilot study was initially performed to determine the feasibility of studying sampling variability between technicians, investigating the effect of pressure exerted on beryllium recovery. Technicians used both Whatman filters (dry) and Ghost Wipes (wet), and despite the small sample sizes used, results indicated a significant relationship between pressure and beryllium recovery when using Ghost Wipes. During the pilot study it was noted that Whatman filters are routinely used at UKAEA. Ghost Wipes are used less frequently. It was concluded that one of the potential reasons that average pressure was not significant for Whatman filters might be because the technicians are more familiar with this sampling medium and rely on their own techniques to get better results, rather relying on pressure alone. Alternatively, no significant relationship may have been picked up by the model due to the granularity of the study which resulted in small subsets of data.

The second study phase, as part of a collaborative robotics and AI programme (RAICo), aims to further the understanding of surface sampling variables pertaining to operator input, including force application, surface types, and sampling protocol. Determining the effect of these factors on sampling results is expected to inform the development of remote sampling techniques through the development of complex swabbing motions for more efficacious and repeatable swabbing using a robotic arm, comparing robotic and manual sampling efficiency using qualitative measures.

Presenting Author: Eilish McKeon
Presenting Author's Affiliation: Culham Centre for Fusion Energy, UK Atomic Energy Authority, UK
Email: Eilish.mckeon@ukaea.uk

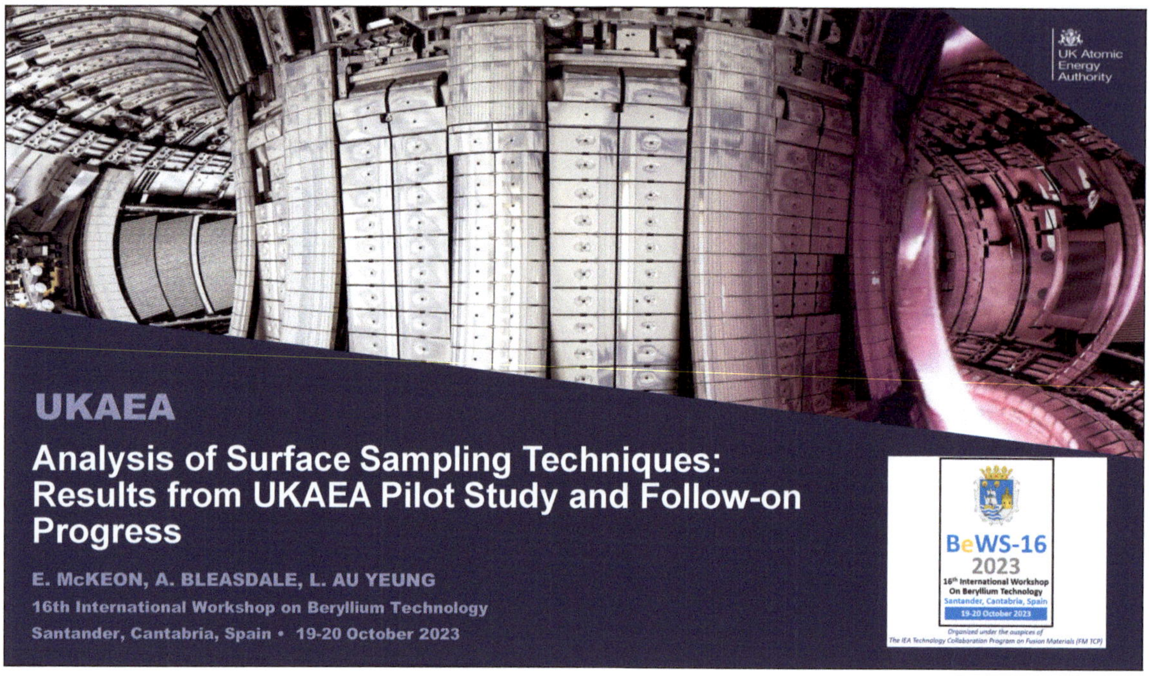
UK Atomic
Energy
Authority
UKAEA
Analysis of Surface Sampling Techniques: Results from UKAEA Pilot Study and Follow-on Progress
E. McKEON, A. BLEASDALE, L. AU YEUNG
16th International Workshop on Beryllium Technology
Santander, Cantabria, Spain • 19-20 October 2023
BeWS-16
2023
16th International Workshop On Beryllium Technology
Santander, Cantabria, Spain
19-20 October 2023

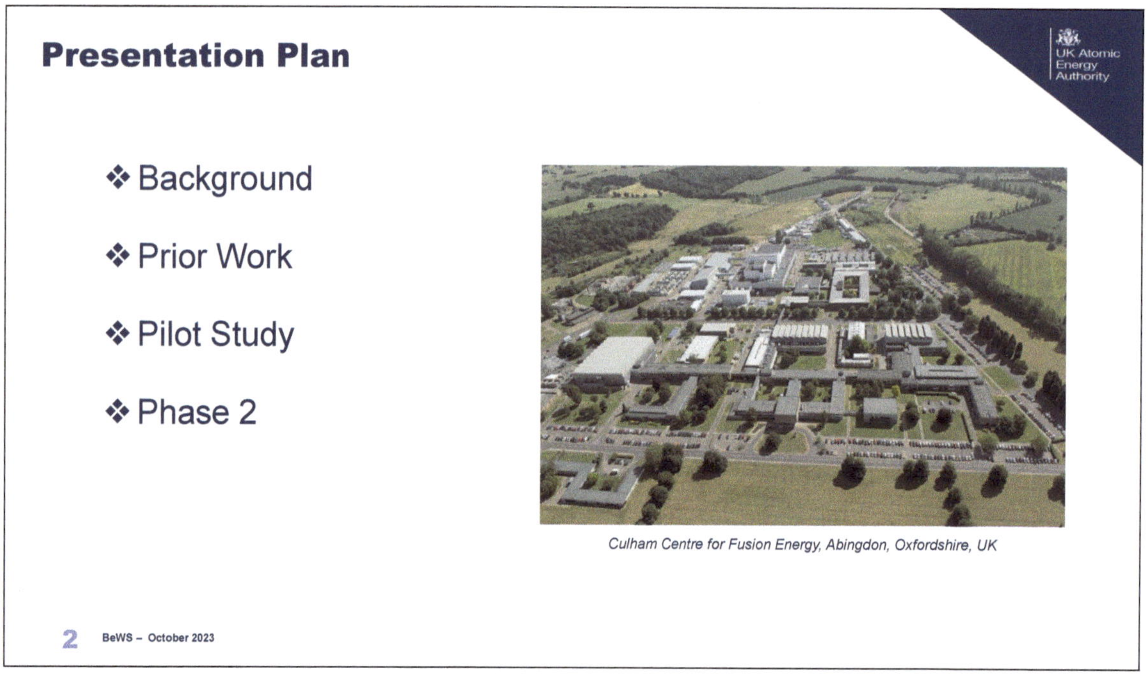
Presentation Plan
UK Atomic
Energy
Authority
❖ Background
❖ Prior Work
❖ Pilot Study
❖ Phase 2
Culham Centre for Fusion Energy, Abingdon, Oxfordshire, UK
2 BeWS – October 2023

Background

UK Atomic Energy Authority

- Beryllium has been in operational use at UKAEA since 1989
- Despite the lack of surface contamination limits, UKAEA foresaw the need to monitor for beryllium surface particulate

- As a result, and based on their experience in radiological protection, UKAEA developed a surface sampling method for beryllium, in conjunction with Sandia National Lab (USA)
  - Dry wiping using Whatman Grade 1 filter papers
  - Analysis of collected samples done on site
- Manual task with natural variation:
  - Person to person
  - Technique to technique
  - Surface to surface

JET Reactor c.2010-11 with Beryllium ITER-Like Wall Installed [1]

3 BeWS – October 2023

Project Aims

UK Atomic Energy Authority

- To aid in the development of robotic sampling methods by determining the typical performance of operators, by:
- Determining the effects of swabbing force on the removal of contamination.
- Determining the effects of sampling efficiency on different surface types.
- Using the physical and chemical data gathered from the previous deliverables to inform robotic smearing techniques, comparing robotic sampling efficiency with human efficiency using qualitative measures.

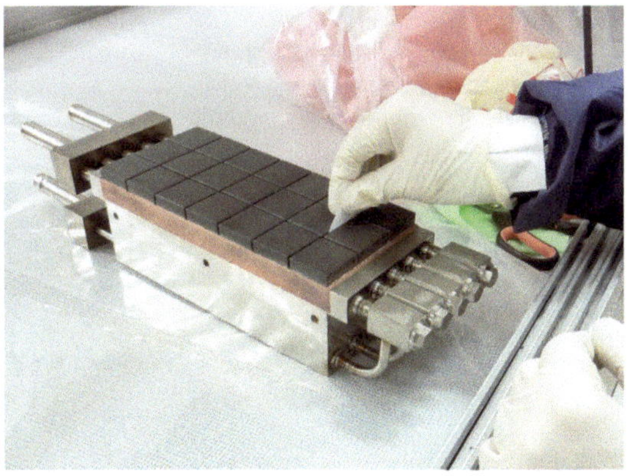

Wipe sampling with dry Whatman filter paper of beryllium tile surfaces

4 BeWS – October 2023

Prior Work

UK Atomic Energy Authority

- **Mirjana Damjanovic, CEA at UKAEA**
  - *"Comparison of Collection Efficiencies of Whatman No.1 Filter Paper and Ghost Wipes for Beryllium Sampling"* (presented at BeWS-14) [2]
  - Study performed with UKAEA's Health Physics technicians and analysis lab.
  - Looked at sampling glass plates and vinyl material.
  - Concluded sampling technique, including pressure exerted, may be an important variable

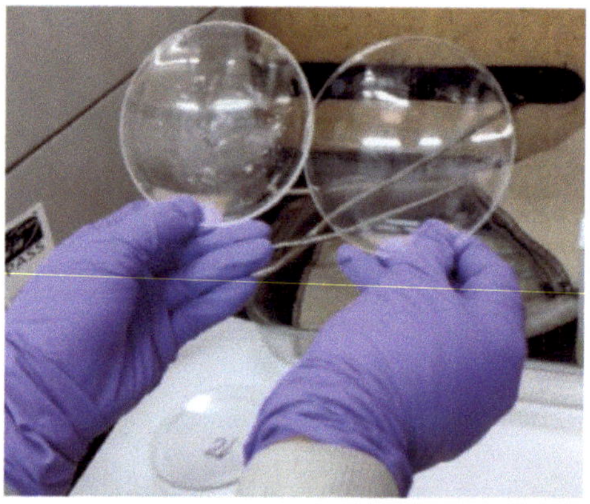

Glass plates spiked with beryllium-containing solution in known amounts were used in the previous study [6]

5 BeWS – October 2023

Pilot Study – Technician Variability

UK Atomic Energy Authority

- **Background**
  - Internally funded by Innovation Dept
  - Pressure rig designed by Materion
  - 4 technicians with experiencing ranging from <2 years to 10+ years
- **Scope of Work**
  - Determining if the results are correlated and if relationship exists between pressure exerted (N/s) and beryllium recovered (%)
  - Assessing the feasibility of pursuing further research into this relationship in a low-risk manner.

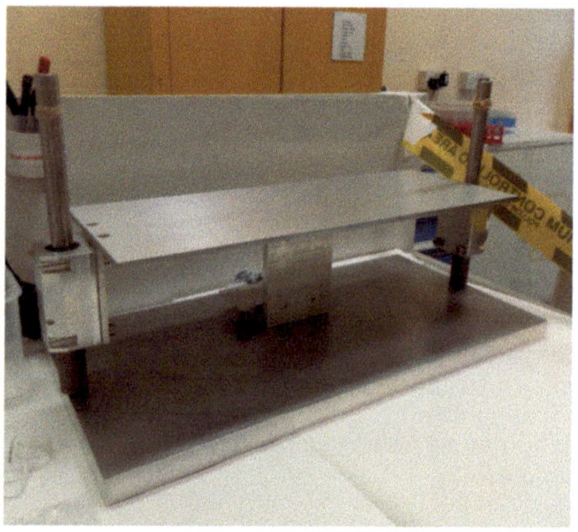

Pressure rig (test fixture) built at UKAEA and used in the study (the force gauge is not installed in this view).
Design credit: Materion Corp., Ohio, USA
Shown here with kind permission [3]

6 BeWS – October 2023

Method

UK Atomic Energy Authority

- Spike a defined surface with a range of known concentrations of beryllium on a plastic surface
- 1ml and 2ml of Beryllium standards at concentrations of:
  - 0PPB (control), 40PPB, 80PPB, 200PPB
- 4x Technicians: HPT A-D
- 2x sample media: Whatman Filters (dry), Ghost Wipes (Wet)
- Measure pressure ranges exerted
- Measure the amount of beryllium detected

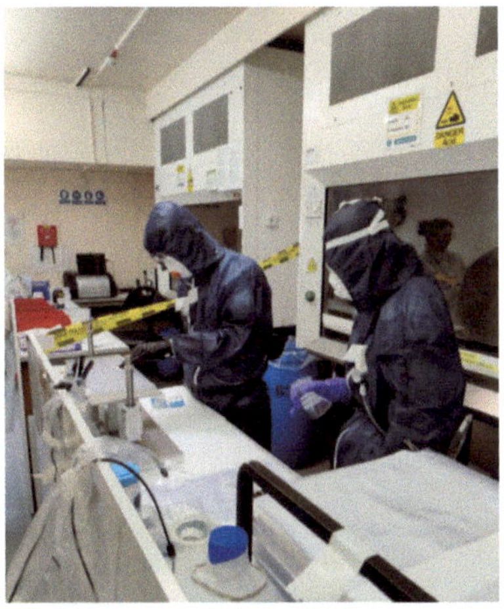

Technicians prepare standardised beryllium samples in a temporary controlled area set up in the Analytical Laboratory at UKAEA expressly for the purposes of this study

7 BeWS - October 2023

Results

UK Atomic Energy Authority

- Despite the small sample sizes results indicated a fairly correlated (+0.33) relationship between average pressure (N/s) and beryllium recovery (%) overall.
- A chi-squared test for independence was used to further check the relationship and found a statistically significant relationship between the variables at 5% significance
  - i.e., the null hypothesis that "*there is no effect of pressure on the percentage of beryllium recovered during sampling*", was rejected
- For subsets of the different combination of technician and sample media (e.g., HPT C and Whatman), no relationship was found between pressure and beryllium recovered.
  - Possibly due to granularity around Be concentrations and small sample sizes.

Min/Mean/Max pressures of technician and sample medium

Technician	Sample Media	Minimum Pressure (N/s)	Mean Pressure (N/s)	Maximum Pressure (N/s)	Mean Be Recovery (%)
Overall	Overall	1.9	5.9	14.4	76.8
A	Dry Whatman	4.2	6.1	7.4	87.6
	Ghost Wipe	5.2	6.2	8.5	67.7
B	Dry Whatman	2.8	3.3	4.0	79.3
	Ghost Wipe	1.9	2.9	3.9	51.0
C	Dry Whatman	2.3	3.2	4.2	94.3
	Ghost Wipe	2.2	3.0	3.6	62.6
D	Dry Whatman	8.1	10.2	12.6	93.9
	Ghost Wipe	9.9	12.3	14.4	81.8

Interpretation of the correlation coefficient

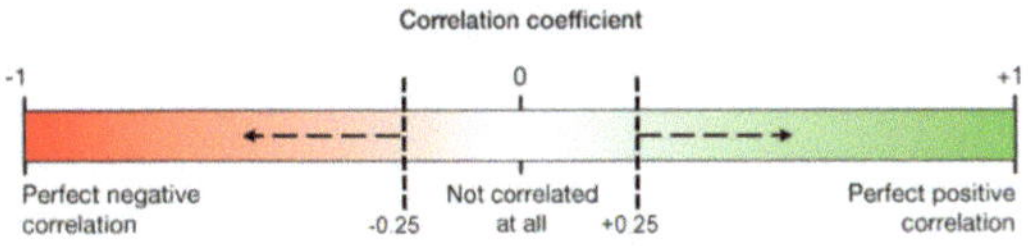

8 BeWS - October 2023

Results (continued)

UK Atomic Energy Authority

- Linear regression of all Beryllium Recovery (%) vs Average Pressure (N/s) data found significance at 10%, 5%, but not 1%.
  - Prob (F-statistic) <0.05
- For subsets of individual technician data, again no significance found
  - Prob (F-statistic) >0.1

Column	Relationship w/ 'Beryllium Recovery (%)' column: Technician	Intercept Coefficient	Column Coefficient	Prob(F-statistic)	Significant at 5%
Average Pressure (Newton/Seconds)	A	73.440	0.648	0.891	No
	B	14.038	16.714	0.240	No
	C	86.923	-2.815	0.800	No
	D	109.962	-1.867	0.246	No

Results table for regressions performed for the individual technicians

OLS Regression Results

Dep. Variable:	Beryllium_Recovery	R-squared:	0.112
Model:	OLS	Adj. R-squared:	0.092
Method:	Least Squares	F-statistic:	5.785
Date:	Fri, 26 Aug 2022	Prob (F-statistic):	0.0202
Time:	16:21:02	Log-Likelihood:	-206.24
No. Observations:	48	AIC:	416.5
Df Residuals:	46	BIC:	420.2
Df Model:	1		
Covariance Type:	nonrobust		

	coef	std err	t	P>|t|	[0.025	0.975]
Intercept	66.9690	5.019	13.342	0.000	56.866	77.072
Average_Pressure	1.6845	0.700	2.405	0.020	0.275	3.094

Omnibus:	3.032	Durbin-Watson:	1.018
Prob(Omnibus):	0.220	Jarque-Bera (JB):	2.234
Skew:	-0.518	Prob(JB):	0.327
Kurtosis:	3.210	Cond. No.	13.9

Regression Results Table

9 BeWS – October 2023

Results (continued)

UK Atomic Energy Authority

- Average pressure and the two sample media types:
  - Not significant for Whatman filters
  - Significant at 1%, 5%, 10% for Ghost Wipes
- Model suggests there is a relationship between average pressure and beryllium recovered for Ghost Wipes, but not Whatman filters.
  - Possibly due to experience in knowing the best ways to manoeuvre the filter on the surface.

Column	Relationship w/ 'Beryllium Recovery (%)' column: Sample Media	Intercept Coefficient	Column Coefficient	Prob(F-statistic)	Significant at 5%
Average Pressure (Newton/Seconds)	Dry Whatman	84.524	0.707	0.473	No
	Ghost Wipe	50.283	2.479	0.000	Yes

Results table for regressions performed for the individual technicians

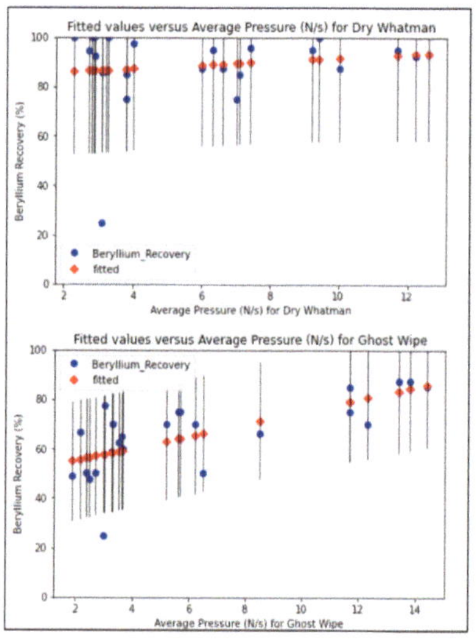

Fitted vs True values of Beryllium recovery for the separate sample media

10 BeWS – October 2023

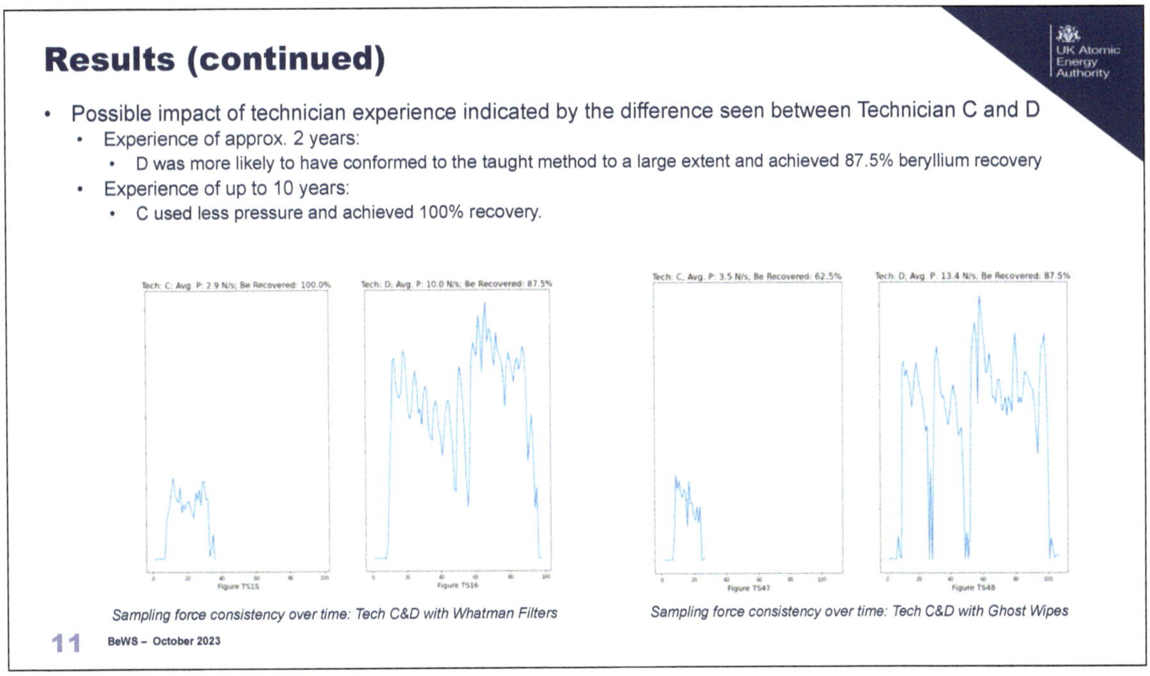
Results (continued)
UK Atomic Energy Authority
• Possible impact of technician experience indicated by the difference seen between Technician C and D
• Experience of approx. 2 years:
• D was more likely to have conformed to the taught method to a large extent and achieved 87.5% beryllium recovery
• Experience of up to 10 years:
• C used less pressure and achieved 100% recovery.
Tech: C; Avg. P: 2.9 N/s; Be Recovered: 100.0%
Figure TS15
Tech: D; Avg. P: 10.0 N/s; Be Recovered: 87.5%
Figure TS16
Tech: C; Avg. P: 3.5 N/s; Be Recovered: 62.5%
Figure TS47
Tech: D; Avg. P: 13.4 N/s; Be Recovered: 87.5%
Figure TS48
Sampling force consistency over time: Tech C&D with Whatman Filters
Sampling force consistency over time: Tech C&D with Ghost Wipes
11
BeWS – October 2023

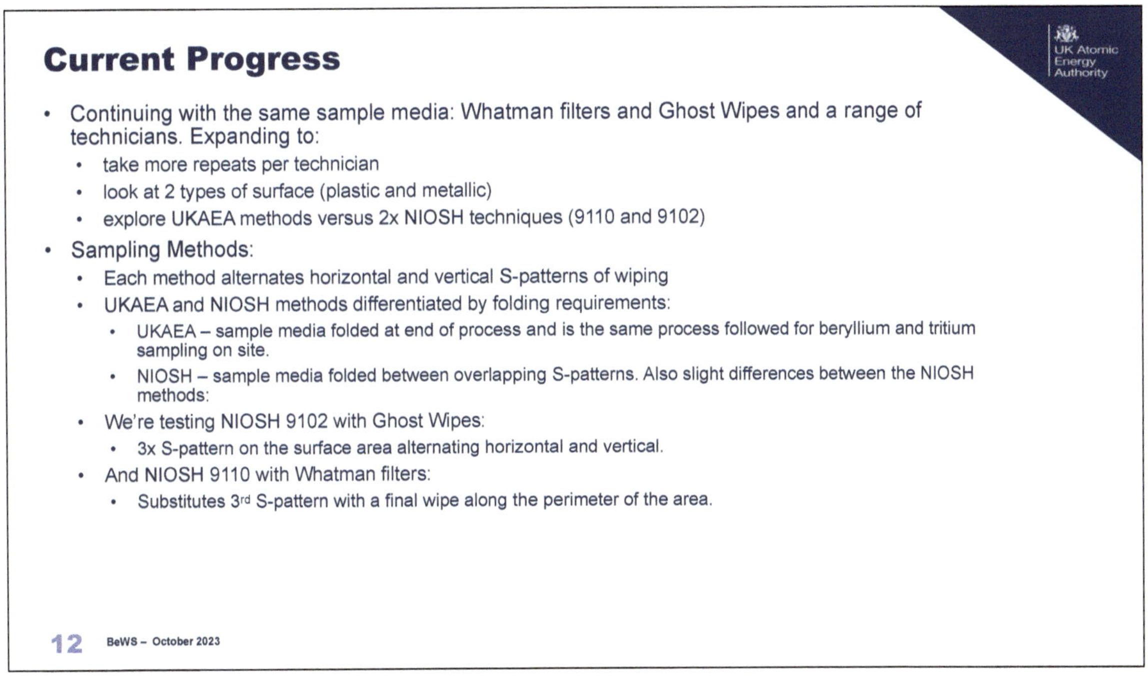
Current Progress
UK Atomic Energy Authority
• Continuing with the same sample media: Whatman filters and Ghost Wipes and a range of technicians. Expanding to:
• take more repeats per technician
• look at 2 types of surface (plastic and metallic)
• explore UKAEA methods versus 2x NIOSH techniques (9110 and 9102)
• Sampling Methods:
• Each method alternates horizontal and vertical S-patterns of wiping
• UKAEA and NIOSH methods differentiated by folding requirements:
• UKAEA – sample media folded at end of process and is the same process followed for beryllium and tritium sampling on site.
• NIOSH – sample media folded between overlapping S-patterns. Also slight differences between the NIOSH methods:
• We're testing NIOSH 9102 with Ghost Wipes:
• 3x S-pattern on the surface area alternating horizontal and vertical.
• And NIOSH 9110 with Whatman filters:
• Substitutes 3rd S-pattern with a final wipe along the perimeter of the area.
12
BeWS – October 2023

What's Next

UK Atomic Energy Authority

- Testing robotic sampling techniques
- Qualitative comparison between robotic sampling and human sampling efficiency using UV powder
- Publication with University of Manchester

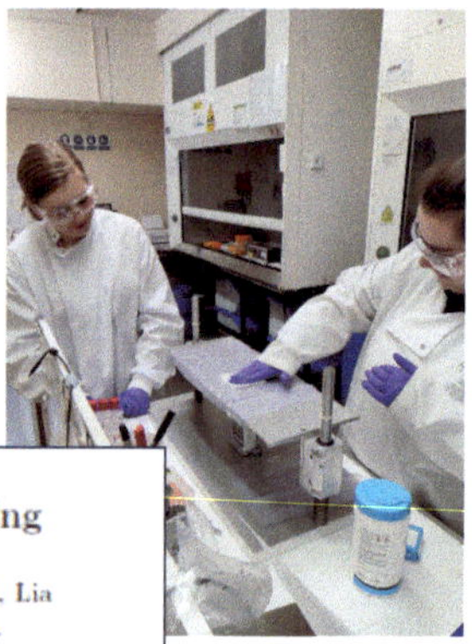

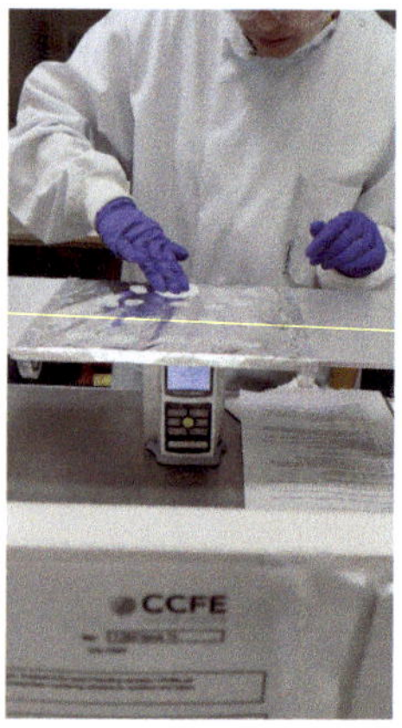

Analysing Human Performance in Wipe Sampling

Thomas W. Johnson[1]‡, Eilish McKeon[2], Amy Bleasdale[2], Lia Au Yeung[2], Simon Watson[1], Thomas Carey[3] and Clint A. Sharrad[1]

1. The University of Manchester, Oxford Road, Manchester, United Kingdom.
2. United Kingdom Atomic Energy Authority, Culham Centre for Fusion energy, Culham Science Centre, Abingdon, United Kingdom
3. National Nuclear Laboratory, Chadwick House, Birchwood Park, Warrington, Cheshire, United Kingdom.

13 Footer

References & Credits

UK Atomic Energy Authority

[1] European Fusion Development Agreement (EFDA) and Joint European Torus (JET). ITER-Like Wall Project, Original URL: http://www.efda.org/jet/, 2009.

[2] Damjanovic, M. "Comparison of Collection Efficiencies of Whatman No. 1 Filter Paper and Ghost Wipes for Loose Beryllium Surface Contamination". *Proceedings of the 14th International Workshop on Beryllium Technology (BeWS-14)*, held in Long Beach, California, U.S.A., 2019, pp. 324-346. Lemmens Medien GmbH, 2021.

[3] Materion Corp., Mayfield Heights, Ohio, USA, 2021.

14 BeWS – October 2023

SESSION 9

BeYOND Industrial Forum

9.1 An Assessment of Beryllium-Based Neutron Multiplier Materials for Breeder Blanket Applications in the Modern Fusion Industry

Aniceto GORAIEB, Christopher DORN

The BeYOND Industrial Forum was first held in 2009, co-founded by Aniceto Goraieb of the Karlsruhe Beryllium Handling Facility (KBHF) and Chris Dorn, then working for Brush Wellman Inc. in the U.S. The name BeYOND is an acronym, which stands for "Beryllium Opportunities for New Developments."

The reason for holding another beryllium-in-fusion-type event in addition to the IEA International Workshop on Beryllium Technology (BeWS) was that it would have a different focus. BeYOND was created with the idea of more emphasis on strengthening the connection between the beryllium research community and the beryllium industry, an aspect not always satisfied by the BeWS. Since BeYOND-V in 2013, the workshop has generally also placed more emphasis on beryllium health and safety than typically found in the BeWS.

After trying various timing relative to other conferences, it evolved that the BeYOND workshops would take place on alternate years from the BeWS and normally held in conjunction with the SOFT Conference (Symposium on Fusion Technology), which is always held in Europe. Meetings to date:

1. BeYOND-I: Karlsruhe, Germany, 2009
2. BeYOND-II: Porto, Portugal, 2010
3. BeYOND-III: Karlsruhe, Germany, 2011
4. BeYOND-IV: Karlsruhe, Germany, 2012
5. BeYOND-V: Barcelona, Spain, 2013
6. BeYOND-VI: San Sebastian, Spain, 2014
7. BeYOND-VII: Berlin, Germany, 2016
8. BeYOND-VIII: Karlsruhe, Germany, 2018
9. BeYOND-IX: Karlsruhe, Germany, 2022

Nowadays, as the driving force in the fusion industry moves from government research institutes to commercial enterprises, it is the commercial companies that will determine the timeframe for achieving a successful, scalable fusion machine. If the realization of that first machine is to succeed, it is clear that collaboration between the traditional institutes and the commercial companies must happen, and that need for working together goes well beyond just what is happening with beryllium. Everyone needs to start thinking bigger.

In June 2023, BeYOND co-founder Aniceto Goraieb participated as a panelist in a satellite session at the Symposium on Fusion Energy (SOFE), which was his original brainchild. That session, which was titled "Next Steps to Global Fusion Partnerships," discussed the opportunities in private-public partnership (PPP) in the fusion energy industry, and it was ultimately supported by Kyoto Fusioneering in the UK, and the Fusion Industry Association (FIA) in the U.S. The presentations and discussion during that session emphasized not only the importance of partnerships to accelerate the development of fusion, but they also showed how many different pathways exist and are already being used for collaboration and partnership.

Based on this recent experience at SOFE, coupled with the large-scale changes in the fusion industry, it seems that now is the time to change the purpose of the BeYOND Industrial Forum. Up to now, BeYOND has been focused on fostering partnerships between government labs and industry for beryllium-related research, but the vision for the future is to significantly broaden the scope of those partnerships. Now, the critical path for accelerating the success of fusion lies in the promise of the afore mentioned public-private partnerships, and those PPPs need to cover all areas of fusion where

LET´S CATCH THE STAR
RESEARCH ON BEHALF OF MANKIND
FUSION
FOR FUTURE
www.fusion-for-future.de
BeWS-16
2023
The Future of the BeYOND Industrial Forum
Aniceto Goraieb, KBHF GmbH
Chris Dorn, Be4FUSION
16th International Workshop on Beryllium Technology (BeWS-16)
Santander, Spain
October 2023

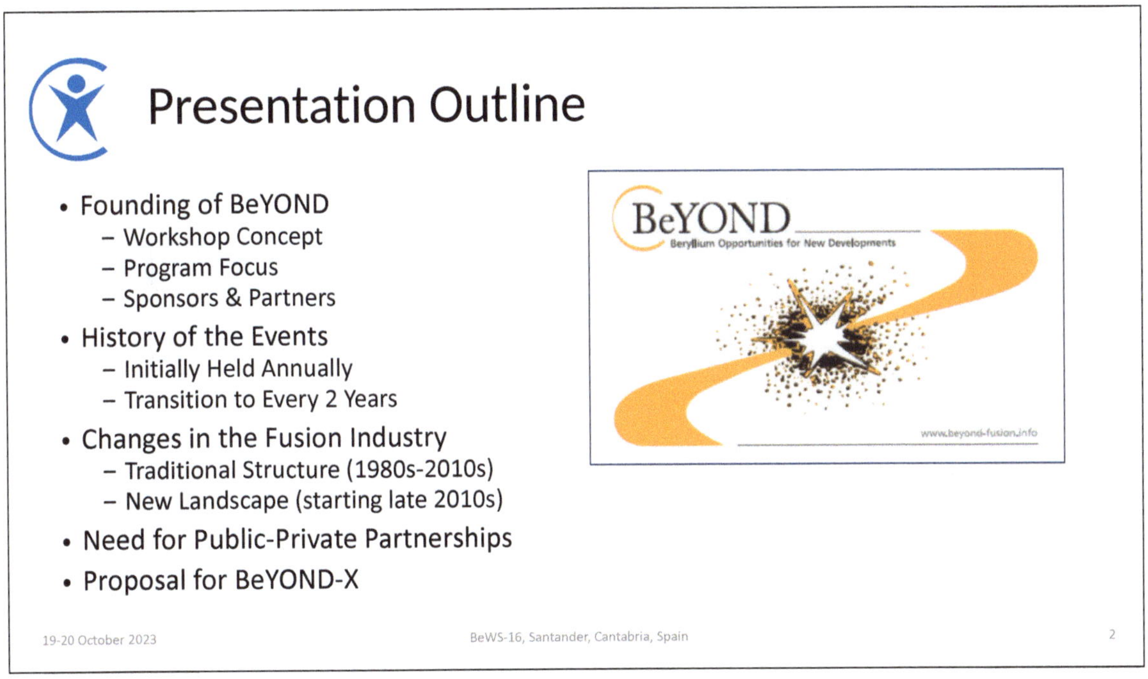
Presentation Outline
• Founding of BeYOND
– Workshop Concept
– Program Focus
– Sponsors & Partners
• History of the Events
– Initially Held Annually
– Transition to Every 2 Years
• Changes in the Fusion Industry
– Traditional Structure (1980s-2010s)
– New Landscape (starting late 2010s)
• Need for Public-Private Partnerships
• Proposal for BeYOND-X
BeYOND
Beryllium Opportunities for New Developments
www.beyond-fusion.info
19-20 October 2023
BeWS-16, Santander, Cantabria, Spain
2

We came a long way through fusion history

- The Start-Up GVT was established 1993
  - Specialized on Solid Breeder Blanket Design
    - Material design, testing & pilot production
    - Work performed on FLiBe, as well
- Nonprofit organization found in 2000
  - "Fusion for Future" by mankindproject e.V.
    - Events for children and adults
    - Science, industry, politics and the public
- KBHF, a Spin-In at KIT established 2009
  - Private infrastructure for Blanket handling
    - Public-Private-Partnership on KIT Campus
    - Transfer company science to industry
- BeYOND started the same year

GVT Goraieb Versuchstechnik

www.versuchstechnik.de

19-20 October 2023 BeWS-16, Santander, Cantabria, Spain 3

Founding of BeYOND

- Workshop Concept
  - Link Research Centers to Industry
  - Greater Emphasis on Industrial Issues
    - Scale-Up from Lab to Production
    - Technology Transfer
    - Environmental, Health & Safety (EH&S)
- Starting the Series
  - Partnership: KBHF & Brush Wellman
  - Funding from KIT & Industry Partners
- Evolution of Program
  - Even-Number Years (opposite BeWS)
  - Timing Coordinated with SOFT
  - Always Held in Karlsruhe, Germany

19-20 October 2023 BeWS-16, Santander, Cantabria, Spain 4

History of the Events

BeYOND
Beryllium Opportunities for New Developments

BeYond

- BeYOND-I: Karlsruhe, Germany, 2009
- BeYOND-II: Porto, Portugal, 2010
- BeYOND-III: Karlsruhe, Germany, 2011
- BeYOND-IV: Karlsruhe, Germany, 2012
- BeYOND-V: Barcelona, Spain, 2013
- BeYOND-VI: San Sebastian, Spain, 2014
- BeYOND-VII: Berlin, Germany, 2016
- BeYOND-VIII: Karlsruhe, Germany, 2018
- BeYOND-IX: Karlsruhe, Germany, 2022

BeYOND-IX
2022
9th Industrial Forum on Beryllium Opportunities & New Developments
Karlsruhe, Germany
16 September 2022

19-20 October 2023 BeWS-16, Santander, Cantabria, Spain 5

Next Steps to Global Fusion Partnerships

- First time in June 2023, at SOFE
  - Symposium on Fusion Engineering in Oxford
  - A brainchild of the BeYOND founders
  - Supported by Kyoto Fusioneering (KF) in UK
  - Also, by Fusion Industry Association (FIA, USA)
  - Organized with the help of UKAEA
    - Discussed the opportunities of PPPs
    - Showed the importance of partnerships to accelerate fusion
    - Many different pathways already exist and are used for this purpose
    - Synergies play a significant role to get ready in time (1 or 2 decades?)
    - Non-Profit Organizations could be an additional key to the success

19-20 October 2023 BeWS-16, Santander, Cantabria, Spain 6

Mankindproject e.V. has just been restarted

- Communication
  - To keep the public informed
  - Connecting Engineers and Visioneers
  - Bridging between science & Industry
  - Student programs for junior stuff
- Event management
  - Support of BeYOND
  - Historical venues to catch up with science
- Co-Financing of research programs
  - Crowd Funding
  - Organization of charity events
  - Tax refunds for donations

19-20 October 2023 BeWS-16, Santander, Cantabria, Spain 7

Merging the concepts at BeYOND-X?

- Global Fusion Partnerships need:
  - New interested and qualified staff
  - Technology Transfer options
  - Partners that enable the possibility of synergies
  - Public and political support
- BeYOND-GFP could help with:
  - Short-Term TT:
    - Finding Technology near state-of-the-art
  - Mid-Term TT:
    - Create research partnerships, train students
  - Long-Term TT:
    - Indirect technology transfer back to science labs
- The Non-profit Organization's role would be:
  - Crowd Funding, State Funding
  - Organization of charity events
  - Tax credits or offsets for donations

19-20 October 2023 BeWS-16, Santander, Cantabria, Spain 8

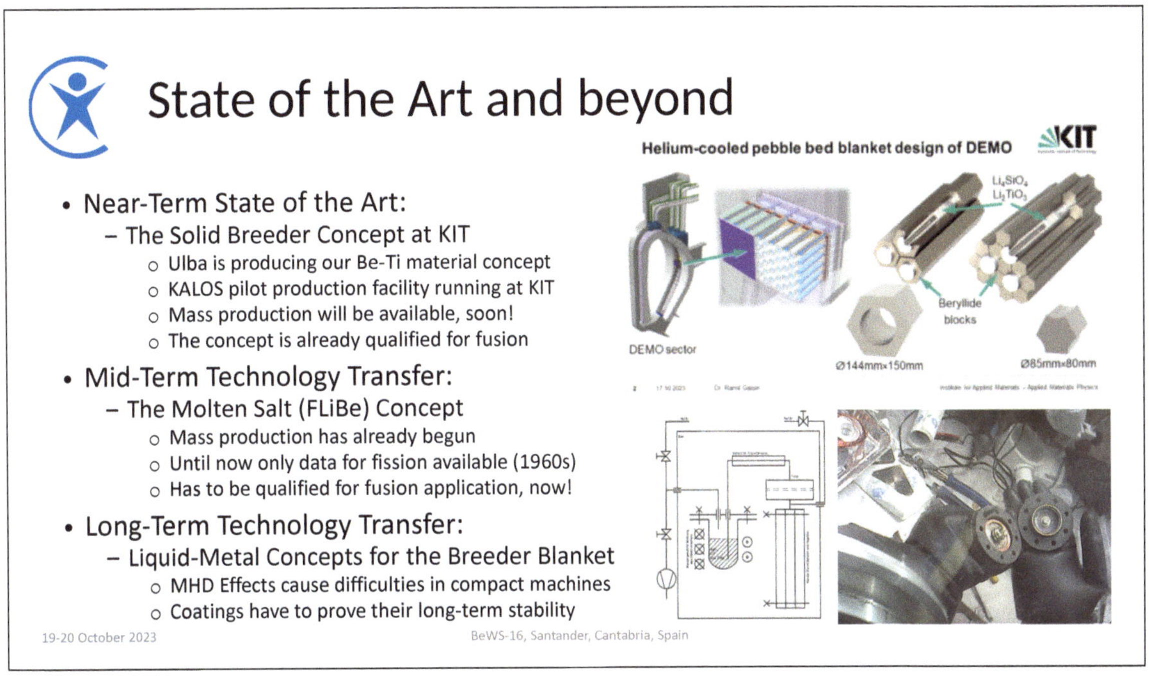
State of the Art and beyond
Helium-cooled pebble bed blanket design of DEMO
KIT
• Near-Term State of the Art:
– The Solid Breeder Concept at KIT
o Ulba is producing our Be-Ti material concept
o KALOS pilot production facility running at KIT
o Mass production will be available, soon!
o The concept is already qualified for fusion
• Mid-Term Technology Transfer:
– The Molten Salt (FLiBe) Concept
o Mass production has already begun
o Until now only data for fission available (1960s)
o Has to be qualified for fusion application, now!
• Long-Term Technology Transfer:
– Liquid-Metal Concepts for the Breeder Blanket
o MHD Effects cause difficulties in compact machines
o Coatings have to prove their long-term stability
DEMO sector
Beryllide blocks
Ø144mm×150mm
Ø85mm×80mm
19-20 October 2023
BeWS-16, Santander, Cantabria, Spain

Thank You Very Much for Your Attention!
FUSION FOR FUTURE
www.fusion-for-future.de
Aniceto Goraieb, KBHF GmbH
Be4FUSION
Specialty Materials & Business Consulting
https://be4fusion.com/
Chris Dorn, Be4FUSION LLC

APPENDIX A

Workshop Photos

Figure 1: Group picture of the BeWS-16 participants. From left to right:

1st row: J. Reimann[1], P. Vladimirov[2], A. Goraieb[3], K. Smith[4], J. Sharp[5], A. Renier[6], B. Walker[7], A. Nicholls[8]

2nd row: M. Klimenkov[9], S. Karliev[10], E. McKeon[11] (partially obscured), T. Knudson[12], Y. Natori[13], R. Gaisin[14], M. Rubel[15], S. Kuksenko[16], Y. Sugimoto[17], S. Nakano[18], T. Hwang[19], J. Taylor[20]

3rd row: M. Damjanovic[21], R.-J. Zabolockis[22], P. Mählmann[23], G. Kizane[24], N. Ikemoto[25], D. Bachurin[26], K. Woller[27], M. Voss[28]

4th row: J. Hinks[29], G. Unquera[30], C. Dorn[31], L. Vandermark[32], C. Gayot[33], H.-C. Schneider[34], A. Frehn[35], J.-H. Kim[36].

Figure 2: Loading on a bus for the excursion to the "Beryllium Facilities" at "Leading", Spain

Figure 3: Visiting the "Beryllium Facilities" at "Leading", Spain: our host Marcos Perez introduces his kingdom to us

Figure 4: Visiting the "Beryllium Facilities": Aniceto Goraieb pictured here with a machine that coincidentally bears his nickname (Cheto)!

Figure 5: Dr. Ramil Gaisin of KIT, Germany, pictured with his trophy, just after he was presented with the Mario Dalle Donne Memorial Award (MDDMA) for BeWS-16

Figure 6: Award presentation at the conference dinner on 19.10.2023

APPENDIX B

BeWS-16 Technical Program

Title:	The 16th International Workshop on Beryllium Technology (BeWS-16)
Location:	Santander, Spain
Date:	19-20.10.2023

BeWS-16: 19.10.23

Start	End	Time	Item	Presenter	Session	Chair	Column1
9:00 AM	9:15 AM	0:15	Welcome of the Cantabria Authorities	to be confirmed	0.1 Opening		
9:15 AM	9:30 AM	0:15	Invest in Cantabria	Angel Gonzalez Madrazo	0.2 Opening		
9:30 AM	9:55 AM	0:25	Present Status of Advanced Neutron Multipliers for JA DEMO activities	Jae-Hwan Kim	1.1 Introductory	Pavel Vladimirov	
9:55 AM	10:20 AM	0:25	Overview of analysis methods for beryllium from controlled fusion devices	Marek Rubel	1.2 Introductory		
10:20 AM	10:35 AM	0:15	Break				
10:35 AM	11:00 AM	0:25	Reaction kinetics in the leaching process of beryllium from beryl	Suguru Nakano	2.1 Fabrication & Recycling	Jae-Hwan Kim, Slava Kuksenko	
11:00 AM	11:25 AM	0:25	Recycling of neutron-irradiated beryllium and titanium beryllide	Vladimir Chakin	2.2 Fabrication & Recycling		
11:25 AM	11:50 AM	0:25	Recycling of Beryllides: Novel Approach for Cost-Effective and Efficient Fabrication	Ramil Gaisin	2.3 Fabrication & Recycling		
11:50 AM	12:15 PM	0:25	Attempts for uranium removing in beryllium using arc plasma	Norihiro Ikemoto	2.4 Fabrication & Recycling		
12:15 PM	12:40 PM	0:25	Method for processing spent products from titanium beryllide TiBe12 after completion of their operation in the DEMO reactor	Sergey Udartsev	2.5 Fabrication & Recycling		
12:40 PM	2:10 PM	1:30	**Lunch**				
2:10 PM	2:35 PM	0:25	FLiBe Molten Salt as a Tritium Breeder/Coolant for Fusion: Fostering Collaboration for Breakthrough Progress	Chris Dorn	3.1 Molten salt breeders	Aniceto Goraieb	
2:35 PM	3:00 PM	0:25	Prototype verification of tritium breeding with liquid immersion blanket technology using flibe	Kevin B. Woller	3.2 Molten salt breeders		
3:00 PM	3:25 PM	0:25	Beryllium Worker Protection Model for the Production and Handling of FLiBe	Theodore Knudson	3.3 Molten salt breeders		
3:25 PM	3:40 PM	0:15	Break				
3:40 PM	4:05 PM	0:25	Screening tests to qualify thermo-mechanical properties of beryllium pebble beds with non-spherical pebbles	Jörg Reimann	4.1 Mechanical properties of Be & Beryllides	Ramil Gaisin	
4:05 PM	4:30 PM	0:25	Study of Mechanical Properties of Beryllium Intermetallic Compounds Fabricated by Plasma Sintering	Taehyun HWANG	4.2 Mechanical properties of Be & Beryllides		
4:30 PM			**ADJOURN**				
Total		**7:30**					
7:00 PM			Conference Dinner				

BeWS-16: 20.10.23

Start	End	Time	Item	Presenter	Session	Chair
9:00 AM	9:15 AM	0:15	"Leading": developments in beryllium area	Ignacio Cobo		
9:15 AM	9:40 AM	0:25	Investigation of radiation damage effects in beryllium: updates on microstructure and properties of proton, neutron and He-ions irradiated samples	Slava Kuksenko	5.1 Irradiation damage & He-implantation	Pavel Vladimirov, Marek Rubel
9:40 AM	10:05 AM	0:25	The swelling property of Neutron multipliers after irradiation with He ions under high temperature	Yutaka Sugimoto	5.2 Irradiation damage & He-implantation	
10:05 AM	10:30 AM	0:25	High temperature helium ion irradiation of Be as a rapid-throughput proxy for neutron irradiation experiments	Joanne Sharp	5.3 Irradiation damage & He-implantation	
10:30 AM	10:50 AM	0:20	Break			
10:50 AM	11:15 AM	0:25	Tritium content in partially oxidized neutron-irradiated beryllium pebbles	Rudolfs Janis Zabolockis	5.4 Irradiation damage & He-implantation	
11:15 AM	11:40 AM	0:25	Microstructural examination of neutron irradiated beryllium and beryllides	Michael Klimenkov	5.5 Irradiation damage & He-implantation	
11:40 AM	12:05 PM	0:25	DFT study of helium behavior in titanium beryllides	Dmitry Bachurin	6.1 Modeling	Michael Klimenkov
12:05 PM	12:30 PM	0:25	Hydrogen diffusion in beryllides	Pavel Vladimirov	6.2 Modeling	
12:30 PM	2:00 PM	1:30	Lunch			
2:00 PM	2:25 PM	0:25	Effect of impurities on resistivity recovery after annealing in beryllium	Cristopher Stihl	6.3 Modeling	
2:25 PM	2:50 PM	0:25	Beryllium facility design	Beth Walker	7.1 Health & Safety	Chris Dorn
2:50 PM	3:15 PM	0:25	Beryllium Stimulated ELISpot as an Alternative Screening Test for Beryllium Sensitization and Chronic Beryllium Disease	Lisa A Maier	7.2 Health & Safety	
3:15 PM	3:40 PM	0:25	Analysis of Surface Sampling Techniques: Discussion of Results from UKAEA Pilot Study and Follow-on Progress	Eilish McKeon	7.3 Health & Safety	
3:40 PM	4:05 PM	0:25	The Future of the BeYOND Industrial Forum	Aniceto GORAIEB	8.1 BeYOND	
4:05 PM	4:20 PM	0:15	Break			
4:20 PM	4:30 PM	0:10	**Closing remarks**	Pavel Vladimirov		

4:30 PM		**Bus transportation to the "Leading" Beryllim Facilities**
5:00 PM	19:00 *	Tour over Beryllium Facilites of Leading
	19:30 *	**Earliest arrival time to the hotel**
	*	*All times are approximate. DO NOT rely on them when planning your return trip.

APPENDIX C

The list of participants

	Last Name	First Name	Institution	Country
1	Bachurin	Dmitry	Karlsruhe Institute of Technology	Germany
2	Barker	Lisa	National Jewish Health	USA
3	Dorn	Chris	Kyoto Fusioneering America Ltd.	USA
4	Damjanovic	Mirjana	UK Atomic Energy Authority	UK
5	Cobo	Ignacio	Leading Metal-Mechanical Solutions	Spain
6	Frants	Yevgenii	Ulba Metallurgical Plant	Kazakhstan
7	Frehn	Andreas	Materion Brush GmbH	Germany
8	Gayot	Charles	NGK Berylco France	France
9	Gaisin	Ramil	Karlsruhe Institute of Technology	Germany
10	Goraieb	Aniceto	KBHF GmbH	Germany
11	Hwang	Taehyun	National Institutes for QST	Japan
12	Hinks	Jonathan	University of Huddersfield	UK
13	Ikemoto	Norihiro	KAKEN Inc.	Japan
14	Karliev	Serdar	New Frontier International	UAE
15	Kim	Jaehwan	National Institutes for QST	Japan
16	Kizane	Gunta	Institute of Chemical Physics University of Latvia	Latvia
17	Klimenkov	Michael	Karlsruhe Institute of Technology	Germany
18	Knudson	Theodore	Materion Brush Inc.	USA
19	Kuksenko	Slava	UK Atomic Energy Authority	UK
20	Madrazo	Angel González	Leading Metal-Mechanical Solutions	Spain
21	McKeon	Eilish	UK Atomic Energy Authority	England
22	Moyret	Pierre	CERN	Switzerland
23	Mroz	Margaret	National Jewish Health	USA
24	Mählmann	Peter	TROPAG GmbH	Germany
25	Nakano	Suguru	National Institutes for QST	Japan
26	Natori	Yuri	KAKEN,Inc.	Japan
27	Nicols	Anthony	Novanta	UK
28	Renier	Angélique	NGK Berylco France	France
29	Rubel	Marek	KTH Royal Institute of Technology	Sweden
30	Sugimoto	Yutaka	National Institutes for QST	Japan
31	Taylor	Jamie	Novanta	UK
32	Schneider	Hans-Christian	Karlsruhe Institute of Technology	Germany
33	Sharp	Joanne	University of Huddersfield	UK
34	Smith	Keith	Materion Inc.	USA
35	Unquera	Gabriel	NGK Berylco France	France
36	Udartsev	Sergey	Ulba Metallurgical Plant	Kazakhstan
37	Vandermark	Lee	General Dynamics Mission Systems	USA
38	Vladimirov	Pavel	Karlsruhe Institute of Technology	Germany
39	Voss	Michael	Tropag GmbH	Germany
40	Walker	Rhiannon Beth	Safer Environment	UK
41	Whalen	Lance	General Dynamics	USA
42	Woller	Kevin	Massachusetts Institute of Technology	USA
43	Zabolockis	Rudolfs Janis	University of Latvia	Latvia

APPENDIX D

Author Index